The Greening of the Automotive Industry

Also by Giuseppe Calabrese

LA FILIERA DELLO STILE E LE POLITICHE INDUSTRIALI PER L'AUTOMOTIVE IN PIEMONTE E IN EUROPA

LO STATO DI SALUTE DEL SISTEMA BANCARIO E IMPRENDITORIALE NEL MEZZOGIORNO

FARE AUTO: La Comunicazione e la Cooperazione nel Processo di Sviluppo Prodotto

Other GERPISA titles:

Bernard Jullien and Lung Yannick (*editors*)
INDUSTRIE AUTOMOBILE: La Croisée des Chemins

Bernard Jullien and Andy Smith (*editors*)
INDUSTRIES AND GLOBALIZATION: The Political Causality of Differences

Elsie Charron and Stewart Paul (*editors*)
WORK AND EMPLOYMENT RELATIONS IN THE AUTOMOBILE INDUSTRY

Jorge Carrillo, Yannick Lung and Rob van Tulder (*editors*)
CARS, CARRIERS OF REGIONALISM?

Jean-Pierre Durand and Nicolas Hatzfeld (*editors*)
LIVING LABOUR: Life on the Line at Peugeot

John Humphrey, Yveline Lecler and Mario Sergio Salerno (*editors*)
GLOBAL STRATEGIES AND LOCAL REALITIES: The Auto Industry in Emerging Markets

Jean-Pierre Durand, Stewart Paul and Juan José Castillo (*editors*)
TEAMWORK IN THE AUTOMOBILE INDUSTRY: Radical Change or Passing Fashion?

Koichi Shimizu (*editor*)
LE TOYOTISME

Michel Freyssenet (*editor*)
THE SECOND AUTOMOBILE REVOLUTION: Trajectories of the World Carmakers in the 21st Century

Michel Freyssenet, Shimizu Koichi and Giuseppe Volpato (*editors*)
GLOBALIZATION OR REGIONALIZATION OF THE EUROPEAN CAR INDUSTRY?

Michel Freyssenet, Shimizu Koichi and Giuseppe Volpato (*editors*)
GLOBALIZATION OR REGIONALIZATION OF THE AMERICAN AND ASIAN CAR INDUSTRY?

Michel Freyssenet, Andrew Mair, Koichi Shimizu and Giuseppe Volpato (*editors*)
ONE BEST WAY? Trajectories and Industrial Models of the World's Automobile Producers

Robert Boyer and Michel Freyssenet (*editors*)
THE PRODUCTIVE MODELS: The Conditions of Profitability

Robert Boyer, Elsie Charron, Ulrich Jürgens and Steven Tolliday (*editors*)
BETWEEN IMITATION AND INNOVATION: The Transfer and Hybridization of Productive Models in the International Automobile Industry

Yannick Lung, Jean-Jacques Chanaron, Takairo Fujimoto and Daniel Raff (*editors*)
COPING WITH VARIETY: Flexible Productive Systems for Product Variety in the Auto Industry

The Greening of the Automotive Industry

Edited by

Giuseppe Calabrese
CNR-Ceris, Moncalieri, Italy

In association with

GERPISA: *Le Réseau International de l'Automobile*
(*International Network of the Automobile*)

Groupe d'Étude et de Recherche Permanent sur l'Industrie et les Salariés de l'Automobile
(Permanent Group for the Study of the Automobile Industry and its Employees)
École Normale Supérieure de Cachan, Paris, France

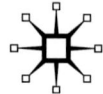

Selection and editorial content © Giuseppe Calabrese 2012
Foreword © Bernard Jullien 2012
Individual chapters © the contributors 2012

All rights reserved. No reproduction, copy or transmission of this publication may be made without written permission.

No portion of this publication may be reproduced, copied or transmitted save with written permission or in accordance with the provisions of the Copyright, Designs and Patents Act 1988, or under the terms of any licence permitting limited copying issued by the Copyright Licensing Agency, Saffron House, 6–10 Kirby Street, London EC1N 8TS.

Any person who does any unauthorized act in relation to this publication may be liable to criminal prosecution and civil claims for damages.

The authors have asserted their rights to be identified as the authors of this work in accordance with the Copyright, Designs and Patents Act 1988.

First published 2012 by
PALGRAVE MACMILLAN

Palgrave Macmillan in the UK is an imprint of Macmillan Publishers Limited, registered in England, company number 785998, of Houndmills, Basingstoke, Hampshire RG21 6XS.

Palgrave Macmillan in the US is a division of St Martin's Press LLC,
175 Fifth Avenue, New York, NY 10010.

Palgrave Macmillan is the global academic imprint of the above companies and has companies and representatives throughout the world.

Palgrave® and Macmillan® are registered trademarks in the United States, the United Kingdom, Europe and other countries.

ISBN 978–0–230–36909–2

This book is printed on paper suitable for recycling and made from fully managed and sustained forest sources. Logging, pulping and manufacturing processes are expected to conform to the environmental regulations of the country of origin.

A catalogue record for this book is available from the British Library.

A catalog record for this book is available from the Library of Congress.

10 9 8 7 6 5 4 3 2 1
21 20 19 18 17 16 15 14 13 12

Printed and bound in the United States of America
by Edwards Brothers Malloy, Inc.

'The awe, the marvel of this reality which imposes itself upon me, of this presence which reaches me, is at the origin of the awakening of human consciousness.'

Luigi Giussani, *The Religious Sense*

Contents

List of Tables	ix
List of Figures	xi
List of Abbreviations	xiii
Acknowledgements	xviii
Foreword by Bernard Jullien	xix
Notes on the Contributors	xxi

Introduction		1
Part I Innovative Design for Alternative Vehicles		
1	Innovative Design and Sustainable Development in the Automotive Industry *Giuseppe Calabrese*	13
2	Manufacturing Capability and the Architecture of Future Vehicles *Takahiro Fujimoto*	32
3	Industrial Designers and the Challenges of Sustainable Development *Monique Vervaeke*	49
4	Towards New R&D Processes for Sustainable Development in the Automotive Industry: Experiencing Innovative Design *Maria Elmquist and Blanche Segrestin*	69
Part II Technological Trajectories in Alternative Vehicles		
5	Balancing a Strong Strategic Intent and an Experimental Approach to Electric Vehicles *Florence Charue-Duboc and Christophe Midler*	89
6	'Sailing Ship Effects' in the Global Automotive Industry? Competition between 'New' and 'Old' Technologies in the Race for Sustainable Solutions *Dedy Sushandoyo, Thomas Magnusson and Christian Berggren*	103
7	Firm Perspectives on Hydrogen *Marc Dijk and Carlos Montalvo*	124
8	CNG Cars: Sustainable Mobility Is within Reach *Andrea Stocchetti and Giuseppe Volpato*	140

9   Institutional, Technological and Commercial Innovations in the Brazilian Ethanol and Automotive Industries   164
    Marcos Amatucci and Eduardo Eugênio Spers

10 New Forms of Vehicle Maker–Supplier Interdependence? The Case of Electric Motor Development for Heavy Hybrid Vehicles   185
    Dedy Sushandoyo, Thomas Magnusson and Christian Berggren

Part III Surrounding Conditions for the Development of Alternative Vehicles

11 Electric Vehicles and Power Grids: Challenges and Opportunities   207
    Ettore Bompard, Elena Ragazzi and Alberto Tenconi

12 Agreements and Joint Ventures in the Electric Vehicle Industry   225
    Giampaolo Vitali

13 Business Model Innovation and the Development of the Electric Vehicle Industry in China   240
    Hua Wang and Chris Kimble

14 Urban Mobility as a Product of Systemic Change and the Greening of the Automotive Industry   254
    Francesco Garibaldo

15 A Forecasting Framework for Evaluating Alternative Vehicle Fuels, Using the Analytic Hierarchy Process Model and Scenarios   263
    Tugrul Unsal Daim and Jubin Dilip Upadhyay

16 Consumer Attitudes Towards Alternative Vehicles   286
    Marc Dijk, Jorrit Nijhuis and Reinhard Madlener

17 The Second Automotive Revolution Is Under Way: Scenarios in Confrontation   304
    Michel Freyssenet

Appendix: The GERPISA International Network   323

Index   327

List of Tables

1.1	Cost and benefit analysis forecast for year 2020	20
1.2	Statistics on electric vehicles, European Union, 2010/11	27
7.1	The mean values of beliefs regarding hydrogen-powered vehicles	130
A7.1	Respondents to the questionnaire surveys	135
A7.2	Respondents to our interview surveys	136
A7.3	Hydrogen-powered car questions	137
8.1	Average CO_2 emissions of cars sold in Europe, by manufacturer, 1997–2010 (g/km)	141
8.2	Octane index and heat of combustion comparison of fuels	144
8.3	Average pollutant emissions by fuels, in relation to Euro 5 standards (NEDC cycle)	145
8.4	Average emissions of different fuels in an urban situation, measured by Euro standards	145
8.5	CO_2 reduction by substitution and by 'methanization'	146
8.6	Numbers of Euro 0–4 cars on the road in Italy by class, fuel type and Euro standard, 2009	147
8.7	Average CO_2 emissions of Italian Euro 0–4 fleet by standard, fuel type and class	148
8.8	Geographical distribution worldwide, known reserves of NG, 2008	149
8.9	Geographical distribution worldwide, known reserves of NG: top 20 countries, 2009	149
8.10	Natural gas production, by geographical area, 2009	150
8.11	CNG pump prices compared to petrol and diesel with equal energy content, various countries, 2011	151
8.12	Natural gas vehicles and number of CNG refuelling stations, selected countries	155
8.13	Natural gas vehicles in the European Union-27	157
9.1	Brazilian economic growth, evolution of external debt and the importance of the oil account, 1972–9	173

9.2	Phases of the use of ethanol as a fuel	181
10.1	Component and architectural knowledge development	195
A10.1	Interviews at Scania	201
A10.2	Interviews at Volvo	202
A10.3	Interviews at Voith	203
A10.4	Interviews with university professors	203
15.1	Criteria to find the best motor vehicle fuel	273
16.1	Trends in vehicle attribute preferences in the USA	290
16.2	Factors mentioned as being important in the purchasing decision	291
16.3	Sizes and characteristics of consumer sub-frames	296
A16.1	Interviews with motor companies/dealerships, 2005 and 2009	301
17.1	Number of cleaner car models on sale across the world, 2011, models to be launched by 2015, and prototypes (non-exhaustive census, June 2011)	305
17.2	Five groups of countries, according to their energy preferences for cleaner cars, with some recent changes in these preferences, as at 2011	307
17.3	Five (shifting) cleaner automobile strategies of some carmakers, giving priority to certain types of vehicle, August 2011	310
17.4	Objectives of numbers of EVs (PHEVs + EVs) on the roads of the most ambitious countries, 2015–50	313
17.5	Electric vehicle start-ups and newcomers in a range of countries (non-exhaustive census, June 2011)	316
17.6	Number of cleaner car models in the world, by types of producer: on sale in 2011, and models to be launched by 2015 (non-exhaustive census, June 2011)	319
17.7	Number of cleaner models, by type of vehicle: on sale in 2011, and models to be launched by 2015 (non-exhaustive census, June 2011)	319

List of Figures

2.1	Design-based comparative advantage	38
2.2	Basic forms of production (structure, function and operation of artefacts)	41
3.1	Diedre Design sketches for the Cybergo vehicle	61
3.2	The Cristal transport system	63
3.3	The Bluecar for the Autolib project	64
6.1	Hybrid car sales, USA, 2000–10	109
6.2	Diesel car sales, Europe and USA, 2005–10	110
6.3	Hybrid car sales in USA, Asia Pacific (mainly Japan) and Europe, 2006–9	111
6.4	Total number of hybrid, diesel and petrol engine patents, 1992–2007	113
6.5	Patents by carmakers related to hybrids, 1992–2007	113
6.6	Diesel patents related to emissions reduction, 1990–2007	114
6.7	Diesel engine patents by major carmakers, 1990–2007	115
6.8	Diesel patents by European carmakers, 1990–2007	115
7.1	Distribution of respondents' present engagement in hydrogen vehicle development	130
7.2	Relationship between frame, willingness and belief-structure regarding a specific technology	133
8.1	Average payback period of a CNG system in a petrol-driven car at various mileages per year	153
9.1	Institutional changes and governance structure mediating the market adoption of a technological innovation	167
9.2	The Brazilian ethanol and automotive innovation system	168
9.3	Brazilian flex-fuel fleet evolution: numbers of cars and light trucks using the technology, 2003–12	178
10.1	Timelines of Volvo's and Scania's hybrid vehicle developments	187
10.2	Patents granted related to power electronics, electric motors/generators and control systems for hybrid and electric vehicle applications	188

10.3	Volvo's parallel hybrid configuration	190
10.4	Scania's series hybrid configuration	192
10.5	Voith's component and architectural knowledge related to electric motors and hybrid drive systems	195
10.6	Technology and application innovations of electric motors relative to suppliers	196
10.7	Technology innovations relative to vehicle manufacturers	196
11.1	Examples of load profiles for Italy	217
15.1	The AHP model	276
15.2	The four scenarios	278
15.3	Scenario analysis and first-tier criteria comparison results	281
15.4	Scenario analysis and choice of fuel results	282
16.1	Trends of average car mass and fuel consumption	289
16.2	Frames for hybrid electric vehicles, 2000 and 2005	293
16.3	Frames for direct injection diesels, 2000 and 2005	295
16.4	Consumer sub-frames: a visualization of Table 16.3	297
A16.1	Sales figures for three Ford models in the Netherlands, 2003–7	300

List of Abbreviations

ABS	anti-lock braking system
AC	alternating current
ACEA	Association des Constructeurs Européens d'Automobiles (European Automobile Manufacturers Association)
ADEME	Agence de l'Environnement et de la Maîtrise de l'Energie (French Environment and Energy Management Agency)
AFDC	Alternative Fuels and Advanced Vehicles Data Center (USA)
Ah	ampere hour
AHP	Analytic Hierarchy Process
ANFAC	Asociación Española de Fabricantes de Automóviles y Camiones (Spanish association of Manufacturers of Automobiles and Trucks)
ANFAVEA	Associação Nacional dos Fabricantes de Veículos Automotores (National Association of Motor Vehicle Manufacturers) (Brazil)
ANP	Agência Nacional do Petróleo, Gás Natural e Biocombustíveis (National Agency for Petroleum, Natural Gas and Biofuels) (Brazil)
ASEAN	Association of Southeast Asian Nations
AV	alternative vehicle
BDVE	Business Development Véhicule Electrique (France)
BEV	battery electric vehicle
BPT	business package team (Volvo)
BRIC	Brazil, Russia, India and China group of countries
CBG	compressed biogas
CCFA	Comité des Constructeurs Français d'Automobiles
CENARGEN	Genetic Resources and Biotechnology Centre for Plants (Brazil)
CEO	chief executive officer
CEPREMAP	Centre for Economic Research and Its Applications (France)
CNG	compressed natural gas
CNGV	CNG vehicles
CNPA	Conseil National de Professions de l'Automobile
CNR/CERIS	Consiglio Nazionale di Ricerche

CNRS	National Centre for Scientific Research (Paris, France)
CO	carbon monoxide
CO_2	carbon dioxide
CTA	Comando-Geral de Tecnologia Aeroespacial (General Command for Aerospace Technology) (Brazil)
CTC	Centro de Tecnologia Canavieira (Sugarcane Technology Centre) (Brazil)
DC	direct current
DI	direct injection (system in diesel engines)
DOE	US Department of Energy
DSO	distribution system operator
EASYBAT	Easy and Safe Battery Switch in an EV
EC	European Community
ECMD	European Centre for Mobility Documentation (The Netherlands)
EDF	Électricité de France
EES	electric energy storage (or system)
EFTA	European Free Trade Association
EFTE	European Federation for Transport and Environment
EHESS	School of Higher Studies in the Social Sciences (Paris, France)
EIA	US Energy Information Administration
ELV	End of Life Vehicle Directive
EMBRAPA	Empresa Brasileira de Pesquisa Agropecuária (Brazilian Enterprise for Agricultural Research)
EPA	Environment Protection Agency (USA)
EPO	European Patent Office
ESC	electronic stability control
EU	European Union
EU FP7	The Seventh Framework Programme of the European Union for the funding of research and technological development in Europe
Euro NCAP	European New Car Assessment Programme
EV	electric vehicle
FAW	FAW Group Corporation (Chinese car manufacturing group)
FCC	Federal Communications Commission
FCEV	fuel cell electric vehicle
FCV	fuel cell vehicle
FDI	foreign direct investment
FE	front end

FFE	fuzzy front end
FFV	flexible fuel vehicle
FIEV	La Fédération des Industries des Equipements pour Véhicules (Federation of Industries of the Equipment for Vehicles)
FUPET	Future Power Electronics Technology
GE	General Electric
GEA	Gustaf Ericssons Automobilfabrik (former automobile manufacturer)???
GEIA	Automobile Industry Executive Group (Brazil)
GEM	Global Electric Motorcars
GERPISA	Groupe d'étude et de Recherche Permanent sur l'Industrie et les Salariés de l'Automobile
GERRI	Grenelle de l'Environnement à la Réunion – Réussir l'Innovation [Green Energy Revolution – Réunion Island]
Gg	gigagram
GGE	Gasoline gallon equivalent
GHG	greenhouse gas
GM	General Motors
GPS	Global Positioning System
GS Yuasa	GS Yuasa Corporation
GSM	Groupe Spécial Mobile
HC	volatile hydrocarbons
HDV	heavy duty vehicle
HEV	hybrid electric vehicle
HMI	human machine interface
HPCU	hybrid power management control unit
HV	high voltage
i-MiEV	Mitsubishi innovative Electric Vehicle
ICE	internal combustion engine
ICT	information and communication technology
IEA	International Energy Agency
INRIA	Institut National de Recherche en Informatique et en Automatique (National Institute of Automatic Control Systems and Information Technology [France])
IPI	Imposto sobre Productos Industrializados (Brazil – federal excise tax on manufactured goods)
IPT	Instituto de Pesquisas Tecnológicas (Technological Research Institute, São Paulo state, Brazil)
KWh	kilowatt hour

LCO	lithium cobalt oxide (battery)
LED	light-emitting diode
LFP	lithium iron phosphate (battery)
LG	LG Corporation
Li-Ion	lithium-ion (battery)
LMO	lithium manganese oxide (battery)
LMP	lithium-metal-polymer (battery)
LNG	liquid (or liquefied) natural gas
LOLP	loss of load probability
LPG	liquefied petroleum gas
LSEV	low speed electric vehicle
LTO	lithium titanate (battery)
LV	low voltage
MAN	MAN SE (Maschinenfabrik Augsburg-Nürnberg), a Munich-based German engineering and manufacturing company best known for its buses and heavy trucks.
M&As	mergers and acquisitions
MEA	more electric aircraft
MMA	Ministério do Meio Ambiente (Ministry of the Environment, Brazil)
MPa	megapascal
MV	medium voltage
NCA	lithium nickel cobalt aluminium oxide (battery)
NCAP	New Car Assessment Programme
NEC	Nippon Electric Company
NEDC	New European Driving Cycle
NERC	North American Electric Reliability Corporation
NG	natural gas
NiMH	nickel-metal hydride (battery)
NMC	lithium nickel manganese cobalt oxide (battery)
NMHC	non-methane hydrocarbons
N_2O	nitrous oxide
NOx	oxides of nitrogen
NPD	new product development
OECD	Organisation for Economic Co-operation and Development
OEM	original equipment manufacturer
OLED	organic light-emitting diode
OPEC	Organization of the Petroleum Exporting Countries
PHEV	plug-in hybrid electric vehicle

PLM	product lifecycle management
PM	particulate matter
PNNL	Pacific Northwest National Laboratory (USA)
QCD	quality–cost–delivery
R&D	research and development
RATP	Régie Autonome des Transports Parisiens (Independent Paris Transport Authority)
RON	research octane number
rpm	revolutions per minute
SAIC	Shanghai Automotive Industry Corporation
SCR	selective catalytic reduction
SFS	software fuel sensor
SHVC	Swedish Hybrid Vehicle Centre
SNCF	Société Nationale des Chemins de Fer Français (French National Railway Company)
SOx	sulphur/sulfur oxides
SUV	sport utility vehicle
TCO	total cost of ownership
TFM	transverse flux machine
THS	Toyota Hybrid System
TWh	terawatt hour
UAE	United Arab Emirates
UITP	International Association of Public Transport
ULEV	ultra low emission vehicle
UN FCCC	United Nations Framework Convention on Climate Change
UNICA	União da Indústria de Cana-de-Açúcar (Sugar Cane and Ethanol Industry Association) (Brazil)
USPTO	US Patent and Trademark Office
VACAR	Virginia and Carolinas, USA
VTLIB	Véolia urban transport
VU log	Web-based service tells its registered users where to find the nearest available electric car. The common pool of small shared electric cars is part of the car-sharing initiative.
V2G	vehicle to grid
VOC	volatile organic compound
w.a.	weighted average
WTW	well-to-wheel (supply path of gas/oil)
ZEV	zero emissions vehicle

Acknowledgements

I would like to thank Jullien Bernard and Tommaso Pardi of GERPISA, and Secondo Rolfo, Director of CNR-Ceris, who encouraged me to edit this book. I would also like to express my gratitude to the authors and colleagues who reviewed the chapters and, in particular, my assistant Enrico Viarisio, who revised the typescripts.

The editor and publishers would like to thank McGill-Queen's University Press for permission granted for the epigraph used in the dedication of this book; the editors of the *American Economic Review* for the permission granted for the epigraph used in Chapter 9; the *Journal of Industrial and Business Economics* edited by Franco Angeli for allowing part-reutilization of the paper by Takahiro Fujimoto, 'Complexity explosion and capability building in the world auto industry: an application of design-based comparative advantage' (vol. 38, no. 2, pp. 25–49) revised for Chapter 2; Diedre Design, Lohr Industrie Group and Autolib for permission to use the illustrations in Chapter 3; Dutch branch organisation BOVAG for permission to use the graph in Chapter 16. Every effort has been made to contact all the copyright-holders but if any have been inadvertently omitted the publishers will be pleased to make the necessary arrangement at the earliest opportunity.

Foreword

GERPISA, the international network for research on the automobile, developed between 2007 and 2011 its fifth research programme, entitled 'Sustainable Development and the Automobile Industry'. Its aim is to understand the extent to which companies and states have taken into account the exigencies of sustainable development, and how they are doing so. More precisely, the intention is to measure the scale and pace of the changes that increasingly important political and social requirements have placed on the industry. These changes are related to different interlinked dimensions: as often stressed by the literature on sustainable development, public and corporate actors have indeed to develop synergies (but also to make trade-offs) between environmental performance, economic performance and social progress.

This is why, since 2007, GERPISA has linked the analysis of the quest for better vehicle performances in terms of emissions with a more general examination of the other constraints that have a simultaneous effect on the development of corporate strategies and of public policy regulations. The 2008–09 crisis has been a brutal reminder of the importance of these other constraints. Environmental sustainability is certainly important, but the economic sustainability of the products sold by the carmakers and of the processes involved in production is still a problematic dimension of automobile production worldwide. This economic sustainability, as GERPISA has always stressed, is linked both to the microeconomic capability of firms to make a profit as well as to develop coherent strategies within their own macroeconomic environment on the one hand, and in relation to employment conditions of workers, on the other.

This collective work edited by Giuseppe Calabrese is the first book published from GERPISA's 5th international research programme. Its main focus is on the issues related to technological innovations aimed at achieving better environmental performance of automobile products. It approaches these questions from a wide variety of points of view. Such an approach reflects not only the value-added generated by the mobilization of an international network to explore such a complex set of questions, but also the interest to the scientific community working on sustainable development of interacting with a network of researchers who have developed a detailed knowledge of the automobile industry and its history. By combining this expertise about the industry with an analysis of the dynamics implied by the integration of the environmental exigencies, the book provides an insightful measure of the changes that are taking place.

It shows in particular that, beyond their technological relevance, these changes have already acquired a new economic and geopolitical dimension: what is at stake today, beyond the reduction of emissions, is the capacity of this industry to make a 'second revolution'; to find, in the context of the extraordinary growth of production and demand in emerging countries, ways to a renewed sustainability. If these ways are still far from clear in 2012, the contributions gathered in this book allow us to grasp those that are emerging and the automotive worlds to which they might lead.

BERNARD JULLIEN
Director, GERPISA International Network
Professor of Economics, École Normale Supérieure de Cachan

Notes on the Contributors

Marcos Amatucci is Associate Dean of Research at Escola Superior de Propaganda e Marketing (ESPM), Brazil and International Management and Innovation Management Full Professor at ESPM-SP (ESPM, central São Paulo, Brazil), researching innovation, internationalization and sustainable innovation in the automotive and correlate industries. His main topics concerning the automotive sector are internationalization strategies and system innovation in transport technology. His most recent publication related to the automotive industry is M. Amatucci and E. E. Spers, 'The Internationalisation of the Automobile Industry and the Roles of Foreign Subsidiaries', *International Journal of Automotive Technology and Management*, 10(1) (2010), 37–55.

Christian Berggren is Professor of Industrial Management at the University of Linköping, Sweden and Director of the KITE research programme, 'Knowledge Integration and Innovation in Transnational Enterprise'. He has written extensively on production systems, and product development and innovation, especially within the automotive, electro-technical and telecommunications industries. Published titles include *The Volvo Experience* (Macmillan, 1992); *The Resilience of Corporate Japan* (Sage, 1997); and *Knowledge Integration and Innovation: Critical Challenges Facing International Technology-Based Firms* (Berggren *et al.*, Oxford University Press, 2011). His current research focuses on the competition for sustainable vehicles in the automotive industry, innovation processes and individual innovators, and the role of regulation in driving innovation.

Ettore Bompard is an Associate Professor of Power Systems at Turin Polytechnic. He is co-ordinating several European and international projects in the field of electricity markets restructuring (generation scheduling, congestion management design, security impacts on the competitive markets, electricity markets simulation). His research interests include power systems security, electrical energy efficiency, smart grids, and electricity market analysis and simulation. He has co-authored more than 100 publications on various topics related to power systems analysis.

Giuseppe Calabrese is a senior researcher at CNR-Ceris (the National Research Council's Institute for Economic Research on Firms and Growth) in Moncalieri, Italy, and teaches as a Visiting Professor of Managerial Economics at the University of Turin. He is co-editor of the *International Journal of Automotive Technology and Management* and a member of the International Steering Committee of GERPISA. His main topics concern new product development and production networks, the role of small to medium firms in the reorganization of the car supply base, R&D organization, and alternative vehicle and car styling. His most recent publication is 'Structure and Transformation of the Italian Car Styling Supply Chain', forthcoming in the *International Journal of Vehicle Design*.

Florence Charue-Duboc is a Senior Researcher at CNRS (the National Center for Scientific Research) and Professor at the École Polytechnique ParisTech, Paris, France. Her research work deals with innovation management in large firms in sectors such as the chemical, pharmaceutical and automobile industries. She investigates the management of major innovations with a strong technical and scientific component and has focused on various phases: the generation of new concepts and competences, the exploration of emerging markets, and the first market introductions targeting the early market and selected customers. The role and specificities of inter-firm co-operation in these exploratory processes is analysed specifically. She has conducted several empirical analyses of new product developments in various firms, and has published several articles and edited two books.

Tugrul Unsal Daim is a Professor and Ph.D. Programme Director in the Department of Engineering and Technology Management at Portland State University (PSU), Oregon, USA. Prior to joining PSU, he worked at the Intel Corporation for over a decade in varying management roles. He has been a consultant to several organizations in sectors ranging from energy to medical device manufacturing. He is also a Visiting Professor with the Northern Institute of Technology at the Technical University of Hamburg, Germany. He has recently been appointed as Extraordinary Professor at the Graduate School of Technology Management at the University of Pretoria in South Africa. He has published over 200 refereed papers in journals and conference proceedings. He is the Editor-in-Chief of the *International Journal of Innovation and Technology Management* and North American Editor of *Technological Forecasting and Social Change*.

Marc Dijk is a Research Fellow at Maastricht University, The Netherlands. He has developed a model for analysing paths of innovation in car mobility. His micro–macro framework with co-evolution of demand and supply emphasizes feedback effects and stakeholder perspectives, combining evolutionary economics with the sociology of technology. He has determined the framework for the case of electric and hybrid-electric engines on the automobile market after 1990. He believes that the main merit of a co-evolutionary, micro–macro approach is the integrated analysis of consumers and firms, and competition between technologies.

Maria Elmquist is an Associate Professor at the Department of Technology Management and Economics and a Senior Researcher at the Center for Business Innovation (CBI), part of Chalmers University of Technology,

Gothenburg, Sweden. She is also an affiliated researcher at the Centre for Management Sciences at MINES ParisTech, in Paris, France. Her current academic research is focused mainly on the management of innovation, organization of R&D and innovation, the relationship between design and innovation, and processes for strategic change. Ongoing research includes projects on the development of innovation capabilities and the management of open innovation. Her research has appeared in journals such as *Research Policy*, *R&D Management* and *Creativity and Innovation Management*.

Michel Freyssenet is Honorary Research Director at CNRS (National Center for Scientific Research) in Paris, France. He is co-founder of GERPISA and currently a member of its international steering committee. His main topics are productive models, national growth models, world productive recomposition, the history of work division, and social relationships theory. His main publication in English was with co-author R. Boyer, *The Productive Models. The Conditions of Profitability* (Palgrave, 2002). His most recent edited book is *The Second Automobile Revolution: Trajectories of the World Carmakers in the 21st Century* (Palgrave, 2009). Most of his texts can be downloaded from the website: http://freyssenet.com.

Takahiro Fujimoto is currently Professor in the Graduate School of Economics at the University of Tokyo, Japan, a position he has held since 1998. He has also served as Executive Director of the Manufacturing Management Research Center at the University of Tokyo since 2003. He specializes in technology and operations management, as well as business administration. He holds a Ph.D. from Harvard Business School (1989), where he served as a researcher following graduation and later as a Visiting Professor (1996–7) and a Senior Research Associate (1997). He has also served as an Associate Professor on the Faculty of Economics at the University of Tokyo, a Visiting Professor at Lyon University, France and a Visiting Researcher at INSEAD.

Francesco Garibaldo is an industrial sociologist by education, a former Director of IPL, and is now a member of the scientific direction of IRES E/R (the Institute of Economic and Social Research, Emilia-Romagna), Bologna, Italy. His latest publications on the automotive industry are a book co-authored with P. Morvannou and J. Tholen (eds) *Is China a Risk or an Opportunity for Europe? An Assessment of the Automobile, Steel and Shipbuilding Sectors* (Peter Lang, 2008) and 'Le politiche industriali', in G. Volpato and F. Zirpoli (eds) *L'auto dopo la crisi*, (Francesco Brioschi Editore, 2011).

Chris Kimble is an Associate Professor of Strategy and Technology Management at Euromed Management, Marseille, is affiliated to MRM-CREGOR at Université Montpellier II, France, and is the academic editor for the journal *Global Business and Organizational Excellence*. Before moving to France, he lectured in the UK on Information Systems and Management at the University of York, and Information Technology at the University of Newcastle. His research interests are business strategy and the management of the fit between the digital and social worlds. His most recent publication is, with Hua Wang, 'Leapfrogging to Electric Vehicles: Patterns and Scenarios for China's Automobile Industry', *International Journal of Automotive Technology and Management*, 11(4) (2011), 77–92.

Reinhard Madlener is Full Professor of Energy Economics and Management at RWTH Aachen University, Co-Director of the energy section of the Jülich Aachen Research Alliance (JARA Energy), and Research Professor at the German Institute for Economic Research (DIW Berlin), Germany. His teaching focuses mainly on energy economics, environmental economics, economics of technical change, and economics of technological diffusion. He is an associate editor of *Applied Economics Quarterly*; *Applied Energy*; *Energy Systems*; *Energy, Sustainability and Society*; the *International Journal of Energy Sector Management*; and *Sustainable Cities and Society*. His main research interests are in energy economics and policy, as well as the adoption and diffusion of technological (energy) innovation under uncertainty.

Thomas Magnusson is an Associate Professor at the Department of Management and Engineering, University of Linköping, Sweden. He is also involved in the KITE research programme, 'Knowledge Integration and Innovation in Transnational Enterprise'. His interests relate to environmental innovation, technology strategy and the management of product development, with a particular focus on complex products and mature industry segments. Recently published articles have appeared in *Technology Analysis and Strategic Management*; *Industrial and Corporate Change*; *Journal of Cleaner Production*; and the *International Journal of Innovation Management*. Currently he is involved in a study of technology and sourcing strategies for the development of hybrid electric vehicles.

Christophe Midler is Research Director at the Management Research Centre and Professor and Chair of Innovation Management at the École Polytechnique, Paris, France. He is a member of two international research networks related to the automobile industry: GERPISA and the International Motor Vehicle Programme (IMVP). His research topics are product development, project management and innovation strategy. With his team, he has explored these topics in various industrial contexts and in particular in the automobile industry. His preferred methodology is long-term interactive researches with firms. His most recent publication is, with R. Beaume, 'Project-based Learning Patterns for Dominant Design Renewal: The Case of Electric Vehicle', *International Journal of Project Management*, 28 (2010), 142–50.

Carlos Montalvo is Senior Scientist on Industrial and Innovation Policy at TNO, The Netherlands. His research output gives support to the European Commission in several key research and technical development (RTD) and innovation actions and policy. Since 2001 he has been the Subject Editor on innovation and the environment for the *Journal of Cleaner Production*. His work on Behavioural Innovation Economics has recently been recognized as pioneering in the literature of innovation studies. He current research interest spreads over: the evaluation of innovation and RTD policy; R&D and structural change; innovation and the environment, innovation and regulation, technology adoption and diffusion analyses, and the application of behavioural dynamic models to explore the interaction between actors influencing innovation and change. He is author of the forthcoming *Analysis of Market and Regulatory Factors Influencing Innovation Patterns in the Automotive Sector* (Brussels: European Commission DG Enterprise and Industry).

Jorrit Nijhuis is an environmental scientist who has worked as a Ph.D. Researcher at the Environmental Policy Group of Wageningen University, The Netherlands, since 2005. He is participating in a research programme on sustainable lifestyles and consumption patterns financed partly by the Knowledge Network on System Innovations (KSI). He is currently finalizing his thesis on the role of citizen-consumers in transition processes to sustainable mobility. One of the case studies focuses on consumer-oriented policies such as environmental labelling and subsidies in automotive production–consumption chains. Since 2009 he has been employed at the Dutch Highways Agency (Rijkswaterstaat) as a consultant in the fields of mobility management and freight transport. His most recent publication is, with S. Van den Burg, 'Consumer-oriented Strategies for Car Purchases. An Analysis of Environmental Information Tools and Taxation Schemes in the Netherlands', in T. Geerken and M. Borup (eds) *System Innovation for Sustainability 2, Case Studies in Sustainable Consumption and Production – Mobility* (Greenleaf, 2009).

Elena Ragazzi is a Researcher at CNR-Ceris (the National Research Council's Institute for Economic Research on Firms and Growth) in Moncalieri, Italy, and teaches as a Visiting Professor of Business Economics at Turin Polytechnic. She leads the Ceris research on the evaluation of training policies, and is a member of several scientific committees on this topic. Her main research interests include the analysis of competitive electricity markets, and policy evaluation (above all, local development and training policies, and regulation of the electricity market).

Blanche Segrestin is a Professor at MINES ParisTech in the Center for Management Science (CGS), Paris, affiliated to the Chair of Theory and Method for Innovative Design. She works on innovation management in close collaboration with industrial partners. Her research interests lie in innovative partnerships, both within and between firms, and corporate governance by articulating the theory of the firm to innovation capabilities. Recent publications include a book on innovative inter-firm co-operation: *Innovation et Coopération interentreprises*, CNRS Editions); and edited a special issue on the governance of the firm ('Quelle norme pour l'entreprise?', *Entreprises et Histoire*). She has published several articles on the automotive industry in *Research Policy*; *R&D Management*; and *Creativity and Innovation Management*.

Eduardo Eugênio Spers is Associate Professor and Researcher on the M.Sc. course in International Management at Escola Superior de Propaganda e Marketing (ESPM), São Paulo, Brazil. He teaches International Marketing and International Marketing Research and was a Visiting Professor at Wageningen University, The Netherlands. His main topics of interest in the automotive sector are ethanol chain production and consumer biofuels purchase behaviour and perceptions. His most recent publication concerned with the automobile industry is, among others: E. Ferrato, R. Q. Carvalho, E. E. Spers and N. K. Pizzinatto, 'Relacionamento interorganizacional e hold-up no setor automotivo: um estudo sob o enfoque da economia dos custos de transação', *Revista de Gestão USP*, 12 (2006), 75–87.

Andrea Stocchetti is Associate Professor of Business Economics and Management at Ca' Foscari University, Venice, Italy. He is involved in several national and international research projects related to the automotive industry. His main research interests include competitive analysis, sustainability management and sustainable mobility. As a result of his empirical and theoretical research activities, he has published two books and around 50 national and international contributions in the form of chapters in books, articles and conference proceedings.

Dedy Sushandoyo is a Ph.D. student at the Department of Management and Engineering, University of Linköping, Sweden. He undertook his undergraduate training in Mechanical Engineering with a specialization in Energy Engineering at the Gadjah Mada University, Indonesia. He then took a Master's degree in Engineering at Chalmers University of Technology in Gothenburg, Sweden. He has had papers published in the *International Journal of Automotive Technology and Management*. His research is focused on strategies for developing hybrid powertrains in the automotive industry.

Alberto Tenconi is currently a Full Professor in the Department of Electrical Engineering at Turin Polytechnic. Between 1988 and 1993, he worked at the Electronic System Division of the Fiat Research Center (CRF) at Orbassano, Italy, where he was engaged in the development of electrical vehicle drive systems. His research activity is documented by more than 150 papers published in international journals and via international conferences. He has participated, both as a designer and a scientist, in many national and European research programmes. He is a reviewer for international journals and has been Associate Editor for the Institute of Electrical and Electronics Engineers' (IEEE) *Transactions on Industrial Electronics*. His current research interests include high-performance electric machines and drives for transport applications.

Jubin Dilip Upadhyay is currently a Ph.D. student focusing on renewable energy in the Engineering and Technology Management Department of Portland State University, Oregon, USA. He has an extensive academic background with multiple degrees at various levels of chemical engineering studies, reinforced with detailed studies in bio-technology, artificial neural networks and biofuels. He has had several years of experience as a business manager (India) and as an analyst with the Computer Sciences Corporation (USA).

Monique Vervaeke is a sociologist and researcher at the Centre Maurice Halbwachs (CNRS), Paris, France. Before studying the professional role of industrial designers in France and the influence of design schools on this process, she did research on housing policies and industrial strategies. Currently her main interests concern the design and innovation strategies of companies. Her research also deals with relationships between engineers and industrial designers, and the link between the cultural economy and industrial design. She is co-author with C. Midler and G. Minguet of *Working on Innovation* (Routledge, 2010).

Giampaolo Vitali is a Senior Researcher at CNR-Ceris (the National Research Council's Institute for Economic Research on Firms and Growth) in Moncalieri, Italy. He is Visiting Professor of European Economics at the University of Turin. His main research topics are industrial economics, local development policy and innovation economics.

Giuseppe Volpato, 1943–2012, was formerly a Full Professor of Industrial Economics at the Ca'Foscari University, Venice, Italy. He was a Research Fellow of the International Motor Vehicle Program (IMVP) of the Massachusetts Institute of Technology (MIT), Boston, USA, member of the steering committee of GERPISA, and a Senior Adviser on the International Car Distribution Programme (ICDP). He was also a member of the editorial board for many national and international academic journals.

Hua Wang is an Associate Professor at Euromed Management École de Marseille, France. His research interests and teaching centre on innovation management, foreign direct investment, clusters and metropolitans, the globalization strategy of Chinese companies, the automotive industry in China, and doing business in China. His most recent publication on the automotive industry is 'Low-cost Strategy through Product Architecture: Lessons from China', *Journal of Business Strategy*, 31(3) (2010), 12–20.

Introduction

Giuseppe Calabrese

The automotive industry has a strong presence in the global economy and employs a significant proportion of the working population. It has been contributing to the growth of modern society by satisfying everyday mobility. However, it has been accused of affecting the environment and public health. In Europe, CO_2 emissions from the transport sector have increased by 29 per cent since 1990, whereas those of other sectors have decreased by 22 per cent. The contribution of the transport sector to the EU's CO_2 emissions now stands at 30 per cent, up from 20.5 per cent in 1990.

This book presents the results of a four-year research programme on sustainable development in the automotive industry. The programme was co-ordinated by GERPISA (Groupe d'Étude et de Recherche Permanent sur l'Industrie et les Salariés de l'Automobile), an international network of researchers in economics, management, history and sociology, all studying the automotive industry.

As Bernard Jullien, Director of GERPISA, stated when presenting the programme, the question of how the automotive industry is integrating the demands of sustainable development is fundamentally tied to the question of how this activity is positioned in societies that produce and/or use automobiles. In particular, the scale on which the position of the automotive sector is currently being renegotiated in society requires us to return to the question of politics, states and the importance of regulation and taxation, as these issues are likely to play a major role in determining outcomes for automotive firms and regions. This can be summarized in three major points:

- The automotive industry presents a degree of unity and continuity that reflects a sectorial community characterized by competition and imitation. Sustainable development has become one of the key trends that concern all firms in the sector.

- Carmakers have experimented specific historical trajectories in terms of strategies and production policies that create trade-offs in the interpretation of the requirements of sustainable development.
- The strategies adopted by firms need to be interpreted more broadly than in terms of competitive analysis and performance. Sustainable development strategies need to be evaluated in order to take up a long-term position in an increasingly global social, political and economic landscape.

In this book, the concept of sustainable development has been correlated with the main theoretical framework of production analysis and managerial economics – that is, manufacturing and architecture theories; the theory of comparative advantage of design location; design-driven and design-thinking theories; concepts of knowledge models; rule-based and innovative design regimes; path dependency theory; literature on breakthrough and disruptive innovations; studies on technology competition; reasoned action and planned behaviour theories; institutional theoretical approaches; firm growth theories; the smart grid paradigm; business model innovation; definition of scenarios through analytic hierarchy process models; and consumer framing. Moreover, the authors have described the concept of sustainable development in the automotive industry by avoiding biases or conditioning, but elucidating their assumptions thoroughly, and converging analyses can be detected.

The contents of the book have been selected from among the contributions presented at the most recent GERPISA conferences, reviewed by the editor and by anonymous reviewers and improved by the authors in the course of the programme. The first conference was held in Moncalieri (Turin), Italy, hosted by CNR-Ceris in June 2008. The second conference was held in Paris, France, hosted by GERPISA in June 2009. The third conference was held in Berlin, Germany, hosted by WZB (The Social Science Research Center) in June 2010. In total, more than 200 contributions were presented. Finally, in June 2011, in the most recent GERPISA conference held in Paris, France, revised and updated versions of the selected papers were presented and discussed further. The final versions of the contributions were received in the autumn of 2011.

The book is divided into three parts and 17 chapters, written by 28 authors from 11 different countries on three continents: Asia, Europe, and North and South America. The first chapter, *Innovative Design and Sustainable Development in the Automotive Industry*, by the editor of this book, attempts to summarize the four-year programme, focusing on the automotive industry's technological trajectories and its race to produce alternative vehicles, and underlining pros and cons. It analyses the role of new strategic players in the automotive industry and reports on some policy issue implications.

Part I: Innovative design for alternative vehicles

This part of the book comprises three chapters and is dedicated to innovation design in sustainable development with regard to vehicle architecture; style and design concepts; and research and development (R&D) organization.

Takahiro Fujimoto, in *Manufacturing Capability and the Architecture of Green Vehicles*, investigates the architectural differences between purely electric vehicles and hybrid electric vehicles, and discusses the limitations of the former. The design of alternative vehicles is described within the wider contest of manufacturing capability and product-process architecture. The chapter provides an explanation of the theory of comparative advantage of design location in the context of theories of organizational capability and architecture. There are substantial differences in the architecture of each car segment, along a continuum from integral types to modular types, and countries and areas tend to enhance their own strengths based on their historical development. Nowadays, requirements and constraints imposed by sustainable development in the automotive industry are becoming stricter every year, and as a consequence, product design is becoming more complex, avoiding the commodification that happened with electronic products. It is therefore difficult to deal with them by using modular-type design concepts, in which functionally complete parts are accumulated, and a new array of integral-type design parts that have been finely adjusted for total optimization is required, so that integral-type architecture is more suitable. On the other hand, electric vehicles are modular-based but, at the time of writing, too expensive.

Monique Vervaeke, in *Industrial Designers and the Challenges of Sustainable Development*, moves the analysis from innovation on technology to industrial design, and to new meanings associated with products or services. The chapter analyses both the debate between eco-design and design thinking, which began in the 1970s, and the role of industrial designers in carmakers' strategies. From a sustainable development perspective, the designers' creative approach includes reflections about the productive process as a whole, and aesthetic research on the properties of lighter materials, as well as the choice of parts, their assembly, and their disassembly for recycling – symbolic aspects of the construction of new markets. The scope of sustainable development in the automotive sector is therefore broader than simply designing cars with reduced exhaust emissions: it also involves looking for new ideas in urban mobility. This chapter puts forward the hypothesis that in the automotive sector, faced with massive social pressure to transform its capital accumulation model, actors other than the dominant carmakers are creating new opportunities and have the chance to earn a place in the value chain. The chapter also presents three case studies of new mobility concepts.

Maria Elmquist and Blanche Segrestin, in *Towards New R&D Processes for Sustainable Development in the Automotive Industry: Experiencing Innovative*

Design, interrelate sustainable development in the automotive industry with complementary classical R&D processes and innovative design strategies. Carmakers must review the values and performance included in their offer, and avoid focusing only on the reduction of vehicle pollutants. To succeed in doing this, new R&D strategies and managerial techniques are needed. In this chapter, some experiences from a collaborative research study with a European car manufacturer are shared. The project used the so-called Knowledge–Concept–Prototype method, based on the Concept–Knowledge theory of design, as a way to try to explore how the cars of the future can at the same time be environmentally friendly and economical, hereafter labelled 'eco-eco'. The case provides an example of how carmakers could operate to complement their work in the R&D process with an innovative design approach, in order to build better prerequisites for a more innovative output, which is how an expansion from rule-based to innovative design in terms of new environmental values can be organized in the automotive industry.

Part II: Technological trajectories in alternative vehicles

The second part of the book comprises six chapters and focuses on the contextualization of the technological trajectories regarding alternative vehicles from the managerial and industrial point of view, in some cases reporting carmakers' attitudes. The most debated alternative vehicles have been considered: pure electric; hybrid electric; fuel cell; natural gas, particularly compressed natural gas; and biofuel, particularly ethanol. Internal combustion engines – petrol and diesel – have also been examined in terms of sustainable development. A chapter is dedicated to trucks and buses as special alternative vehicles with peculiar characteristics different from those of passenger cars.

Florence Charue-Duboc and Christophe Midler, in *Balancing a Strong Strategic Intent and an Experimental Approach to Electric Vehicles,* examine electric vehicles in terms of environmental breakthrough innovations. They contextualize their analysis by broadening the scope of traditional projects: product usage contexts have to be redefined; users have to learn new practices; business models need to value collective environmental externalities; and traditional value-chains must be deeply extended and reshaped. Therefore, managing such innovations requires both a consistent strategic intent and a capability to learn. Based on the case of the Renault Electric Vehicle Programme, the chapter focuses on the role of pilot deployments in managing such dynamics. It highlights the way that pilot deployments support the structuring of local ecosystems, the development of collective learning capability, and the creation of recoverable assets having a prescriptive power. They provide leverage for building a micro-local ecosystem, an aspect emphasized in the literature on environmental initiatives but not often mentioned in project literature, and connections between micro-local pilot projects and the generalization towards a broader ecosystem. The authors clearly identify the difficulties, uncertainties and possible options as they arise.

Dedy Sushandoyo, Thomas Magnusson and Christian Berggren, in *'Sailing Ship Effects' in the Global Automotive Industry? The Competition Between 'New' And 'Old' Technologies in the Race for Sustainable Solutions*, focus on the near future and on the competition between the well-established but still vital diesel technology and its new contender, petrol-electric hybrid vehicles. The comparison is based on technological product development and process competence. This is followed by an analysis of the different innovations involved, architectural versus modular, of how they relate to the integration capabilities needed in the future, and of the sourcing and supply strategies used by the main actors involved in this competition. A general problem encountered by studies of technology competition is their tendency to reach obvious conclusions: the new technology wins, even though it may take a long time. According to the authors, depending on fuel prices and future regulations, diesel hybrids stand a good chance of offering a cost-competitive and fuel-efficient vehicle alternative for the next 5–10 years. Thus the old diesel technology will continue to evolve, partly as a result of being challenged by other technology options. The chapter also investigates competition among technological trajectories, and the major firms involved.

Marc Dijk and Carlos Montalvo, in *Firm Perspectives on Hydrogen*, look into the framing of firms by studying their belief systems and actual engagement. The authors pose two main questions: Which underlying beliefs drive engagement significantly in the development of hydrogen technology? And are there differences among firms? Hydrogen technology is described through the eyes of carmakers, and their technological frames are explained through their attitudes towards hydrogen engines, the perceived social pressure to implement hydrogen models, and their perceived level of control over the implementation process. The outcomes suggest that different firms deal with hydrogen vehicle development in different ways. Three types of firms can broadly be distinguished in relation to their engagement with hydrogen technology: uncertain, unwilling, and optimistic. Furthermore, the authors find that the firms' level of technological and organizational capability seem to be related most strongly to their actual engagement in hydrogen technology development. Finally, this contribution addresses the relationship between actor beliefs and actor framing, concluding that a carmaker will wait before launching a new product, above all fuel-cell vehicles, until a profitable outlook can be envisaged.

Andrea Stocchetti and Giuseppe Volpato, in *CNG Cars: Sustainable Mobility Is Within Reach*, describe the opportunities related to the diffusion of compressed natural gas (CNG) as an alternative fuel for vehicles, which is seen as the most effective option to achieve sustainability targets in road transport quickly.

CNG is less costly and less polluting than conventional fuels; moreover, it can be used in common petrol engines with relatively inexpensive adaptations. Many carmakers are already offering one or more CNG bi-fuel vehicles, and in countries that already have a network of methane pipelines, the needed infrastructure can be developed in a short time and with a fairly low level of investment. The world's known reserves of gas will be able to satisfy demand for the next 50 years at least, and in the long run CNG might come to be included among renewable resources, since it will be obtained from biogas. According to the authors, incentives to equip vehicles with CNG propulsion devices would be even more effective than supporting the purchase of new cars against the scrapping of old ones. The technology associated with this kind of fuel is able to be developed further, producing interesting innovations from the economic, productive and environmental points of view. Moreover, the authors estimate the reduction of CO_2 that would occur in the Italian car fleet if the oldest cars were converted to use CNG.

Marcos Amatucci and Eduardo Eugênio Spers, in *Institutional, Technological and Commercial Innovations in the Brazilian Ethanol and Automotive Industries*, shed light on the Brazilian approach to sustainable development strategies and on the role of both the governance structure and the national innovation system in the transition to find alternatives to fossil fuels. The authors describe how ethanol-related innovations have spread throughout the automotive industry and the ethanol value chain, by looking at the rapid growth of the ethanol sugar cane industry, the emergence of the ethanol engine and its infrastructure, and flex-power technology. They also present a model, based on technological, commercial and institutional innovations for ethanol and the automotive industry, which considers the evolution from technological innovation to market adoption with institutional changes governing the process. These developments have led to an interaction between automotive carmakers and the sugar cane agribusiness. The chapter also explores innovations and path dependence in the automotive chain; that is, the ethanol vehicle and the flex-fuel engine. To conclude, the authors provide some remarks on the lessons that can be learnt from the Brazilian experience.

Dedy Sushandoyo, Thomas Magnusson and Christian Berggren, as indicated in the title of their contribution *New Forms of Vehicle Maker–Supplier Interdependence? The Case of Electric Motor Development for Heavy Hybrid Vehicles*, provide an analysis of commercial vehicles – trucks and buses – as special alternative vehicles with peculiar characteristics different from those of passenger cars. The aim of this chapter is to investigate key issues for heavy vehicle makers in their collaborations with electric motor suppliers, by presenting an in-depth study of hybrid product development carried out by two heavy vehicle makers together with their respective suppliers of electric motors. The volume of heavy hybrid vehicles is expected to be very low, so collaborations with suppliers are essential because diseconomies of scale are likely. These two vehicle makers used different types of electric motors and different types of hybrid architecture for their hybrid powertrains. By analysing similarities and differences in these two collaborations, the chapter underlines three essential issues: the selection of suppliers with complementary technological knowledge; the development of an internal technological knowledge base to create sufficient overlap with suppliers; and the adaptation of product development collaborations to the knowledge contributions of the organizations involved.

Part III: Surrounding conditions for the development of alternative vehicles

The third part of the book comprises seven chapters and looks at the conditions for the development of green vehicles, in particular electric vehicles, which represent a more disruptive technology when compared to the current regime. The aspects analysed are: the impact of electric vehicles on the grid; the agreements and joint ventures between carmakers and suppliers specializing in electric vehicles; and the role of China – the main national player in electric vehicles development. The last four chapters in the book contextualize and outline the development of alternative vehicles for what concerns the mobility structure, the definition of a forecasting framework, possible consumer attitudes, and a description of likely scenarios.

Ettore Bompard, Elena Ragazzi and Alberto Tenconi, in *Electric Vehicles and Power Grids: Challenges and Opportunities*, pose a fundamental question about the diffusion of electric vehicles: the complex management of the electricity supply. In particular, three aspects are considered. The first is the evaluation of generation capacity: will it be enough in case of a huge and rapid diffusion of electric vehicles? Second, an evaluation of distribution network constraints is provided: will the grid support wide and dispersed connections for recharging batteries? Finally, an integrated vision of electricity supply and vehicle energy efficiency is indispensable for a correct evaluation of the environmental impact. In this chapter, the authors illustrate the basic categories of electric vehicles and refuelling options, discussing their impact on the electricity system. Then they introduce the future structure of electricity grids, discussing two aspects of electric vehicles, where they are able to interact actively with the network: when they are connected to the grid they can be used to store electricity, which can later be discharged to support the grid. They also outline possible advantages and disadvantages of a massive diffusion of electric vehicles for the electric power systems and distribution grids.

Giampaolo Vitali, in *Agreements and Joint Ventures in the Electric Vehicle Industry*, describes the growth strategies of the companies involved in the emerging industrial sector of electric vehicles. To deal with this challenging scenario, firms need to be very flexible as far as their growth strategies are concerned. They cannot afford to follow a high-risk strategy, or a high-sunk-costs strategy, because the technology scenario might change in a few years, depending on changes in technology. This is why strong relationships between carmakers and suppliers specializing in electric vehicle components have been established to develop new technologies and new components. Partnerships can be set up between manufacturing firms and public institutions, as well as between manufacturing firms and electricity producers. This chapter describes some agreements and joint ventures agreed by firms in the electric vehicles industry and its downstream area, in order to shed light on the importance of growth strategies based on agreements and joint ventures. There are significant implications not only for companies but also for policy-makers who intend to promote this emerging industry.

Hua Wang and Chris Kimble, in *Business Model Innovation and the Development of the Electric Vehicle Industry in China*, examine China's capacity for business model innovation rather than technological innovation. Their assumption is that the development of new energy vehicles is a significant technological challenge, but there are numerous examples to show that the successful adoption of technology involves more than merely producing a technologically elegant solution. While business models from emerging economies have not been the focus of many studies, the authors believe there are two reasons why an improved understanding of their development might be of interest: these reasons are not sufficient but are necessary for the success of the business models, and they have the potential to be disruptive and to challenge Europe and America's fascination with the internal combustion engine. The authors present a case study on one of the largest producers of low-speed electric vehicles in China. In terms of business models, this is an example of a product that is still in the process of being defined; in terms of business model innovation, the case study illustrates a market that has grown without the support of the central government and outside the boundaries of the mainstream automobile industry.

Francesco Garibaldo, in *Urban Mobility as a Product of Systemic Change and the Greening of the Automotive Industry*, places the technological trajectories of alternative vehicles within the context of the car-driven mobility crisis. Some dimensions and features are selected and highlighted in this chapter. They are all deeply interconnected in a dynamic way, but two drivers of change in particular are illustrated: the networked social dimension of all kinds of mobility systems; and the role of utility value. In the author's opinion, what is relevant is not the technical means to ensure mobility, but mobility as a social asset and an individual right to be safeguarded. The focus is on designing both the urban context and infrastructures for mobility, with cars being just a part of a broader system of sustainable development. This implies the necessity for vertical integration between the traditional automobile sector and the emerging urban design and management sector. The automobile industry needs a new industrial perspective capable of ensuring mobility both within and between cities. The author supposes the creation of a new segment, the city car, based on clean powertrains, which is of the utmost importance for the future of the car industry as well as for urban mobility.

Tugrul Unsal Daim and Jubin Dilip Upadhyay, in *A Forecasting Framework for Evaluating Alternative Vehicles by Using the Analytic Hierarchy Process Model and Scenarios*, begin their analysis from the assumption that the greening of the automotive industry is ascribable to constant uncertainties concerning oil prices, and the chapter examines the reasons behind oil price fluctuations. The Analytic Hierarchy Process model is then presented. The model has different levels, like those of a tree. The lowest level includes all the existing motor fuels, each of which is briefly described. Issues such as cost and pollution impact are assessed for each of these alternatives. In the next level of the model, the criteria are introduced, including the factors that affect the adoption of each alternative, such as economic, cultural, environmental, sustainability and development time criteria. The topmost level of the tree model presents the goal of this study: finding the best motor fuel for the future. To prove the robustness of each alternative, four different scenarios are applied to a 40-year time span (status quo; environmental challenge; economic challenge; and catastrophe) and quantitative results are presented for each scenario.

Marc Dijk, Jorrit Nijhuis and Reinhard Madlener, in *Consumer Attitudes Towards Alternative Vehicles*, wonder if the current sale trends in alternative vehicles truly reflect a more fundamental change in demand for car engines or are merely the result of stricter environmental regulations on car emissions and environmental subsidies for eco-efficient vehicles. While several studies have analysed consumer perspectives on automobiles in general, very few researches have been carried out specifically on the perception of car engines. The authors consider demand to be more than simply sales levels, as it also encompasses consumer framing. A frame is the way in which an innovation is described or interpreted by consumers, and it is the structure of relevant beliefs, knowledge, perceptions and appreciation that underlie consumer attitudes. Nevertheless, the authors do not find any evidence that consumer preferences are leaning more towards alternative vehicles, despite information about the fuel efficiency of vehicles being much easier to find now than it was only a few years ago. The conclusions reached here are the same as those of most other studies: environmental factors do not currently seem to play a major role in consumers' car choices.

Michel Freyssenet, in *The Second Automotive Revolution Is Under Way: Scenarios in Confrontation*, affirms that three of the four conditions have been met that make a new car revolution probable: the urgency of solving the crisis of the 'door to door' transport system; the race for innovations and the transfer of new technologies from other sectors; and the creation of coalitions of public and private actors to make the energy required available in any place and at any time. The fourth condition – the existence of economic policy decisions that enable the diffusion of new, cleaner cars, will be needed only when one of the three possible transition scenarios has imposed itself. The scenario of diversity is currently in progress but, according to the author, paradoxically it has few supporters and is therefore unlikely to win. The scenario of progressiveness, which the author sees as the only reasonable and realistic one, might, however, suddenly turn into its exact opposite. The scenario of rupture seems to be the riskiest but, according to the author, it is not the most improbable, with the immediate adoption of plug-in hybrid vehicles, or, even more radically, of all-electric vehicles.

Part I
Innovative Design for Alternative Vehicles

1
Innovative Design and Sustainable Development in the Automotive Industry

Giuseppe Calabrese

Finding new methods of propulsion for automobiles is currently a subject of intense debate, driven by the issue of global warming and more generally by the demand for sustainable development in the automotive industry.

The most likely situation in the foreseeable future is the coexistence of a portfolio of co-developed technologies to satisfy different user segments, in terms of vehicle performance largely skewed towards conventional vehicles (EUCAR, 2009). It might be expected that cars will become less general-purpose and that certain technologies will be suited only to certain needs or functions; for example, zero emission vehicles mainly for city users; hybrid vehicles mainly for commuters; range-extender vehicles mainly for long distance users, and so on. As for the distant future, all scenarios are uncertain and debatable; studies generally tend to overestimate the diffusion of alternative vehicles (AVs)[1] and fail drastically in forecasting.

The evolution of powertrains is influenced by the path dependence of countries and carmakers' trajectories. Three sources of path dependency can be detected: business models, consumer attitudes and policy regulations:

- Carmakers' business models are generally characterized by risk aversion and by return optimization through continuous improvement and cost cutting. Overproduction leads manufacturers to offer incentives in order to increase demand, thus return to capital is low and often negative in economic downturns. Despite consistent investment in the reduction of CO_2 emissions, new cars have become 13 per cent cheaper on average in real terms compared to 2003 (EFTE, 2011), absorbing the predicted marginal costs completely.[2] In other words, the current automotive business model is characterized by a lack of profitability (Nieuwenhuis and Wells, 2003), given that profits come mainly from the sale of automobiles and not from the use of them (Ceschin and Vezzoli, 2010). This is the main reason why it is economically not very attractive to invest a great deal in a new, still immature technology, such as that of most AVs. The traditional automotive business model should be changed and the

relationship between producers and users should not end after purchase but continue over time (Ceschin and Vezzoli, 2010) through the offer of services.
- Most consumers are satisfied with the fact that the internal combustion engine (ICE) performs as they expect it to and at a predictable cost. Those who prefer clean and fuel-efficient engines and are willing to pay slightly higher purchase prices represent only a niche market. In sum, consumers favour ICE innovations over AVs, and in particular over electric engines (Dijk and Kemp, 2010). The authors of Chapter 16 in this volume have not discovered any evidence that consumer preferences are inclining towards AVs, despite information about the fuel efficiency of vehicles being much easier to find now than it was just a few years ago. The conclusion they reach is the same as that of most other studies: environmental factors do not currently seem to play a major role in consumers' car choices. Consumers care a great deal about fuel consumption but very little about vehicle emissions.
- As for the European Union (EU), measures to develop AVs have followed the usual path, with the institution of regulatory requirements that increasingly restrict the sale of new vehicles, the setting of specific limits on emissions, and the provision of direct support for basic research. The focus has been on new vehicles, while the reduction of pollutants from vehicles currently in use has barely been considered. In the EU, 34 per cent of the vehicles in use are more than ten years old. Step by step, Euro I to VI regulations have mainly favoured incremental innovation of the ICE in spite of radical innovations (Oltra and Saint Jean, 2009). Over the course of time, R&D financing has moved from fuel cell vehicles (FCVs) to the other types of AVs, because the timeline for introducing FCVs is still largely undefined and their success is still very uncertain in relation to costs, infrastructure, hydrogen generation and storage (see Chapter 7). In addition, the regulatory impact of the policy response to sustainable development is unclear, and the commitment of member countries has been limited, fragmented and disorganized.

The transition to safer and more eco-friendly automobiles often evokes a new vision of mobility and a new structure of the automotive industry (see Chapter 14). The question is whether the changeover will culminate in the coexistence of the current ways in which automotives are used or in the formation of a new strong coalition, such as the one between carmakers and oil companies that led to the triumph of the ICE standard at the start of the twentieth century (Freyssenet, 2009). A new major role could be taken by the producers of biofuels, electricity or hydrogen.

This chapter has a further four sections. The first is dedicated to current technological trajectories, highlighting the costs, opportunities and industrial repercussions of each option. The second section illustrates the race for AVs in the automotive industry, focusing in particular on carmakers' strategies. The third section considers new strategic players and, above all, corporate alliances between carmakers and battery manufacturers. The final section presents some conclusions regarding industrial policy implications.

Discontinuous innovation and technological trajectories

Innovation is frequently associated with the level of newness introduced. It can be described in terms of continuity from incremental to radical change (Tidd *et al.*, 2005), in terms of discontinuity when disruptive technologies bring about new market perspectives (Christensen and Overdorf, 2000), or in terms of systemic innovation that requires complementary goods and competences (Teece, 1984). The typical model used in the economic literature to describe the dynamics of innovation focuses mainly on the processes of creation, management and the enhancement of scientific and technological knowledge, in which innovation processes generally flow from basic research to product launch via applied and pre-competitive research (Malerba, 2000).

This model has been largely adopted in the automotive industry. In recent decades, carmakers have profoundly redefined the ways in which their products are designed, developed and constructed, reducing the time required from concept to production, changing product strategies, and introducing simultaneous design methods. The strategic decisions taken to reach these objectives have included integrative mechanisms and the widespread introduction of overlapping activities, the creation of strategic teams for project management, and the extensive use of information technology (Calabrese, 2001).

Nevertheless, the automotive sector is a mature industry, technologically dominated by all-steel bodies and ICE. These technical aspects have resulted in an industry with high fixed costs, necessitating mass production to achieve low per-unit prices in the market (Wells, 2010). Thus innovation is, for the most part, incremental, conservative and process-oriented (MacNeill and Bailey, 2010). However, as Vervaeke states in Chapter 3, the scope of sustainable development in the automotive sector is broader than simply designing cars with reduced exhaust emissions, as it also involves searching for new ideas in urban mobility. The design-driven approach overcomes the search for innovation based on problem-solving, market pull or technology push (Calabrese, 2010). Moreover, these days requirements and constraints imposed by sustainable development on the automotive industry are becoming increasingly strict and, as a consequence, product design is becoming more complex, avoiding the commodification that happened with electronic products (see Chapter 2).

Rising oil prices, concerns over global warming and, more recently, the widespread financial crisis have led governments and consumers to seek fuel economies, thereby presenting major challenges to the industry in a possibly disruptive scenario. Yet the technological state of affairs regarding AVs is far from clear, and improvements are feasible for all types of propulsion. Each AV technology can be placed within one of the traditional R&D phases. The expertise acquired thus far on natural gas technologies, liquefied petroleum gas (LPG), and compressed natural gas (CNG) might be considered to be almost consolidated. Some improvements can still be placed within the pre-competitive research phase, since possible upgrades are arising from incorporating CNG into hybrid electric vehicles (HEVs). While the Toyota Prius has been a great success and many carmakers are launching electric vehicles, HEVs and fully electric vehicles (EVs) are still in the industrial research phase, and the basic research phase for fuel cell vehicles (FCVs) has not yet been completed.[3]

Nevertheless, these technologies are not being developed separately, but rather they are following a technological trajectory that comprises a flow of related innovations. The various technological trajectories co-evolve/compete, and the established paths are self-reinforced through learning processes (Dosi, 1982). The best solution has not yet arisen and technological co-evolution can favour investment sharing.

ICE performance is essential in CNG and biofuel vehicles, in HEVs, and in hydrogen vehicles without fuel cells. Gas storage systems are similar in LPG, CNG and hydrogen vehicles, and can be used both on board the vehicle and at filling stations. Electric storage system batteries are common to electric, hybrid and fuel cell vehicles. Power electronic systems are common to many AVs and, since fuel cells are energy generators, they are also being developed for other industrial purposes. HEV technologies and recent developments in batteries are potentially more disruptive, since they might change both the technology and the architecture of vehicles (Aggeri et al., 2009), whereas ICE developments do not radically change the identity of vehicles. Moreover, carmakers have to focus on the organization of innovative design activities around the issue of eco-innovation as a complement to the established R&D processes (see Chapter 4). These various options all have different costs and opportunities, as well as different industrial repercussions, and require considerable investment flows.

Recent studies have shown that the potential for further improvements on ICE-related technologies (for example, the catalyst technology and powertrain control systems) might satisfy future emission standards (see Chapter 7) and postpone the launch of vehicles with electric batteries. On the other hand, some technologies have proved to be fuel efficient but are still hindered by high costs[4] or difficulties in mass production.[5]

The price and performance of CNG vehicles are comparable to those of petrol vehicles. They greatly reduce local pollution, cost less to operate as long as governments keep excise duties down, and there is a great deal of technical expertise in this area, but these vehicles do require a distribution network, though to a lesser extent for LPG than for CNG (see Chapter 8).

However, natural gas reserves are far more spatially extensive than oil reserves, and if a country is crossed by a gas pipeline, these distribution problems might be overcome rapidly and without significant cost. Where no pipelines are available, distribution might be achieved using gas carrier ships or through local production of biomethane via the exploitation of biomass. The next technological issues concern turning the fuel from a gaseous to a liquid state by direct injection, combining double fuel feeding with diesel rather than petrol, and mixing it with hydrogen.

Looking at a wider scenario for ICEs, which includes biofuels, Brazilian experience with ethanol and biodiesel has shown that ethanol has reached a mature level of sustainability, whereas biodiesel still relies on institutional initiatives in order to develop it fully (see Chapter 9). Biodiesel is becoming popular because it has a low environmental impact and requires only minor modifications to existing engines. However, despite its ecological advantages, certain negative elements prevent it from becoming more widespread, such as the price, and the origin and accessibility of both the final product and the raw materials.

With regard to synthetic fuels, whether from diversified renewable energy, or from coal, natural gas and biomass, the final product is a liquid fuel with characteristics equivalent to those of petroleum-based fuels, thus it can be distributed through the existing fuel network and used in ICE vehicles.[6] The great advantage of this new generation of biofuels is that the amount of CO_2 generated during combustion is the same as that absorbed by plants as they grow.

In terms of CO_2, considering all the costs from well to wheel, the most promising biofuel seems to be biomethane (CBG, compressed biogas) produced from organic waste or through biomass conversion processes. It prevents methane molecules from being released into the atmosphere, which are 23 times the greenhouse effect of CO_2.

Batteries are the most crucial components affecting the basic performance of EVs. Their energy density determines the driving distance of a vehicle and their cost represents a major portion of the overall cost, up to 30 per cent of the total future cost of EVs. As for their technical features, competing lithium-ion technologies can be compared along seven dimensions: safety, lifespan, performance, specific energy and power, charging time, and cost. Choosing a technology that optimizes one dimension inevitably means compromising on others. At the time of writing, there are five different types of lithium-ion batteries.

Safety is the most important criterion for electric vehicle batteries. The main concern is avoiding thermal runaway, by the use of robust battery boxes, cooling systems and cell-discharge balancing. As far as lifespan is concerned, cycle stability is essentially under control, whereas difficulties persist in overall age. The performance of batteries degrades if they are engineered to function over a wider range of temperatures than those between the levels needed for cold-weather and warm-weather tyres. The specific energy of batteries continues to limit the driving range of EVs, despite their specific power being managed relatively well by current technologies. Fast charging systems add cost and weight, and require enhanced cooling systems on board the vehicles. In addition, to improve driving distance, EVs must fulfil electrical specifications. Automobiles feature numerous components and optional extras that absorb significant amounts of energy, such as heating, air conditioning, lighting, electronic devices and so on.[7] In ICE vehicles, the reduction of energy consumption has rarely been taken into account, while in EVs it is a compulsory target that requires the redesign of all components.

Regarding business issues, high costs remain the major hurdle. The key challenge in the future will be to reduce manufacturing costs through scale and experience effects as market volumes increase. Yet most of these costs are variable, essentially depending on materials and, to a much more limited extent, on quantities (see Chapter 2). Operating costs will depend on battery maintenance and are not yet quantifiable. The distribution network is less problematic than for other alternative fuels, even though a trade-off exists between battery leasing and fast charging. The former requires a common battery standard, while the latter adds cost and weight, and requires enhanced cooling systems on board the vehicle. Both need high public infrastructure investment.

The electronic management systems and optimization of the entire system, particularly the batteries, all have room for improvement. In particular, lithium batteries built for cars must meet far more stringent standards than those for electronic devices. Furthermore, batteries designed for hybrid vehicles have different requirements from those built for EVs. The latter need more specific energy to provide greater autonomy, whereas the former require a particular specific power. In hybrid vehicles, charging and discharging are ultrafast and intense (approximately 1 million cycles of charging and discharging), while EV batteries require only 2,500 cycles. EVs can therefore contribute to a small part of the development needed for hybrid batteries. But their technological paths are quite different.

One important advantage is related to regenerative braking, which contributes significantly to decreasing fuel consumption. As an example, in urban driving, regenerative braking allows for up to a 15–20 per cent reduction in fuel consumption in comparison to conventional cars with the same power-to-weight ratio.

The long-term scenario for hydrogen-fuelled cars cannot easily be hypothesized, since there is no standard technology in place and the existing alternatives are not economically feasible. Moreover, at present none of them are without an environmental impact, apart from the production of hydrogen (which is not found in nature) from solar energy or in new hydro-electric, aeolian (wind-powered) and geothermal plants (see Chapter 7).

Well-to-wheel emissions of fossil fuel-driven ICEs are lower than those of fossil-based hydrogen-driven fuel cells. Hydrogen vehicle technology is only meaningful when the production of hydrogen is from renewable materials. In addition, the catalytic action of fuel cells is obtained by using platinum, a rare natural resource with a very inelastic curve. The cost of platinum is approximately €5,000 for any fuel-cell vehicle. A further technological question for the large-scale use of hydrogen is inherent in its chemical and physical characteristics, particularly in its low energy density per unit volume. This implies the need to increase pressure in the storage system up to as much as 70 MPa. A further increase in density per unit volume is possible via its transformation into liquid hydrogen, which requires a constant temperature below −253°C, consequently requiring heavy and expensive isolation of the storage containers. In the best case, liquid hydrogen, with the same stored energy, occupies a volume nearly four times greater than that of petrol, even though its weight is almost three times less.[8]

A new frontier has been detected in the form of solar cars using photovoltaic panels. Some carmakers, such as Toyota and Fiat Group Automobiles, are investing in prototypes quite different from the demonstration vehicles and engineering exercises we are used to seeing.

The race for alternative vehicles in the automotive industry

The technological trajectories described in the previous section prove that carmakers are mainly committed to improving the ICE, in both its petrol and diesel versions (see Chapter 6; Dijk and Kemp, 2010). Gains are achievable via turbocharging, direct fuel injection, cylinder deactivation, and advances in engine timing, either on individual components or by redefining engine systems completely. Carmakers are also committed to finding systems that can reduce emissions: electric steering, stop and start features, dual clutch transmission, braking regeneration, downsizing, weight reduction, low rolling resistance tyres, LED and the recent OLED headlamps, aerodynamics, mission control, electrification, CO_2 sequestration and so on.

As noted in the Introduction, in recent years traditional vehicles have become cheaper in real terms, and a direct relationship between investments to reduce CO_2 and price increases has not been proven (EFTE, 2011). Indeed, incremental improvements in mature technology and production methods usually ensure a steep fall in costs, particularly for new components.

Furthermore, the improved performance of batteries, from nickel-hydride to lithium-ion or polymer technology, has revived purely electric-powered cars and perhaps discredited Fréry's (2000) myth of EVs as 'eternally emerging' technology. EVs are no longer considered to be technologically less efficient (Beaume and Midler, 2009), but among generalist carmakers only the Renault–Nissan Alliance seems to pursue this business line resolutely, encouraging new strategies in co-operation patterns and innovation management[9] (see Chapter 5). A similar path has also recently been followed by GM.[10] However, the success of these initiatives is still uncertain. The crucial issues to be dealt with are the social acceptance of such a radical innovation, the fact that it is still very much dependent on changes in the business and political environment, and the rising level of competition from other EV models as well as from other plug-in vehicles or other kinds of improved vehicles. Renault is managing a double process: enlarging the scope of usage from product to service, and extending the definition of users and stakeholders.

The difference in price between hybrid electric vehicles (HEVs) with nickel-metal hydride batteries (NiMH) and similar non-electric cars has been reduced to a few thousand euros, but for cars equipped with lithium-ion batteries (Li-Ion) the gap is about 15 per cent, and in the case of newly launched plug-in HEVs it is over 60 per cent.

Table 1.1 shows the results of research (Deutsche Bank, 2008) on the cost and benefit analysis forecast for the year 2020. The incremental costs and payback in years were calculated for HEVs, plug-in HEVs and EVs in comparison to ICEs. With the US retail price of gasoline at around US$4 per gallon at the time of writing, which is an all-time peak, and using lithium-ion batteries, the payback of plug-in HEVs and EVs in comparison to ICEs is, respectively, 7.4 years and 8.1 years in the USA, and 3.9 years and 4.2 years in Europe. The payback remains too high for the US market, but there might be much greater opportunities in Europe, where fuel prices are almost double US prices. The price gap between EVs with lithium-ion batteries and similar non-electric cars will be US$11,000 and this represents a significant incremental upfront cost.[11]

According to Chanaron and Teske (2007), hybrid vehicles are a temporary step between the traditional technology based on petrol and diesel engines and the introduction of full electric vehicles, given that the latter, as described above and even more so in the case of hydrogen powered fuel

Table 1.1 Cost and benefit analysis forecast for year 2020

	Incremental costs (US$)		US fuel costs (payback in years)		European fuel costs (payback in years)	
	NiMH	Li-ion	NiMH	Li-ion	NiMH	Li-ion
Hybrid electric vehicles (HEVs)	2,700	2,400	5.1	2.8	2.9	1.6
Plug-in hybrid electric vehicles (HEVs)	16,400	8,000	15.1	7.4	7.9	3.9
Electric vehicles (EVs)	28,400	11,000	20.9	8.1	10.8	4.2

Source: Adapted from Deutsche Bank (2008).

cells, require breakthrough innovations. At the time of writing, with the exception of the GM Volt (the Opel Ampera in Europe) and the Volvo V60 models, carmakers are focusing more on parallel structures rather than series layout (range-extender). The latter shows a more positive impact on fuel economy but needs larger batteries to provide for peaks in power demand, though less than EVs.

Moreover, in the short term, demand for lithium batteries will be relatively low, as carmakers and suppliers are still validating products and gearing up for large-scale production. The leading companies also have limited experience in producing these batteries on an automotive scale, hence carmakers have a long way to go to make EVs affordable.

LPG and CNG vehicles also have a major position in the strategies of carmakers, directly involved increasingly in offering dualfuel feeding, whereas previously this was limited exclusively to the aftermarket business segment. Gas systems are developed jointly with suppliers and validated by the car manufacturers.

In contrast, biofuels are suffering a negative response concerning both their impact on the prices of agricultural products and their potentially damaging effects on health, even though this widespread concern has not been confirmed (see Chapter 9). Looking at a wider scenario for ICEs, which includes biofuels, Volkswagen's strategic approach seems very interesting, since it supports synthetic fuels from either diversified renewable energy, or coal, natural gas and biomass.

The Honda FCX Clarity is probably the longest-running and best-known hydrogen fuel-cell car currently being tested. It is available for a limited lease (US$600 per month) throughout North America. Both Honda and Toyota have set the year 2015 as their goal for large-scale production of FCVs.

The mass availability of environmentally friendly cars presents an opportunity not only for manufacturers but also for citizens and workers. However, this opportunity comes with its own set of obstacles, such as the need to convert the fuel distribution network and to define common standards. Furthermore, there is no consensus on energy cost comparisons for these different methods of propulsion. Some experts feel that the main problem at present is urban pollution, which can be solved immediately by using CNG vehicles (see Chapter 8). On the other hand, if the entire production cycle is considered, some studies conclude that hybrid vehicles are more ecologically sound than those powered by hydrogen. Others believe that future developments of traditional engines (see Chapter 9), together with the use of biological or synthetic fuels, will lead to a 30–40 per cent reduction in fuel consumption, and to the use of renewable sources of energy.

ICE supporters affirm that hybrid engines are barely an improvement on the current state of affairs and ultimately too expensive to be widely applied. Conversely, fans of hybrid engines pin their hopes on the development of battery-run EVs, to enable them to be used for local travel. In agreement with the promoters of EVs, they highlight potential progress in terms of battery bulk, weight, power, safety and charging speed.

Comparing the performance structure of EVs with the typical criteria of ICEs would always give the advantage to the latter (Beaume and Midler, 2009). Consequently, according to EV supporters, performance assessment should be shifted towards new value content, such as quietness, ease of driving, less stress to the environment, collective benefits and so on. This belief is based on a statement by Bower and Christensen (1995), claiming that successful disruptive technologies do not generally outperform the existing one on established performance parameters but generate new breakthrough performance on new value parameters.

Until now, carmakers have not pursued optimal technical solutions, but have rather concentrated on a particular profit strategy (Freyssenet, 2009) or brand image. In the past, PSA Peugeot Citroën achieved excellent performance from EVs, but they were abandoned in 1998 to concentrate on improving the environmental performance of existing ICEs. Fiat followed a similar path, moving from EVs to CNG vehicles. Both Toyota and Honda developed a new specific model with a hybrid driving system, but their other vehicles show higher emissions. BMW did the same, using hydrogen to power a conventional ICE. Nissan and Renault decided to support the new generation of batteries and push ahead with cars with a medium-distance driving range, to be sold in less densely populated areas. To sum up briefly, carmakers' strategies seem to be conditioned by the minimization of risk and an aim to appear both innovative and environmentally friendly.

The turbulence that the automotive industry will face depends on which standards will ultimately prevail. If the changes only involve fuel types or hybrid petrol/electricity engines, they will not be particularly disruptive. However, they will be more radical if electricity is the only future source of fuel. Yet, from an electric energy systems perspective, a massive diffusion of EVs may represent a serious problem as well as a new, challenging opportunity. The electricity needed to feed EVs has to be made available at the distribution level and there are huge differences in terms of global impact, depending on whether it is produced locally from renewable sources or centrally through traditional fossil-fuel-fired power plants (see Chapter 11). In this case, new actors are bound to arise, who will attempt to appropriate a significant proportion of the value produced in this sector and perhaps even control its distribution: producers of new fuels, electricity producers, battery makers, fuel cell manufacturers, manufacturers of new driving systems, or companies that design and produce the new electronic power systems, and the functions of control and regulation (Freyssenet, 2009).

However, it should be noted that vehicle electrification is by no means a permanent solution to CO_2 emissions and the automotive industry's sustainability problems. Zero emissions require not only clean cars but also the complete decarbonization of electricity generation (see Chapter 11).

New strategic players and relationships

The increased use of AV technologies has also led new players to enter the automotive industry as important contributors to the design and manufacture of components and sub-systems (see Chapter 12). At one end of the value chain are the chemical companies and battery component producers; while at the other are mobility operators and power companies. In a similar way to the situation during the previous electronics revolution in cars, the automotive industry is developing interfaces to tap into different knowledge bases, which requires organizational structures to manage the new flow of information (Lenfle and Midler, 2003).

The management of these interfaces is a major challenge for carmakers in their bid to gain a competitive advantage and survive in the market place (see Chapter 12). The way forward is arguably to develop co-operative business models between 'traditional' automotive actors and the new players. This outlook means that, more than ever, 'traditional' automotive actors will have to operate as system integrators and interfacing actors to secure an effective symbiosis of knowledge and technologies that are governed on either side of the boundaries of their own firms (Kamp and Tozun, 2010).

At the time of writing, the fiercest competition is between carmakers and battery manufacturers working towards the development of high-performance rechargeable batteries. In this context, corporate alliances or an exchange of stakes between carmakers and battery manufacturers are appearing. These relationships give carmakers exclusive access to the know-how, technology and production capacity of cell manufacturers, and allow the carmakers to differentiate their vehicles in terms of a chosen battery technology. However, relationships of this kind can limit a carmaker's ability to react quickly to the technological advances achieved by other cell manufacturers. Furthermore, exclusivity can limit scale effects and delay manufacturing-based cost reductions.

Some tier one suppliers are also teaming up directly with cell manufacturers. Relationships of this kind allow tier one suppliers to apply automotive-integration expertise to the battery business and give cell manufacturers access to a range of car manufacturers through established relationships. As for the carmakers, this model yields less control and a less detailed knowledge of battery technology, but it allows them to benefit from the scale effects of leveraging a cross-carmaker supply base. It also reduces their up-front costs and the potential cost of switching to an alternative technology, should one emerge. This scenario will be of the greatest benefit to carmakers if pack-level standards emerge and allow for flexibility in battery technology.

Interactions between carmakers and battery manufacturers can be classified into pure relationships of supply and demand; partnerships for the joint development of batteries; and partnerships including the setting up of venture companies. Among these, the last type of relationship is essential to boost the carmakers' competitive edge, so they will avoid being totally dependent on external battery suppliers. It will be natural for carmakers to link up with particular battery manufacturers in a bid to secure battery technologies and control both technology development and manufacturing operations. The main Japanese carmakers have founded joint venture companies with battery manufacturers, and these alliances support the above concept well, while US and European carmakers are followers.

Another possible scenario sees companies that control rechargeable batteries also controlling the new-generation automotive market. For example, the BYD Group is a Chinese battery manufacturer that has diversified into the automotive sector and is now on its way to producing EVs. Other examples, among many, are Tesla Motors of Silicon Valley, California, and the US–Indian EV manufacturer Reva, whose G-Wiz EV is marketed in Europe, the USA and Japan (MacNeill and Bailey, 2010). Bolloré is another battery producer that has launched its own EV (Calabrese, 2010). Tesla has signed an agreement with Lotus to manufacture in the UK, and Daimler has recently purchased a 10 per cent stake in the company in order to integrate Li-Ion battery technology into the electric Smart car, which has a drivetrain developed, manufactured and fitted by the UK firm Zytek.

Other examples include electricity suppliers, which have been collaborating with carmakers for years; aircraft manufacturers such as Dassault, which has launched an EV; or tyre producers such as Michelin, which has been testing tyres fitted with electric motors. By supplying tyres, Michelin might eventually be able to supply a drive system and perhaps ultimately an entire vehicle.

These examples illustrate how new players can develop innovations but may need to enter into joint ventures with established assemblers to help them to reach a wider market. Such ventures trigger changes in the territorial organization of the industry since, as in some of the examples above, the players may be located outside 'traditional' areas. In addition, the new actors are developing research-based knowledge with a largely codified content. However, this needs to be allied to the established knowledge of the major companies and linked to their networks of upstream supply and downstream logistics and marketing. At this stage, knowledge networks may still be a long way off.

Conclusions: policy issue implications

At the European level, three major policy areas impact on sustainable development in the automotive industry. They involve common standard regulations on emissions (Euro 4; Euro 5, late 2009; and Euro 6, 2015) and safety (Euro NCAP crash tests), which are compulsory in all member states. The third important area concerns recycling and the End-of-Life Vehicle (ELV)

Directive, which came into force at the beginning of 2007. This barrage of increasingly stiff regulations is driving substantial changes and presents a challenge for the global automotive industry, though regulatory regimes for cars around the world remain variable. Some markets have specific safety tests and others distinct cycles for the testing of emissions (Ryan and Turton, 2007; Sperling and Cannon, 2007). Even more pronounced are the differences in the fiscal regimes.

In 2010, Germany, France, Spain and Portugal exhorted the EU to support transnational development projects on electric mobility and to define standards for the charging system. The European Union is clearly missing political leadership, and the case of the 2003 biofuel directive is a typical example. Indeed, only Sweden has adopted the directive to replace fossil fuels with biofuels and support flexi-fuel vehicles.

At the European national level, policies have been for the most part non-interventionist and concentrated on improving the business environment (Bailey and Driffield, 2007). In the manufacturing arena, national policy has shifted away from sector-specific support to general support to all manufacturing sectors, and from vertical to horizontal industrial policy measures. Much of the automotive-specific effort has been directed through programmes to fund collaboration among companies and research organizations to reduce life-cycle emissions (MacNeill and Bailey, 2010).

Countries such as Denmark, France and Israel, which are now establishing attractive incentive schemes for EVs, could potentially generate a huge competitive edge for their domestic automotive and power industries. But unless other governments act promptly to provide adequate incentives for consumers to purchase these cars, and for investors to provide the necessary infrastructure at affordable prices, AVs may be off to a false start. The isolated and top-down experience of California is particularly significant. California implemented a legislation that made it compulsory for carmakers to sell at least 2 per cent Zero Emission Vehicles by 1997, rising to 15 per cent in 2003, but the mandate was gradually reduced and it disappeared in 1998.

Nevertheless, most of the considerations related to the development of AVs and their successful placement on the market undoubtedly depend on fuel prices and state intervention. In July 2008, gasoline reached an all-time peak of about US$150 per barrel in the USA, but five months later its price was 75 per cent lower. High price variability is a hindrance in any market and makes it difficult to define company strategies, in particular those of carmakers aiming to improve the production plans for AVs (see Chapter 15). It is obvious that, if oil prices are low, customers will tend to buy conventional vehicles, but where other conditions are met (battery prices decrease, public utilities provide suitable infrastructures, and the EU sticks to its 95g/km CO_2 emissions target for 2020), the future of AVs will be much brighter. In this context, the proposal to add extra excise taxes on oil prices to reach a lasting and fixed level seems sensible. According to calculations by the Boston Consulting Group (2009), hybrid vehicles are more attractive than petrol vehicles when the price of oil reaches around US$70 per barrel, and more appealing than advanced diesel vehicles when the price reaches about US$170 per barrel. However, EVs remain relatively unattractive unless they are subsidized, or battery costs drop sharply (US$500 per KWh, and oil price at about US$120 per barrel).

One of the most popular schemes pursued by industrial policies is to launch fleet renewal programmes, including market incentives and car scrapping schemes. Scholars and practitioners have different opinions on the matter, however. Above all in Europe, these incentives are seen as a measure intended to modify customer requirements and distort the market, leading only to limited, short-term benefits, because they do not actually encourage people to buy more cars but just to purchase new vehicles earlier than they would normally do. Moreover, in contrast with the common belief, it seems that cost saving is not the main driver motivating people to buy AVs (Nadin *et al.*, 2009). So the decision to purchase an AV is more often a consciousness choice, and only marginally for cost-saving reasons. Hence a more efficient policy might consist of developing communication actions to support the moral and psychological value of this pro-environmental behaviour in terms of responsibility and consciousness.

Evidence on the latest fleet scrapping schemes is discordant, of course. Those opposing fleet renewal programmes are backed by Italian evidence. Significant incentives for AVs in Italy notably increased the percentage of green vehicles over total new passenger car registrations from 3.8 per cent in 2007 to 22.1 per cent in 2009 but, as soon as the incentives stopped, sales dropped to previous levels. In contrast, it has been noted that in Germany a significant number of people who traditionally drove used vehicles are now purchasing new cars for the first time. Moreover, supporters of fleet renewal schemes highlight many other positive effects:

- more than 80 per cent of all cars sold in the EU are produced within the EU, and scrapping incentives can help to maintain jobs in production facilities, component manufacturing and dealerships, saving member states expenditure on unemployment benefits, job search assistance, retraining and other social welfare costs; and
- as the schemes involve cars that are at least 9 or 10 years old, their replacements benefit from increased safety and a decrease in noise levels. Old, polluting vehicles are scrapped rather than being exported to Central and Eastern European countries, and better equipment, such as ABS, ESC, airbags and navigation systems, make vehicles safer.

The main question is how much these incentives reduce both the price gap between AVs and conventional vehicles, and the institutional conditions that foster the current technological paradigm.

This shows why government subsidies have not boosted EV sales in Europe significantly. The discrepancies highlight the apparently weak influence of incentives on purchasing decisions even when the incentives are significant (see Table 1.2). For example, the Mitsubishi i-MiEV, one of the best EVs available, costs more than €36,000, whereas a comparable conventional car costs less than €10,000. Institutional conditions that can influence EV sales are, among others: the degree of urban geography, market maturity, charging infrastructure, the ability to use bus lanes, and free city-centre parking.

Another aspect that must be considered in the comparison between ICEs and EVs is usage cost. The price of energy used shows apparent competitiveness between electricity and petrol, but the gap does not yet seem to be appealing enough. For example, in the case of the Smart model, the EV version consumes 12.2 KWh/100 km and the unleaded petrol version consumes 4.4 litres/100 km. Using the prices quoted in Europe's Energy Portal (www.energy.eu) on 15 October 2011, the usage cost of the EV Smart is 24.3 per cent that of the ICE version in France; 27.5 per cent in the UK; 35.5 per cent in Italy; 43.2 per cent in Spain; and 47.9 per cent in Germany.[12] Unfortunately,

Table 1.2 Statistics on electric vehicles, European Union, 2010/11

Country	A	B	C (€)	D (%)	E (%)
Austria	347	96	2,571	57.3	27.7
Belgium	85	34	10,907	60.0	26.0
Czech Rep.	43	4	271	53.3	17.7
Denmark	283	15	20,588	57.1	55.7
France	953	133	5,000	60.1	24.7
Germany	1,020	185	380	59.5	43.8
Ireland	36	17	5,000	53.8	1.01
Italy	103	40	1,200	56.5	27.7
The Netherlands	269	87	4,936	61.6	25.8
Norway	850	353	17,524	n.a.	27.2
Portugal	93	18	9,442	57.7	36.3
Romania	2	0	3,700	49.3	20.2
Spain	122	76	6,500	49.3	19.4
Sweden	111	167	470	60.0	34.6
Switzerland	239	5	0	n.a.	n.a.
UK	599	90	6,400	60.5	4.8
Others	67	16	n.a.	n.a.	n.a.
Total	**5,222**	**1,336**			

Notes and sources:
A: Sales (first half of 2011), adapted from Jato Dynamics (2011).
B: Sales (2010) adapted from Jato Dynamics (2011).
C: Incentives (in €), adapted from Jato Dynamics (2011).
D: Shares of unleaded taxes, adapted from Europe's Energy Portal (www.energy.eu; accessed 11 November 2011).
E: Shares of taxes on electricity for household consumers, adapted from Eurostat (2010).

the prices of electricity and petrol include different levels of excise duties (see Table 1.2) to which the coffers of the various states are highly sensitive. If governments imposed on electricity the same level of taxes collected from petrol, the percentages would be: France 66.5; UK 61.8; Italy 66.4; Spain 57.1 and Germany 79.6. The percentages speak for themselves.

With regard to industrial policy, it is worth pointing out that, at the end of the twentieth century, France and Italy were the only countries involved in the AV market. In the last few years, however, other countries have concentrated their efforts based on their own specific carmakers and energy structures.

The most likely short-term scenario seems to be one of diversity. Italy is focused mainly on CNG vehicles and the same is true of Russia, which is counting on its large reserves of natural gas. Brazil is the traditional leader in agrofuels. The large amount of nuclear power it produces has led France to concentrate on EVs, whereas Germany has moved from LPG to hybrid cars and biofuel solutions. Nevertheless, other scenarios can be hypothesized: one of progressiveness, involving a smooth transition from ICE to FCV, or one of rupture, with the immediate adoption of plug-in hybrid vehicles, or even more radically, all-electric vehicles (see Chapter 17).

Exponential growth in both car production and sales enabled China to overtake the USA in 2009 and to focus on sustainable growth, less dependency on fossil fuel, and a reduction in pollution (see Chapter 13). As a consequence, the Chinese government has placed EVs among its main priorities in its automotive industrial policy, with the goals of achieving worldwide leadership and avoiding the long and expensive acquisition of ICE technologies. The Chinese leapfrogging strategy can count on abundant reserves of essential materials for the production of batteries, on a technological architecture that has not yet been defined and is simpler in comparison with the mechanical content of the ICE, and on a large number of projects being planned by the companies.

Unlike their US and Japanese counterparts, Chinese carmakers have only a few major difficulties to deal with. Beside additional costs and unwillingness to sell at a loss, the main obstacle is their dependence on hefty infrastructural investments to foster green technologies, which only state planning can afford (Volpato and Zirpoli, 2011).

For this reason, in the short run the most promising AVs able to reduce local pollution seem to be the CNG type, in particular when old vehicles are equipped with CNG devices (see Chapter 8). Policy-makers should intervene by implementing regulations (exploiting environmental and safety benefits as opposed to supporting traditional vehicles); through technology (improving energy performance, and incorporating CNG into hybrid cars); and by supporting demand (not adjusting excise duties and promoting the conversion of cars already in use). The fundamental issue involves expanding the distribution network, which is limited in Italy and almost non-existent in other European countries. Recently, Italy has allowed CNG filling through the home network. This will break the vicious circle that has been created between CNG distributors, who do not want to expand the network because of low demand, and consumers who are not willing to buy CNG cars because of the lack of filling stations.

Finally, and more generally, policy response to sustainable development should aim at the implementation of measures capable of (Ceschin and Vezzoli, 2010):

- encouraging companies to shift their business models by adopting use-oriented (for example, leasing, sharing, pooling) and result-oriented (for example, pay per service unit schemes, integrated mobility schemes) services;
- changing agents' behaviours (for example, public procurements, consumer awareness);
- supporting demonstrative pilot projects (for example, promising business models without direct market pressure); and
- involving universities and research centres in supporting knowledge transfer and the dissemination of information.

In this context, a number of noteworthy public policies are being put forward by local authorities. Increasing numbers of city councils are promoting electric urban mobility systems, renewing their fleets with EVs, and installing charging stations. To name just a few: E-mobility in Berlin, London and Milan; Car2go in Ulm, Austin, Texas and a number of other cities; Connected car in Galicia; Autolib in Paris, and so on.

The most promising initiative is the Better Place project based in Palo Alto, California, and its innovative business strategy. The focus here is not on the cars, the batteries or the sources of electrical power; rather, the project provides the important components of an all encompassing scheme to supply the required supporting infrastructure, most notably charging points, for the roll-out of plug-in EVs (Andersen et al., 2009).

Notes

1. In this chapter, alternative vehicles include alternative fuels (natural gas and biofuels) and alternative engines powered by electric batteries.
2. According to EFTE (2011), the studies conducted in 2001 by AEA Technology and in 2006 by TNO overestimated that reducing CO_2 emissions to 140g would make cars €2,400 and €1,200 more expensive, respectively, compared with the 1995 and 2002 baselines. However, it might also be argued that prices could have been lower without these additional costs.
3. The few prototypes that have been built serve more as research laboratories than as vehicles similar to the ones actually in use, to the extent that universities and research centres have the most significant role in this area.
4. As in the case of small diesel engines designed to fulfil the Europe 2020 strategy.
5. As in the case of the electromagnetic valve train, cylinder deactivation and controlled auto-ignition.

6. The industrial process was proposed by Fischer and Tropsch in 1920 and widely used in the Second World War to supply fuel for army vehicles.
7. Test drives made by trade journals show that autonomy for urban use is on average one third lower than that declared by EV manufacturers.
8. An alternative is to use hydrogen to power ICEs. Its main benefits are the elimination of carbon dioxide emissions and the reduction of nitrogen oxide. However, thermal efficiency remains essentially the same as that of ICEs. This solution was followed only by BMW and recently abandoned.
9. Renault/Nissan is investing 4 billion euros in EVs and aims to sell 1.5 million cars a year by 2016.
10. As well as the Volt/Ampera models, GM is working on two EVs: a two-seater vehicle and a battery-powered minicar.
11. The price gap for the EVs currently on the market is certainly higher. Please see 'Conclusions: policy issue implications', below.
12. At the time of writing, with the two Smart models, in France, for every 1,000 km travelled, €17.0 are spent on electricity and €70.0 on unleaded petrol; in Italy, the figures are €24.9 and €70.1, respectively; in Germany €32.5 and €68.0; in Spain €24.9 and €57.6; in the UK €19.3 and €70.3.

References

Aggeri, F., Elmquist, M. and Pohl, H. (2009) 'Managing Learning in the Automotive Industry – The Innovation Race for Electric Vehicles', *International Journal of Automotive Technology and Management*, 9(2): 123–47.

Andersen, P., Mathews, J. and Morten, R. (2009) 'Integrating Private Transport into Renewable Energy Policy: The Strategy of Creating Intelligent Recharging Grids for Electric Vehicles', *Energy Policy*, 37(7): 2481–6.

Bailey, D. and Driffield, N. (2007) 'Industrial Policy, FDI and Employment: Still "Missing a Strategy"', *Journal of Industry, Competition and Trade*, 7(2): 189–211.

Beaume, R. and Midler, C. (2009) 'From Technology Competition to Reinventing Individual Ecomobility: New Design Strategies for Electric Vehicles', *International Journal of Automotive Technology and Management*, 9(2): 174–90.

Boston Consulting Group (2009) *The Comeback of the Electric Car?*, FOCUS, 1/09, rev 2.

Bower, J. L. and Christensen, C. M. (1995) 'Disruptive Technologies: Catching the Wave', *Harvard Business Review*, 73(1): 43–53.

Calabrese, G. (2001) 'R&D Globalization in the Car Industry', *International Journal of Automotive Technology and Management*, 1(1): 145–59.

Calabrese, G. (ed.) (2010) *La filiera dello stile e le politiche industriali per l'automotive in Piemonte e in Europa*. Milan: FrancoAngeli.

Ceschin, F. and Vezzoli, C. (2010) 'The Role of Public Policy in Stimulating Radical Environmental Impact Reduction in the Automotive Sector: The Need to Focus on Product–Service System Innovation', *International Journal of Automotive Technology and Management*, 10(2–3): 321–41.

Chanaron, J. and Teske, J. (2007) 'Hybrid Vehicles: A Temporary Step', *International Journal of Automotive Technology and Management*, 7(4): 268–88.

Christensen, C. M. and Overdorf, M. (2000) 'Meeting the Challenge of Disruptive Change', *Harvard Business Review*, 78(2): 66–76.

Deutsche Bank (2008) *Electric Cars: Plugged in. Batteries Must Be Included*, FITT Research.

Dijk, M. and Kemp, R. (2010) 'A Framework for Product Market Innovation Paths – Emergence of Hybrid Vehicles as an Example', *International Journal of Automotive Technology and Management*, 10(1): 56–76.

Dosi, G. (1982) 'Technological Paradigms and Technological Trajectories: A Suggested Interpretation of the Determinants and Directions of Technical Change', *Research Policy*, 11(3): 147–62.

EFTE (European Federation for Transport and Environment) (2011) *How Clean Are Europe's Cars?* Brussels: EFTE.

EUCAR (European Council for Automotive R&D) (2009) *The Automotive Industry – Focus on Future R&D Challenges*. Available at: www.eucar.be.

Europe's Energy Portal. Available at: www.energy.eu; accessed 11 November 2011.

Eurostat (2010) *Electricity Prices for Second Semester 2010*. Available at: epp.eurostat. ec.europa.eu/statistics_explained/index.php/Electricity_and_natural_gas_price_ statistics.

Fréry, F. (2000) 'Les produits éternellement émergents: le cas de la voiture électrique', in D. Manceau and A. Bloch (eds), *De l'idée au marché. Innovation et lancement de produits*. Paris: Éditions Vuibert, pp. 234–64.

Freyssenet, M. (2009) 'The Second Automotive Revolution – Promises and Uncertainties', in M. Freyssenet (ed.), *The Second Automotive Revolution: Trajectories of the World Carmakers in the 21st Century*. London: Palgrave Macmillan, pp. 443–54.

Jato Dynamics (2011) *Jato Consult CO_2 Report*. Available at: www.jato.com.

Kamp, B. and Tozun, R. (2010) 'Automotive Industry and Blurring Systemic Borders: The Role of Regional Policy Measures', *International Journal of Automotive Technology and Management*, 10(2–3): 213–35.

Lenfle, S. and Midler, C. (2003) 'Innovation in Automotive Telematics Services: Characteristics of the Field and Management Principles', *International Journal of Automotive Technology and Management*, 1(2): 144–59.

MacNeill, S. and Bailey, D. (2010) 'Changing Policies for the Automotive Industry in an Old' Industrial Region: An Open Innovation Model for the UK West Midlands?', *International Journal of Automotive Technology and Management*, 10(2–3): 128–44.

Malerba, F. (2000) *Economia dell'innovazione*. Rome: Carocci Editore.

Nadin, G., Savorgnani, G. T. and Besana, M. (2009) 'The Environmental Awareness of Car Owners: The Case of Natural Gas Vehicles in Italy', *International Journal of Automotive Technology and Management*, 9(2): 209–28.

Nieuwenhuis, P. and Wells, P. (2003) *The Automotive Industry and the Environment*. Cambridge, UK: Woodhead Publishing.

Oltra, V. and Saint Jean, M. (2009) 'Sectoral Systems of Environmental Innovation: An Application to the French Automotive Industry', *Technological Forecasting and Social Change*, 76: 567–83.

Ryan, L. and Turton, H. (2007) *Sustainable Automotive Transport: Shaping Climate Change Policy*. Cheltenham, UK/ Northampton, MA: Edward Elgar.

Sperling, D. and Cannon, J. S. (eds) (2007) *Driving Climate Change: Cutting Carbon from Transportation*. Burlington, MA: Academic Press Elsevier.

Teece, D. J. (1984) 'Economic Analysis and Strategic Management', *California Management Review*, 26(3): 87–110.

Tidd, J., Bessant, J. and Pavitt, K. (2005) *Managing Innovation, Integrating Technological, Market and Organizational Change*. Chichester: John Wiley.

Volpato, G. and Zirpoli, F. (2011) 'The Auto Industry: From Unfettered Expansion to Sustainable Development. Challenges and Opportunities', *Economia e Politica Industriale*, 15(2): 5–24.

Wells, P. (2010) 'Sustainability and Diversity in the Global Automotive Industry', *International Journal of Automotive Technology and Management*, 10(2–3): 305–20.

2
Manufacturing Capability and the Architecture of Future Vehicles[1]

Takahiro Fujimoto

Based on the framework of design-based comparative advantage, in which the dynamic fit between organizational capabilities regarding *genba* (manufacturing sites) and architectures of products and processes affect the competitiveness of the sites and industries, this chapter describes and analyses various issues within the world automotive industry. The current US automobile crisis is seen as a long-term consequence of the gap between the division-of-labour type capability and the integral-type architecture of small cars. Japanese competitiveness in small cars in the late twentieth century is illustrated as the fit between path-dependent accumulation of co-ordination-type organizational capability and stricter safety–energy–environmental constraints imposed on vehicles in advanced nations. The possibility of the commoditization of automobiles is discussed from the modularization point of view. The modular nature of locally-designed Chinese vehicles, as well as the integral nature of Indian low-cost vehicles, are considered. The architectural differences between purely electric and hybrid vehicles are emphasized as the chapter discusses the possibilities and limitations of the former type. Toyota's recent recall problem, discussed later in the chapter, is seen as a problem of product complexity overwhelming the company's organizational design capability.

The broad concept of manufacturing management

The broad concept of manufacturing refers to the total range of industrial activities that embodies the creation (development) of design information providing utility to users (customers), the transcription of this into media (production), and its transmission to the market (sales). In other words, 'manufacturing theory' can also be about contemplating the nature of what 'design' means to the economy (Clark and Fujimoto, 1991; Fujimoto, 1999).

Manufacturing management rearranged from the perspective of design information comprises three pillars: organizational capability in manufacturing; a capability-building environment; and the product-process architecture.

- Organizational capability in manufacturing refers to a system of organizational routines for carrying out the processes of creating, transcribing and transmitting design information to customers in a way that is more accurate (high-quality), more efficient (low-cost), and faster (reduced lead time) than that of competitors (Nelson and Winter, 1982; Fujimoto, 1999, 2007). In other words, an organizational capability to conduct the simultaneous achievement and improvement of quality–cost–delivery (QCD). This is closely connected to the ability of the site to engage in development, purchasing, production and sales.
- A capability-building environment refers to the environmental factors (conditions such as workforce, capital, materials, design information, market needs and government) that played an important role in building the organizational capability of a country's firms and *genba*. Generally, it is considered to be the experience of prolonged periods of rapid growth that determines the organizational capability or the organizational climate of a firm. Therefore, the capability-building environment before and after prolonged periods of growth is important. For example, as Japan's rapid economic growth began unexpectedly early because of the Cold War, businesses grew amid chronic shortages of labour, capital and materials, leading to growth based on the long-term employment and long-term trading that secured these resources. As a result, the organizational capability of 'integrated-type manufacturing' based on multi-skilled teamwork became disproportionate in Japanese firms. A typical example of this is the Toyota Production System. The key to the success of this system is minimizing 'time not spent on the creation and transcription of design information' and creating 'good flows' of design information to customers that do not become stale.
- Product-process architecture refers to the formal pattern of dividing and connecting the design information of artefacts. Generally, the elements that are the target of design activities are product function (required specifications and so on), product structure (parts and so on), and production processes (equipment, tools and so on), and while the domain of 'specific technologies' deals with the concrete causal relationship between individual elements, it is the theory of architecture that argues the forms of divisions and connections of these elements (Suh, 1990; Ulrich, 1995). The forms of divisions and connections of elements of product function and elements of product structure are called 'product architecture', while the forms of divisions and connections of elements of product function and elements of production process are called 'process architecture'.

Defined in this way, product-process architecture can be divided into two basic types: the 'modular (mix and match) type' in which there is a one-to-one correspondence between the functional elements, structural elements and processual elements; and the 'integral type' in which there is a many-to-many correspondence between the elements. Actual product line-ups are deployed on a spectrum somewhere between these two types (Ulrich, 1995; Baldwin and Clark, 2000).

Organizational capability and architecture

Manufacturing capability can be divided broadly into two types: 'division-of-labour type capability', which emphasizes the specialization of each processing step of design information; and 'integrated-type capability', which emphasizes the teamwork (the ability to adjust) of multi-skilled workers. From a historical perspective, I and others assert that the USA – a nation of immigrants – was of the 'division-of-labour type', which tended to make immediate use of existing individual talent, whereas Japan – a nation that experienced rapid economic growth amid a chronic shortage of labour – had a disproportionate tendency towards the 'co-ordination (integration) type' based on long-term employment and long-term trading, as noted above.

On the other hand, the formal aspect of design information – in other words, the abstract relationship between the design elements of function, structure and process – is called 'architecture (design conception)'. Architecture can be divided broadly into two types: the simple 'modular (mix and match) type' that is close to the one-to-one correspondence between the functional elements, structural elements and processual elements; and the complex 'integral type' in which the relationship between the design elements is intricate (that is, they are deployed sequentially on the spectrum). Of these, integral-type products involve solving simultaneous equations between the functional elements, structural elements and processual elements; in the case of complex products, what is required above all is the close mutual adjustment of design parameters through teamwork.

The comparative advantage of design location

Therefore, as a result of historical evolutionary processes, it appears that countries that have accumulated a disproportionate tendency towards co-ordination-type organizational capability (the capability to adjust) – Japan, for example – tend to have a comparative advantage in design cost with products that require several co-ordinated abilities; in other words, products with an 'integral-type' architecture. Moreover, if design sites are left for products that have a comparative advantage in design, will this not influence the location of a production site (factory); that is, the industrial structure of each country? The above is a hypothesis of the comparative advantage of design in which the organizational capacity disproportionate in countries and the appropriateness (fit) between the architectures of product and process bring about a comparative advantage in design. By applying the engineering principle of 'design' to the industrial structure theory and trade theory, we can endeavour to gain a new insight into the comparative advantages of twenty-first-century Japan (Fujimoto, 2007).

However, there is no architecture inherent in each product category (for example, 'automobiles'). As the user environment and the evolutionary pathway of design information are different in Japanese and Chinese firms, for example, even with similar compact cars, it is well known that there are substantial differences in the architecture. Japanese cars tend towards the integral type, while Chinese vehicles tend towards the modular type – that is, even artefacts (products) that have identical functions, because of differences in such things as the discipline of the user environment, the standards of functions required by users, and the strictness of regulations controlling designers, the architecture is likely to differ by country, firm and time.

Constraints and architectures

Generally, in the event that the required standards regarding design function are high or the constraints relating to design structure are strict, if we consider other conditions as being constant, then the architecture of the artefact (for example, a product) can be supposed to tend towards the integral type. If the conditions are strict, then there is a need to identify and optimize strictly both the functional and structural parameters. In this case, solving the functional and structural simultaneous equations corresponding to an artefact makes it easy to reach a specific solution for the product (this does not mean that the method of solution is specific, of course, but that the value of the design parameters is product-specific). In other words, if in general the functions that customers require or the constraints of society (such as environmental and safety measures) become more sophisticated or complex, I predict that dealing with these through modularization will become difficult, and the product architecture will become something that is both integral and complex.

Therefore, the recognition that 'the twenty-first century is the era of globally-shared environmental constraints', which continues to emerge as another characteristic of this century, tends to have important implications. Needless to say, the present age is one in which constraints relating to the global environment are becoming stricter. With issues such as the fierce international competition for scarce energy resources, controversy over global warming, the movement of environmental pollutants across national borders, and the international adjustment and increasing rigour of environmental regulations, energy consumption regulations and safety regulations in response to these, it has become an era in which both users and designers of artefacts share strict constraints that transcend national borders.

Competitive advantage in integral-type products

On the premise of the above observations, I will provide an explanation of the theory of comparative advantage of design location from the theories of organizational capability and architecture.

The organizational capability of the *genba* of firms is imbued with the history of those fields and organizational co-operation through competition in capability building between firms and across *genba*. On the other hand, the architecture of products also evolves through the unique influence of technology and market needs. Moreover, in the event that the fit (compatibility) between the organizational capability of the *genba* of certain firms and the architecture of the relevant products and processes is good, the competitiveness of that *genba* ('deep-level competitiveness' such as productivity, initial costs, lead time and the rate of defects) increases, which becomes linked with the competitiveness ('surface-level competitiveness' such as price and product strength) of products in the marketplace.

In other words, the organizational capability of firms is disproportionate in each of the world's main regions and it is the theory of comparative advantage of design location that considers this 'regionally disproportionate organizational capability' and 'compatibility' in architecture of the relevant products as in no small part influencing the industrial competitiveness of each region.

What would happen if, for example, this framework were to be applied to manufacturers in Japan? Clearly, each firm has its own characteristics; however, we can consider those firms up against the same capability-building environment as tending to build similar organizational capacities. For example, as noted above, as Japan in the second half of the twentieth century encountered a capability-building environment of rapid growth amid a chronic shortage of resources for production, it had a tendency to build 'integrated-type manufacturing' organizational capacities based on multi-skilled teamwork through competitiveness in capability building that had as its base long-term employment and long-term trading. This was disproportionate in Japan.

What was good about this type of organizational capability and compatibility was the products of 'integral-type' architecture listed earlier. As a result, through these types of products, a strong tendency could be observed in Japanese industrial competitiveness. This is one hypothesis relating to Japanese industrial competitiveness that is predicted by the comparative advantage of architecture. This tendency to comparative architecture was established in the second half of the twentieth century and is a tendency I see continuing into the first half of the twenty-first century.

Industrial architectural geopolitics

With regard to the countries and areas other than Japan, while at this stage these are impressionistic hypotheses from case studies, I forecast (Fujimoto, 2006) the following (my analysis focuses mainly on Asia, with only a rough analysis of Europe):

- Japan, with its disproportionate 'co-ordinative (integrative) capability of *genba*', has its strength in integral-type products that emphasize hidden competitiveness as a result of the historical development of its post-war economy.
- The strength of continental Europe, with its disproportionate 'expression capability', is in integral-type products that emphasize hidden competitiveness (brand design) as a result of its history of manoeuvring between medium-scale countries.
- The USA, with its disproportionate 'conceptualization capability', has its strength in knowledge-intensive modular-type products as a result of its historical background as a nation of immigrants.
- South Korea, with its disproportionate 'concentration capability', has its strength in capital-intensive modular-type products as a result of its history of growth based on conglomeration without the dissolution of conglomerations.
- China, with its disproportionate single-skilled 'mobilization capability', has its strength in labour-intensive modular-type products as a result of its history of suddenly moving away from a Soviet Union-type model to an open economy based on its coastal areas.
- Compared with China, a section of the ASEAN members that in general have a high labour-retaining capability has the potential to be strong in labour-intensive integral-type products.
- I also get the impression that, with the noteworthy exception of the Bangalore information industry, Indian manufacturers also tend more towards 'integral-type' than do those in China.
- Taiwan, located on the so-called 'intersection' of the US and China 'modular-axis' and the Japan/ASEAN 'integral-axis', has the ability to change its business partners swiftly in response to product architecture.

Thus, if we consider the theory of comparative advantage in architecture originating from design, we can anticipate that both the types with disproportionate organizational capability and those that excel in organizational capability will differ by region. This is because each country and region carries its own differing history. Moreover, this can be a basic framework in which to analyse the global economy in the twenty-first century and the comparative advantages of the industries that become evident in it.

Organizational capability and architecture evolve

The framework of the 'design-based comparative advantage', which predicts that the fit between the 'co-ordinated capability of the design sites' and the 'co-ordinated workload of the product design', is a dynamic theory that should embrace evolution if viewed from a broader perspective. The organizational capacity of the design field and the product architecture will evolve together and the fit between these will be a dynamic adaptation.

As shown in Figure 2.1, the organizational capability of the *genba* evolves while being influenced by the capability-building environment of regions, the capability-building competitiveness of industry, and the capability-building ability of firms (Fujimoto, 2007). On the other hand, the architecture of products and processes also evolves while being influenced by the requirements of users as well as social and technological constraints that designers encounter. Therefore, in evaluating the types of designs that Japanese industry will have to deal with in the future, an extremely long-term level of insight is necessary to predict how the environment surrounding *genba* and products will change in the future.

What we should pay attention to here regarding architecture is that, as previously stated, there is no architecture specific to each product category (for example, 'automobiles'). Even with similar compact cars, there are substantial differences in the architecture of Japanese cars that tend towards the integral type, and Chinese cars that tend towards the modular type; that is, even artefacts (products) with identical functions can have different architectures. There is no architecture specific to products.

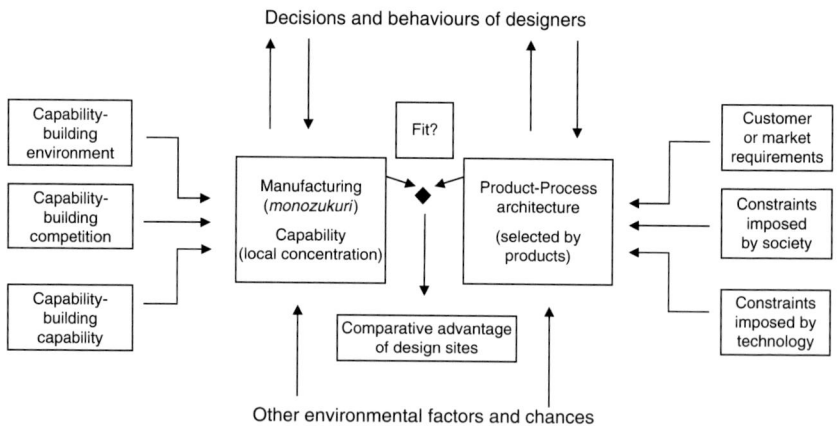

Figure 2.1 Design-based comparative advantage
Source: EHESS website: http://ffj.ehess.fr/index/article/265/takahiro-fujimoto.html.

Generally, in the event that the required standards regarding design function are high, or the constraints relating to design structure are strict, if we consider other conditions as being constant, then the architecture of the artefact (for example, a product) can be supposed to tend towards the integral type. If the conditions are strict, there is a need to identify and optimize strictly both the functional and structural parameters. In this case, solving the functional and structural simultaneous equations corresponding to an artefact makes it easy to reach a specific solution for the product. If, in general, the functions that customers require or the constraints of society (such as environmental and safety measures) become more sophisticated or complex, then the compilation of already designed products through modularization will become difficult, and the product architecture will become something that is both integral and complex.

As automobiles are 'heavy objects that travel at high speeds through public space', they have a negative impact on society, by being involved or implicated in a wide range of problems, such as traffic accidents, air pollution and global warming. In fact, requirements and constraints imposed by society on automobiles are becoming stricter every year.

In general, when constraints become stricter, the product design becomes more complex. What is meant by design is the vision to connect the functions and structure that an artefact should have, but if the required functions and constraints become stricter, it is then difficult to deal with them by using modular-type design concepts (architecture) in which functionally-complete parts are accumulated, and a new array of integral-type design parts that have been finely co-ordinated for total optimization is required.

Thus the design of automobiles in industrial nations has become more complex. For example, compared with the start of the 2000s, the ratio of common parts in the average car in Japan has decreased, the ratio of electronic control has increased, and the volume of new product developments has increased. While engineers have made efforts to simplify design, the selective environment (that is, society) does not permit the persistence of modular-type designs, and products as a whole have become integrated. This is the difference between the design evolution of automobiles and personal computers.

Assuming this, the perception of the 'twenty-first century as an era of environmental constraints' has important implications. Even after there has been a recovery from the global recession, we shall still face a century beset with issues such as the fierce international competition for scarce energy resources, the controversy over global warming, the movement of environmental pollutants across national borders, the international adjustment and increasing strictness of environmental regulations, and energy consumption regulations and safety regulations in response to these; it will be an era in which both users and designers of artefacts share strict constraints that transcend national borders.

Commodification and loss of competitiveness in Japanese industry

At the time of writing, with the intensification of global competition, Japanese firms and *genba* have lost the competitive advantage across a range of products. In many cases this occurred when, on the one hand, a large part of the global market began to attach importance to price rather than functional difference, while on the other hand firms and production fields in developing nations began to provide extremely cheap products, despite some degree of a reduction in function. To a large extent in the market, Japanese products lost ground in their branding because of 'excessive design, excessive quality and excessive function', and were defeated. This phenomenon is generally known as 'commodification'.

While commodification is close to the phenomenon known as standardization in the product cycle hypothesis, at present this is a weak point for Japanese industry. Japanese products, which pursued extreme performance that became excessive design, lost their markets repeatedly to the low-priced products from developing nations in Asia that emulated Japanese goods, and researched and simplified them. Many of these were digital data goods such as CD media, in which the emergence of Taiwan meant a significant drop in prices, overwhelming Japan, and DVD players that started as Japanese products but ultimately the main production moving to China as Japan watched helplessly.

Thus, because of these repeated cases of Japan – a country renowned as a nation of technology – losing its market share in high-tech products, it became linked to a growing pessimism in the Japanese economy itself. Though exports of luxury goods from Japan, such as automobiles, were booming during the US economic bubble, this too disappeared, and once again the world was shrouded in pessimism about Japanese manufacturing.

However, if household electrical goods are commodified, then is this not also the case with automobiles? Initially, there is a need to observe actual products open-mindedly and to analyse the fields of their design, production and consumption.

The design, production and consumption of artefacts

Here, let us try to reinterpret commodification from the theory of design (Fujimoto, 2007). First, *'genba'* refers to the place where people become involved with artefacts. Designers design artefacts, workers operate them, and users gain function and utility from them (see Figure 2.2).

As shown in the figure, an 'artefact' is a designed object; in other words, it refers to individual objects that have had design information (elliptical boxes) transcribed into media (rectangular boxes). Moreover, what is meant by 'design' is the conception that forms connections between the function

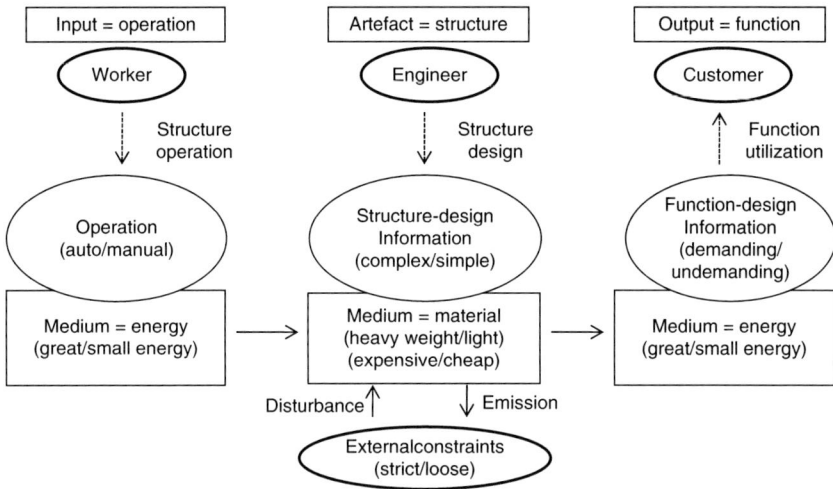

Figure 2.2 Basic forms of production (structure, function and operation of artefacts)
Source: Revised from Fujimoto (2007), Figure 2.

and structure of an artefact. Structure design transcribes objects, whereas function design transcribes energy.

The people connected with the field are workers who operate the structure of artefacts, users who use function, and designers who design structures. I will not go into detail about the processes here, but in the case of material goods consumption, the owner of an artefact = user = worker; in the case of production, artefact = manufacturing facilities; and in the case of services, user ≠ worker. All of these are *genba*.

At any rate, design is by nature information, and media are material and energy. In other words, from the perspective of fields, we must ascertain the characteristics of the design information of products and processes, and the characteristics of the material and energy of the media. The basis of industrial competitiveness analysis is to concentrate on *genba* observation, putting aside for a moment the government's industrial classification, industry groups and competent authorities.

Design is also the activity of simulating consumption and production in advance. The designers of consumable goods simulate the consumption process in advance, while the designers of production goods simulate the production process in advance, tracing the structure from the required functions of users, operating environment and operating content.

For example, in the case of automobiles, designers trace the structural parameters (such as dimensions and configuration) of such things as the body, chassis, engine, transmission and control system from the required functions and constraints, including fuel consumption, emissions, safety, horsepower, ride quality and attractive style. We can consider that this is a task of solving simultaneous equations from functional and structural elements.

Generally, when the required functions or the constraints of consumers and producers become stricter, then modular-type design becomes difficult. There is a many-to-many entanglement of structural elements and functional elements, and designers come under pressure to solve complex simultaneous equations. The solving of equations becomes specialized (in other words, parts become product-specific) and the search for the optimum solution becomes problematic, and modular-type design is not carried through. As stated previously, this type of complex design format is called integral-type architecture.

Thus we understand that the commodification that has given Japanese firms a difficult time has, in many cases, accompanied modularization. In fact, the reality that Japanese firms, products and *genba* lose the competitive advantage when structural parts and production facilities become functionally complete, body and connecting parts and interfaces become standardized, and the mixing and matching of these are able to ensure customer satisfaction, has been made clear by many researchers, both inside and outside Japan.

The conditions of commodification

However, does this mean that all products eventually become modularized and commodified? I do not feel that this is the case. What becomes necessary is the level-headed confirmation of *genba* (manufacturing site) and *genbutsu* (actual product).

As previously stated, as product design is a simulation of consumption, we cannot decide whether the design is simplified or the product is commodified without investigating individually the strictness of functional requirements and constraints that occur in the field of that product's consumption. If the constraints become lax, there is the immediate possibility of products becoming modularized, though if the constraints are strict, there is the possibility that modularization will not occur.

Returning to the artefactual analysis of the *genba*, let us now consider the characteristics (of information, objects and energy) of products that are easy to modularize or commodify.

First, when the functional requirements that customers expect in a product are not rigid, then that product is easy to modularize and commodify. This is because these less precise requirements can be dealt with by the mixing and matching of parts that have already been designed.

Second, for the same reason, those products on which society places fewer constraints are also easy to commodify.

Third, in the event that the mass (size) of the media of the product (artefact) is small, there are fewer constraints relating to safety problems. Digital items that are being reduced in size fall into this category.

Fourth, in the event that the function of products developed through operating does not have a large input or output of energy, then the energy saving constraints that have become an inevitable part of contemporary society can be avoided. Typical of these are electronic digital products.

Fifth, in the event that the media – in other words, the materials used in the making of a product – are cheap, then the mass production of such products and parts becomes linked directly with cost cutting, and thus commodification is easy. In this regard, we can say that silicon and iron are examples of low-cost media that exist in abundance on the Earth.

Therefore it is also easy to simplify the design information for products with a small mass, low input and low output, comparatively relaxed constraints and customer requirements, and that have cheap media costs. In other words, they are easy to commodify. Modular-type digital products, such as computers, which Japanese firms have struggled with in recent years, are undoubtedly close to these conditions.

However, it is also easy to complicate the design information for products with a large mass, high input and high output, strict constraints and customer requirements, and that have high media costs. Compact cars are a typical example of this.

Will automobiles commodify in developed nations?

In this respect, automobiles are 'heavy objects that move through public spaces at high speed'. They have a large mass, input and output. Customers have high standards; they require multiple functions such as movement, leisure, self-expression and a substitute for the home, though these are accompanied by the strict constraints that society imposes on automobiles because they are associated with increasing traffic accidents, air pollution, global warming and the depletion of resources. They are objects that came into being with the original sin of being harmful to society, and efforts to offset these negative aspects are endless. Society monitors them strictly with regard to such matters, and public regulations are also moving towards becoming even tougher. Moreover, the discerning eyes of the consumers who pay large amounts of money for cars are also critical. This is the definitive difference between automobiles and digital commodities driven by electrons, which can disregard the existence of mass.

As previously stated, if the conditions of consumption are strict, then the relationship between the required functions and structural elements becomes one of complex many-to-many correspondences. One part takes on many functions and many parts support one function. Fuel efficiency as a function is supported by many other functions, including the body, chassis, engine, transmission, tyres and auxiliary items. The body as a part contributes to many functions, including the safety of the occupants, the safety of pedestrians and other road users, fuel efficiency, external appearance, the quality of the interior, sound insulation, heat insulation, ride quality and the ensuring of visibility. Moreover, the need to adjust the parameters of these becomes increasingly important.

On the other hand, if safety and environmental regulations and the functional demands of customers become lax, then automobiles are quickly commodified and are no longer products that workplaces in Japan, whose strengths lie in multi-skilled team production planning, have a comparative advantage over. At this point, as with many digital products, it may be necessary to accept the reality of transferring fields overseas; in other words, the Japanese automobile industry as a collection of workplaces in Japan might lose ground.

The dissemination of electric vehicles is not easy

However, if nothing else, the design information of automobiles in developed nations is becoming more complex. In extreme cases, there are 10 million lines of code in control software, 100 chips in microcomputers, and 30,000 single parts. These are put together with 4,000 bolts and 4,000 welding points. The ratio of common parts has decreased since the early 2000s and there are an increasing number of key tasks and labour hours required in development.

Though the main forms of power are petrol in Japanese, US and European automobiles, in recent years some 50 per cent of new European automobiles have been powered by diesel, as are large commercial vehicles. All are driven by internal combustion, and hybrid and electric vehicles are in the minority. A keynote of the next few decades will be the diversification of engine types, though it is inconceivable that in the near future internal combustion engines will disappear, and thus it will be difficult for firms that cannot develop a good combustion engine to prosper. When automobiles become more complex, the key to survival is 'how to make complex components at a lower cost' and not 'simplifying products to make them cheaper'.

Though details of electric vehicles and Chinese and Indian local automobiles will not be discussed here, they can be predicted to a certain extent from the framework of the artefactual analysis highlighted above.

First, the design of electric vehicles, at present a popular topic for discussion, has a high possibility of modularization in the future, but this will not be connected directly with commodification. This is because automobiles are moving objects with a large mass, large input and large output that will not change. Electric vehicles that rely only on electrical energy to power them require battery materials with a high energy density, which are extremely expensive. At present, Mitsubishi's electric vehicle, one of the best electric vehicles available, is fitted with a cutting-edge lithium-ion battery that costs around ¥2,000,000 (more than US$20,000), weighs 200 kg and has a realistic operational distance of 100 km of continuous driving. There is no sign of any other type of revolutionary battery material being in development that will replace this.

Because of this, there is no choice but to reduce the absolute amount of the battery for each vehicle to a fraction of the current amount. This is what has happened with hybrid vehicles. As purely electric vehicles, we can consider a battery-sharing system, although this will mean limiting their use to short distances within cities. Whatever the case, the system of selling electric vehicles needs a co-operative solution to be found by national and local government, the industry and other players, but it is inconceivable that we shall reach a stage in the near future where this will replace ICE-type, privately-owned automobiles through stand-alone sales (though it is possible with commercial vehicles, such as taxis). Keeping in mind the economic principle that the cost of battery materials is high, the material cost is a variable one, and that variable costs will not decrease even if a million vehicles are produced, we need to engage in level-headed discussions rather than being swept away by possibilities. The articles one often sees proclaiming that the age of the electric vehicle is at hand lack supporting theories regarding the fields of production, design and consumption, and are thus unreliable.

Between increasing complexity and commodification

Level-headed discussions are also needed regarding low-cost Chinese and Indian automobiles. Many of the low-cost automobiles made by the Chinese firms are a miscellany of commercial parts, copied and modified parts, and self-manufactured parts, and are thus do not meet the functional standards required by the markets and societies of the developed automobile nations. The low-cost automobiles developed by India's Tata Motors are much better than Chinese automobiles with respect to fulfilling the required functions and constraints, though they have a completely different architecture from typical Chinese automobiles.

In short, electric vehicles are modular but expensive; China's low-cost cars do not fulfil the required functions of the global market; and while India's low cost cars – for example, Tata Motors' Nano – are more sophisticated functionally, they have many specially-designed parts, and tend towards the integral type rather than the modular type. Whatever the case, the reason that commodification that occurs frequently in household electrical appliances and electronic products does not do so on any great scale in the world of automobiles is suggested from an industrial analysis of the architecture.

In general, if they are products that require a diverse range of functions and models, do not have strict standards required by customers, are formless, have few limitations on capacity, are light, are not used in social places (so as not to cause inconvenience to others), and do not put much pressure on energy and the environment, they can be modularized and commodified relatively easily. However, the compact automobiles on sale in the markets of developed nations do not satisfy these conditions. Therefore, even if designers make the strong efforts to create a modular architecture, following these efforts is likely to remain on the integral side for the foreseeable future because of constraints related to safety, the environment and energy use becoming stricter.

As things stand at the start of the twenty-first century, it is difficult to envision automobiles suddenly having architectures that are modularized in a similar way to those of digital electronic devices. On the other hand, there is no shortage of fields that confront the long-term issue faced by firms of 'the growing complexity of artefacts', including automobiles that transport people safely, medical devices that target the human body, production equipment that processes delicate products, industrial instruments that process difficult-to-handle materials, electronic devices that require extreme miniaturization, and embedded software that requires the complex control of real time. The twenty-first century has been characterized to date by environmental constraints, to which many industries and firms have no choice but to respond simultaneously and in line with increasing complexity and commodification, against a backdrop of rapidly changing design, production and usage conditions surrounding artefacts.

Conclusion: the crisis in the automobile industry and the future

The 2008 global recession that had its roots in the US financial crisis also landed a major blow to the Japanese economy. Until then, the demand for luxury automobiles and home electrical appliances originating in the US bubble benefited Japan directly or indirectly. However, with the US economy a shambles from 2008, the Japanese export and commodity industries and firms that relied on exports to the USA experienced a drastic decrease in production and a worsening of its financial performance in 2009.

From relying on the US luxury sport utility vehicle (SUV) market, Japanese automobiles, particularly those of Toyota, increased in size, became heavier and more complex. This brought huge profits to Toyota from the USA; however, with the collapse in demand after the US bubble burst, there was a sudden reduction in sales volume and a worsening of the company's finances. These were exacerbated by the major strategic mistake of its overly slow construction of a US truck factory. Product development inclining towards luxury automobiles and the latest ecologically-friendly automobiles to meet demand from developed nations such as the USA meant that there was a delay in detailed product development aimed at the markets of developing nations, and the development of competitive low-cost automobiles. On the other hand, in the US and European markets, Toyota could not keep up with building capability in line with the rapid increase in complexity of products and operations, causing substantial problems in quality.

In other words, because of the over-reliance on the US market for large, luxurious and complex automobiles, and their own strategic mistakes, Toyota found itself crisis-ridden. However, assuming that the strengthening of social constraints regarding safety, the environment and energy saving, and the increasing sophistication of the functional requirements of customers in developed nations will continue in the future, in the long term this will not necessarily be a disadvantage for Japanese firms such as Toyota, who excel in maintaining co-ordination-type organizational capability, and in integral products that become complex. In the future, despite the battle with design complexity in developed nations continuing, and in the short term the recurrence of crises is also foreseen, Japanese firms should see the recovery of markets in developed nations such as the USA and begin to revive their own business performance.

Concurrently, the challenge for Japanese firms, particularly Toyota, is on the one hand to tackle the increasing complexity of products and on the other to move forward with product development and market cultivation in the medium-to-low price zone (the so-called volume zone) in newly-developing nations. With regard to the cultivation of markets in developing nations, taking Japan and Korea as examples, there is a tendency for Korean firms to transcend Japanese firms, and thus there is much that Japanese firms can learn from this. On the one hand, they need to deal with the problem of design quality in developed nations – in other words, the problem of under-engineering – and, on the other, to deal with the problem of over-engineering of Japanese automobiles aimed at developing nations. The formulation and implementation of this double-sided strategy will not be easy. This challenge is not only for Japanese firms; it is likely become an important issue for all automobile manufacturers aiming at the global market in the first half of the twenty-first century.

Notes

1. This chapter is a revised and shortened version of T. Fujimoto, 'Complexity Explosion and Capability Building in the World Auto Industry', *Economia e Politica Industriale*, 38(2): 25–49.

References

Abernathy, W. J. (1978) *The Productivity Dilemma*. Baltimore, MD: Johns Hopkins University Press.
Baldwin, C. Y. and Clark, K. B. (2000) *Design Rules: The Power of Modularity*. Cambridge, MA: MIT Press.
Clark, K. B. and Fujimoto, T. (1991) *Product Development Performance*. Boston, MA: Harvard Business School Press.
Fujimoto, T. (1999) *The Evolution of a Manufacturing System at Toyota*. New York: Oxford University Press.

Fujimoto, T. (2006) 'Architecture-based Comparative Advantage in Japan and Asia', in K. Ohno and T. Fujimoto (eds), *Industrialization of Developing Countries: Analysis by Japanese Economics*. Tokyo: National Graduate Institute of Policy Studies, pp. 1–10.

Fujimoto, T. (2007) 'Architecture-based Comparative Advantage: A Design Information View of Manufacturing', *Evolutionary and Institutional Economics Review*, 4(1): 55–112.

Jones, G. (2005) *Multinationals and Global Capitalism*. Oxford: Oxford University Press.

MacDuffie, J. P. and Fujimoto, T. (2010) 'Why Dinosaurs Will Keep Ruling the Auto Industry', *Harvard Business Review*, June: 601–3.

Nelson, R. R. and Winter, S. G. (1982) *An Evolutionary Theory of Economic Change*. Cambridge, MA: Harvard University Press.

Ricardo, D. (1971 [1819]) *On the Principles of Political Economy, and Taxation*. Baltimore, MD: Penguin.

Samuelson, P. A. (1948) 'International Trade and the Equalization of Factor Prices', *Economic Journal*, 58(230): 165–84.

Suh, N. P. (1990) *The Principles of Design*. New York: Oxford University Press.

Ulrich, K. T. (1995) 'The Role of Product Architecture in the Manufacturing Firm', *Research Policy*, 24(3): 419–40.

Womack, J., Jones D. T. and Roos, D. (1990) *The Machine that Changed the World*. New York: Rawson Associates.

3
Industrial Designers and the Challenges of Sustainable Development

Monique Vervaeke

The challenge of sustainable development and new mobility facing the automotive sector implies a change in strategies among car manufacturers, government intervention (at the local or national level) and compliance with new regulatory frameworks. New mobility also has broader implications for society in general. The automotive sector's profit model, founded on the internal combustion engine, needs to change to make room for alternative models (Freyssenet, 2009). New models are now in the experimental phase, and are based on a new conceptual approach to innovation and the social organization of mobility. The public sector is stimulating the private sector in setting up and encouraging new modes of transport.

In economics and business management fields, research on innovation is too often focused on technology, mass-market globalization and scale economies. This approach fails to consider the strategic role of industrial design that, over and above aesthetic ideals, takes into consideration the cultural meanings attached to products and services (Lorenz, 1986; Manzini, 1986 and 1990; Calabrese, 2010). Thinking of innovation as a *design driven* process implies extending the limits imposed by 'techno push' or 'market push' approaches. A design driven process is based on exploratory studies formulating future socio-cultural scenarios to determine the new meanings that will be associated with a product or service (Brown and Katz, 2009).

The *design thinking* approach also permits the study of experiments conducted by small companies, alternative players to the major carmakers operating globally. This chapter puts forward the hypothesis that, in the automotive sector, faced with massive social pressure to transform its capital accumulation model, actors other than the dominant carmakers are creating new opportunities and have the possibility of taking a place in the value chain. New competitors with a sustainable strategy have initiated experiments providing interesting case studies. They could open up new perspectives by redefining the principles governing the supply of products and services.

Industrial designers have engaged in a debate criticizing industrialization that fails to take environmental impact into account. The first part of this chapter deals with the issue of eco-design and design thinking as debated and experimented between industrial designers (Papanek, 1972; Manzini, 1986; Lotti, 1998; Manzini and Vezzoli, 1998; Brown and Katz, 2009).

The second part examines the role of industrial designers in the policies of automakers: the reformulation of transportation from the perspective of 'product/service'; aesthetic research on the properties of lighter materials; new digital techniques contributing to a reduction in the number of prototypes; and symbolic aspects of the construction of new markets. Industrial designers co-operating with engineers become increasingly integrated with the design process.

The third part presents three case studies of new mobility: the Cristal project under development by the supplier LOHR in the field of public transportation; the Cybergo project, developed by the electronics company Induct in collaboration with Diedre Design; and the Bolloré–Pininfarina Bluecar project chosen by the mayor of Paris for the electric-car-sharing system 'Autolib'. The eco-design issue opens up thinking on multi-mobility systems and new partnerships between firms and public policy initiatives.

Design thinking as a new approach to mobility and eco-design theory

Henry Ford once said: 'If I'd asked my customers what they wanted, they'd have said a faster horse', a remark indicating that modes of mobility are subject to cultural relativism. The Fordist industrial revolution brought about a change in society affecting the way automobiles were manufactured and consumed (Ford and Crowther, 1922; Brown and Katz, 2009).

Imagining the future of mobility implies rethinking or reimagining the way we travel. It implies more than simply limiting a vision of the future to the improvement of existing cars. Improving the performance of the internal combustion or hybrid engine is a medium-term strategy. Prospective research should lead to the elaboration of alternative solutions to counter the inevitable exhaustion of fossil fuels. The actors preparing future transformations in mobility will be in a position to impose their innovative solutions and capitalize on future changes to their advantage.

Design alters the concept of the product as a single tangible object to a product-system integrating a combination of products, services and communication systems. In France, the distance travelled between home and workplace is on average 35.4 km per day and, for 50 per cent of employees, less than 5.8 km per day (Baccaïni et al., 2007). Automobiles are often an oversized means of transport, costly in energy and harmful to the environment. Designers' researches into aesthetics are driven by cultural considerations and their implications regarding the products that constitute our material environment. Design introduces societal issues into company culture (Manzini, 1990).

In publishing *Design for the Real World* (1972), Victor Papanek[1] stated that designers share the responsibility for pollution. His book was published in the same year as the Club of Rome report entitled *The Limits to Growth* (Meadows *et al.*, 1972), which emphasized the devastating effects of the existing development model in terms of pollution, the downgrading of cultivated soils and the depletion of energy sources. Papanek called a chapter of his book 'Design for Survival and Survival for Design'. Designers are assigned the task of seeking the 'collective good', working on a quest for beauty that introduces an aesthetic aspect into the day-to-day environment. Designers share a social and critical responsibility. Papanek criticized any design that proposes the use of wasteful materials or objects. As long ago as 1972, Papanek stated that the new generation of designers should work on products such as compact electric cars, battery-powered means of travel, non-polluting industrial complexes and the use of biodegradable materials.

He wrote his book 15 years before the publication of *Our Common Future*, also known as the Brundtland Report (Brundtland Commission, 1987). By defining the notion of sustainable development, this report went beyond the established fact that industrialization has pernicious effects on environmental equilibrium. It set out the principle that each inhabitant of the Earth has an equal right to access natural resources, and that each successive generation has the responsibility to use natural resources without endangering future generations.

Some industrial designers are developing a critical standpoint concerning the development of market segmentation driven by high product renewal rates. A great many products have no basic utility or cultural value. According to Ezio Manzini, one of the issues underlying aesthetic research by designers is giving aesthetic form to sustainable development. The aesthetic aspect is a crucial factor of the changes that accompany a mode of production and consumption promoting sustainable development. The renewal of aesthetics can only emerge if it is based on a system of values (Manzini, 1989).

From the 1980s onwards, Manzini elaborated a framework of ideas going beyond friendliness to the environment. He introduced the notion of a connection between innovation as a cultural project and the social relationships of the future. His introduction of products as a service incorporating cultural codes and multi-modal sensory integration was reformulated by Giovanni Koenig through his vision of *'design per la comunità'*. According to this author, the creative approach regarding the shape of objects incorporates an analysis of the purpose and use of objects. Effective use of the tools of representation enables designers to give new meaning to a product with instantly perceived values. Product meaning depends on its symbolic and social value characteristics. Design processes that highlight the way products are made, used and perceived decrease the importance of technological constraint (Koenig, 1991).

According to Giuseppe Lotti, companies became interested in environmental issues in the 1990s when consumer movements initiated green marketing, and political organizations such as the European Community (EC), set eco-objectives. In terms of 'greenhouse' gases and product recycling, the environmental involvement of firms concerns their market strategies. This involvement is linked to the impact of companies' environmental strategy on their balance sheets. The environmental involvement requires significant organizational changes, initiatives and co-ordinated planning (Lotti, 1998).

From a sustainable development perspective, designers' creative approaches include reflection on the productive process as a whole, the choice of parts, their assembly, and their disassembly for recycling. Designers seek to promote the emergence of new alternatives and design products whose manufacture and usage tend to minimize environmental resources (Lotti, 1998; Manzini and Vezzoli, 1998; Kazazian, 2003).

Stemming from a study of designers' use of materials, Manzini considered design as being oriented towards creating a new combination between products and services, respecting usage value and using low material-energy intensity associated with the use of recycled materials (Manzini 1986; Manzini and Vezzoli, 1998). According to Manzini, the deterioration of the environmental equilibrium and the limitation of natural resources require radical changes in the industrial culture. The new 'ecology of the artificial' consists in making products with special care being taken over details and the life-cycle of objects in relation to human life and the environment (Manzini, 1990).

Eco-design challenges production, consumption and distribution modes. Product strategy based on sustainable development is associated with broader prospective reflections on mobility as a service. The scope of sustainable development in the automotive sector is therefore broader than simply designing cars with reduced exhaust emissions. Design thinking enables the search for new ideas in urban mobility compatible with sustainable development objectives. The designer Roberto Giolito (head of design at Fiat, and designer of the new Fiat 500) expresses his views on city mobility and public space as follows:

> The city must be regarded as resulting from diverse interventions between transportation means and their infrastructures as well as with communication technologies, which evolve dramatically and offer the opportunity of being less dependent on cars and individual vehicles in general. I do not think that the city will evolve in accordance with cars and I do not feel I am a heretic in expressing such opinion as I work for a carmaker, but I believe that it has a great influence on my work. In the gradual transformation of the city, space must be considered with respect

to pedestrians. This also implies improvement of the traffic by reorganizing roads and parks so as to make vehicles lighter, more environmental-friendly and more transparent than they are currently. (Finizio, 2005)

Design thinking is a way of questioning human practices and proposing a human-centred approach to innovation. Instead of targeting a clientele according to existing market organization, mobility is observed and interrogated to encourage behaviours and practices more compatible with environmental protection. It announces the transformation of mobility systems.

A product strategy based on sustainable development is associated with broader prospective reflections on mobility as a service. Different design theoreticians, and in particular Manzini and Thierry Kazazian, have led thinking about the sensory, aesthetic and symbolic properties of new materials that could be introduced in an eco-design approach. These studies have provided theoretical possibilities to enable the rethinking of the design of energy saving vehicles, not only in terms of industrial processes, but also in terms of usage and life-cycle.

Research on form, materials and colours is correlated closely with the values, social practices and cultural representations of society at a given time. Certain behaviours are the implicit expression of expectations that anticipate future changes. Design thinking implements methodologies to observe mobility behaviour, social customs and practices. The data collected is then used to elaborate scenarios and learn from them in order to design a system. The design driven approach to innovation conceives a product as a system integrating the values to be communicated, the message that will be perceived by the consumer from the initial design phase and throughout the development process (Lorenz, 1986; Maldonado, 1991).

Since the 1980s, numerous concept cars have involved the development of low-pollution engines, initiating a major change towards a sustainable future. The architecture of cars will be transformed through the installation of batteries, new engines and electronic commands. Concept cars are preparing for the mutation of the car market towards vehicles using energy sources other than petrol and preparing future users to change their representations and behaviours associated with cars. But this is only one aspect of the contribution to contemporary changes in transport systems provided by industrial designers through their ways of working. Design thinking examines mobility as a social behaviour and service offer.

Solicited by the bicycle parts manufacturer Shimano, Tim Brown, CEO of the design firm IDEO, used an alternative approach to design research. Rather than approaching the problem from the company's traditional top-market niche, the IDEO team explored the problem from a human-centred perspective. Their observations on the relationships of adults with the bicycle showed that it was often a nostalgic reminder of childhood, generally abandoned as an adult. The sophistication of bicycle design with professionally oriented distribution channels, complex maintenance and insufficient roadside urban cycle tracks guaranteeing safety had resulted in adults abandoning this type of mobility. Inspired by the past, when cycling was considered as a simple, fun activity that made life easier, designers developed innovative components that were nevertheless easy to use. In collaboration with local governments and cycling organizations, the team produced a public relations campaign including a website that identified safe places to ride (Brown and Katz, 2009). The IDEO team contributed to setting up the industrial and social environment and creating partnerships between companies and associations encouraging multi-mobility combining the use of bicycle-and-bus or bicycle-and-train. This constitutes an alternative approach to innovation no longer based on 'problem solving', 'market push' or 'technology push'.

Complementary systems of mobility could evolve combining rail transportation, public transportation and individual means of transportation for short-distance urban mobility. Small electric urban vehicles made available through car-sharing schemes could contribute to new ways of thinking about multi-mobility.

Mobility as a service offer: from style to design

In the nineteenth century, the first car designs were influenced by the forms of carriages with high floorboards and large diameter wheels. The driver was seated like a carriage driver, with the steering wheel in a low position. Pierre Francastel noted that innovations introduced to the market were often based on old aesthetic models. As an example, he cites the first iron columns introduced in architecture that were moulded in the classic style, with Gothic and Renaissance style decorations (Francastel, 1956).

In 1899, Camille Janatzy broke a speed record by reaching a top speed of over 105.98 km/h in an electric car named the *'La Jamais Contente'* ('The Never Satisfied'). The body-maker, Rothschild (Rheims Auscher's successors), used sheet aluminium. The bodywork, with its aerodynamic shape, small wheels and lower position of the deck, attracted attention in that it totally deviated from the cultural codes embodied by the 'hippomobiles'.

In terms of mobility, the automobile imposed itself as the dominant mode of transport because of the numerous advantages it offered compared to the hackney carriage (speed, no smell of horse dung in city streets, for example) and the train (independent travel). At the end of the nineteenth century, talented body-makers who had mastered the art of moulding curves and volumes associated with craftsmen (woodworkers, saddlers, goldsmiths, upholsterers) to design custom-built bodywork for automobiles. This craft tradition was maintained in the French luxury car sector until the 1950s.

Before society was prepared to accept a mass-produced car, Henry Ford had to overcome a number of prejudices. Compared to a train weighing hundreds of tons, the lightweight cars were initially perceived as being dangerous. Heaviness associated with safety compared to lightness perceived as dangerous was already a subject of public debate (Ford and Crowther, 1922). The Model T Ford was ground-breaking because of the standardization of parts that rationalized the manufacturing process. It represented the transition from luxury custom-made bodywork to industrialized, mass-produced bodywork. Its cubic form, however, remained close to that of the traditional carriage.

At General Motors (GM), the creation of a style department (the 1927 Art and Color studio, which became the Style Section in 1937) marked a breakthrough in automobile design. Alfred Sloan, CEO of General Motors, sought to stimulate sales by accelerating the obsolescence of existing models by rapidly changing their style. Harley Earl, who directed the General Motors design department, greatly influenced automobile style in the 1950s and following decades. He innovated by generalizing the use of modelling clay to sculpt scale mock-ups, creating a new repertory of forms inspired by the aviation industry. The aerofoils and chrome bumpers turned the car into a symbol of modernity and speed (the Buick Y-Job model, for example). Styling and restyling permitted the creation of surface diversity, differentiating and renewing models without involving large-scale investments in the research and development of new ranges of models.

Raymond Loewy (1951) criticized the styling incorporating chrome features that loaded down the car with decoration. He gave much thought to the connection between the product's appearance, its components and its use. For this industrial designer, the key to design aesthetics was simplicity. His approach marked the transition from stylistic design limited to external appearance, to industrial design that relates form to the overall product system. He defended the principle that each component must be designed skilfully in order to economize on raw materials. As the multiplicity of elements creates confusion, the designer must attempt to simplify by eliminating or combining. His approach consisted of reducing form to its essentials, textures and colours, and eliminating noise and parasitical vibrations to create a harmonious whole. He advocated that fuel gauges and speedometers be easily legible. He considered that a well-designed car, and in particular his Studebaker models, had a shape that ensured good driver visibility, thus protecting the life of the passengers. Loewy had worked for the Pennsylvania Railroad Company and his work is emblematic of a sleek, aerodynamic style: the 'streamline' also used in his designs for the Studebaker. According to Loewy (1951), a car must look fast, whether it is moving or parked, and look as if it were leaping forward. His reassessment of style brought about an evolution in automobile design moving towards an increasingly integrated approach. His creative approach made the car an icon of modernity, an object of desire on which a growing number of families focused their hopes and aspirations.

The industrialization process that permitted the mass production of cars redefined the division of labour between the carmakers (engine and chassis specialists) and body-makers (design specialists). The rationalization of the design and manufacturing process that standardized vehicle components progressively eliminated small, independent body-makers and carmakers (Bellu, 2007). Design as an activity asserted itself in association with the industrialization of bodywork. The different players along the value chain reorganized themselves around the carmakers that employed designers in newly integrated style departments. Style conferred elegance to the industrial object and was a mark of differentiation and competitive advantage.

The small and popular automobile models democratized the automotive industry (Herbert Austin launched the Austin Seven in 1922; the Fiat 500 'Topolino' in 1936; the Volkswagen Beetle was industrialized after 1946; and the Fiat 600, designed by Dante Giacosa, in 1956). The Citroën 2 CV, designed by Flaminio Bertoni, the first studies for which dated back to 1935, was launched in 1948, and was to remain on the market for four decades.

In contrast to French coachbuilders at the beginning of the nineteenth century, Italian coachbuilders survived as independent stylists/designers working for the carmakers. By preserving their distinctive stylistic signatures, they evolved progressively to become independent engineering and design companies, intervening in R&D or outsourcing resources for the creation of concept cars or computer-assisted design (Calabrese, 2010).

The carmakers' design teams were built up progressively and collaborated with renowned stylists. From the 1950s, regular collaboration between French carmakers and Italian stylists from Turin (Giovanni Bertone, Carrozzeria Ghia, Battista Pininfarina) gradually transformed luxury custom-built bodywork to standardized bodywork, and brought about the diversification and market segmentation of models. The 1973 and 1979 oil crises led to restructuring within the automotive industry (Michelin's management of Citroën transferred to Peugeot; Peugeot acquired European Chrysler subsidiaries). Matra finalized a series of prototypes that included plastic materials.

The 1980s were marked by the intensification of Japanese competition (Freyssenet *et al.*, 2000). Increasing productivity and speeding up product development became a major challenge for carmakers. To introduce a greater number of vehicles with a high differentiating value on the market as rapidly as possible, they reinforced their capacities in the exploration of new services and innovative technologies (Midler, 2007). The influence of in-house industrial design in the product development process increased.

In 1987, Patrick Le Quément was recruited by Renault chairman, Raymond Lévy to manage the 'style centre', which was later to become the Department of Industrial Design. Design finally acquired the same corporate status as the 'Études techniques' or 'le Produit' (technical research or product departments). This organizational change was a recognition that design played a strategic role within the company. It illustrated the increasing influence of design in the product development process, and in particular, the design of the Twingo (Midler, 1993).

Electric concept cars began to appear at motor shows during the 1990s. Nicolas Hayek, who completely transformed watch design with the launch of the Swatch, formed a team to design a small, two-seater electric car for rental. In 1994 he began a partnership with Mercedes, but abandoned the collaboration in 1998 after the launch of the petrol-engined Smart car, as he considered it to be a deviation from his original project.

During the 1990s, carmakers encouraged automotive equipment suppliers to create in-house design departments to liaise with automotive equipment engineers and carmaker design departments (Vervaeke, 2010). These co-design partnerships led equipment manufacturers to design complete modules for subsequent assembly by the carmakers.

Confronted by increasing international competition, carmakers were forced to extend and rapidly renew their vehicle ranges, and in particular Renault and PSA–Peugeot–Citroën reinforced their in-house design departments. Massive investments were allocated to the creation of new design centres, bringing together research and development (innovative project teams), creation (design studios) and the building of prototypes (milling machines, paint and assembly workshops and so on); the Renault Technocentre was inaugurated in 1998, and in 2004 the Peugeot Automotive Design Network. On these sites, service-provider employees, sub-contractors and suppliers collaborate with Renault and PSA–Peugeot–Citroën teams.

In the pre-project phase, industrial designers work in close collaboration with engineering teams to facilitate exploratory research and the elaboration of new solutions (Midler, 1993; Hirt, 2004). This increasingly integrated design process is able to assess the industrial constraints (quality, cost, deadlines) in the pre-development phase and confirms the feasibility of designers' ideas.

Carmakers wanted more and more control over their design and brand identity. The concept cars presented at motor shows (Paris, Geneva, Frankfurt, Detroit), demonstrate the creative research that will orient the development of automobile ranges in the more-or-less long-term or enable them to test their new models among consumers. Certain companies specialize in representational techniques permitting the virtual realization of mock-ups, prototypes and concept cars. They offer specialized design services that are complementary to the carmakers' design centres. These new digital techniques reduce the number of prototypes that need to be built. They also permit the creation of virtual scenarios simulating mobility services according to hypothetical travel modes and their utility for consumers and local authorities. In September 2007, Dassault Systems, world leaders in 3D and product lifecycle management (PLM), strengthened their in-house design team by appointing Anne Asensio as vice president of design experience. Previously head of ranges at Renault, she then joined General

Motors as executive director of advanced design where for seven years she created concept cars in collaboration with Bertone in Italy and automotive equipment suppliers in the United Kingdom and Italy (Asensio, 2008). Asensio's mission in Dassault System group is to develop the full potential of 3D and PLM for collaborative design platforms. This provides a seamless product development chain from design concept to realization.

By internationalizing production, carmakers diversified their product range, renewing models more rapidly and reinforcing their brand identity (Midler *et al.*, 2002). To meet the challenge of cost competitiveness, the principal carmakers rationalized design and production by creating design and manufacturing platforms. New digital methods permit the adoption of a 'simultaneous engineering' workflow that facilitates communication between designer teams working on vehicle design and its technical architecture. The platform policy multiplies the number of differentiating alternatives that can be tested while controlling costs and development cycle time.

Major design department contributions were mainly for model differentiation in line with brand strategy. Design departments were mobilized to sustain this strategy by designing concept cars and prototypes. The scheduling of new models was based on market segmentation aimed at extending the supply of existing ranges to a wide range of market segments. At the beginning of the 2000s, with the exception of Toyota, who launched the Prius, carmakers' R&D was essentially oriented towards extending model ranges, reducing energy consumption, safety features and luxury interior design. Sustainable development was not a strategic orientation of the carmakers.

New ideas of mobility provide designers with numerous creative perspectives. PSA–Peugeot–Citroën had already studied the possibility of an electric-car-sharing project in 1995. The Individual and Public Urban Transport Programme (*Transport Urbain Libre Individuel et Public* – TULIP) imagined a car rental subscription service with electric recharging points and stations. The two-seater concept car consisted of five main elements assembled by heat fusion. Fifteen years later, in 2010, PSA–Peugeot–Citroën competed with the Mitsubishi i-MiEV model in response to the 'Autolib' call for tenders launched by the city of Paris. This illustrates the lack of continuity between prospective design research and the development of a mass-produced electric vehicle. The concept car 'Tulip' supported a brand image policy instead of a long-term sustainable policy. With the BB1 concept car (Frankfurt motor show 2009), PSA-Peugeot-Citroën abandoned the small urban car segment to develop a motorized quadricycle. The steering wheel has been replaced by handlebars. Passengers are seated astride, as on a motorcycle, and the front seat back rests fold down to allow access to the back of the vehicle. Motorization is supplied by Michelin's Active Wheel (10Kw electric motors lodged in the rear wheels).

Hayek's forward-thinking ideas (as a businessman involved in wristwatch production) resurfaced in 2008. Shai Agassi, former manager of the software company SAP, manager of Project Better Place, proposed a partnership with Renault to launch an electric car in Israel and Denmark. This collaboration removed the infrastructure and maintenance obstacles that limited electric vehicle diffusion. Project Better Place will manage a subscription service, the battery recharging station infrastructure, and network and battery maintenance.

Renault is currently developing four electric car projects. A model based on the Megane, named 'Fluence' will be proposed with a traditional petrol combustion engine or an electric version, initially destined for the Israeli and Danish markets. It was conceived for production on a manufacturing platform. A standard car body, designed to carry an internal combustion petrol engine, will be used to reduce design and production costs. The basic architecture of this derivative will nevertheless require modification because of the constraints imposed by the installation of on-board batteries. The Kangoo Express Z.E. is a lightweight commercial van. The Twizy is a two-seater tandem.[2] This range will be completed by the five-seat Zoe.

Cost competitiveness has generalized the platform policy. Initially introduced between models produced by the same carmaker, this strategy led to alliances between manufacturers (Fiat and Chrysler, Renault and Daimler). Renault, Nissan and Daimler made such an alliance public in 2010. Components will be shared between Renault models, Class A and Class B replacements and the future Smart. The development costs for small car mass production (new Smart, Twingo and Twizy models) will be mutualized on a common manufacturing platform (Edison Project). The designer of the Renault Logan, Kenneth Melville, is collaborating in research on electric vehicles programmed for this platform. Creating a car for emerging markets breaks the spiral of ever more luxurious, and thus expensive, cars. Certain cost-reducing principles adopted for the Logan can be transferred to the design process for a mass-produced electric car.

Melville explains just how difficult it is to build a US$6,000 car, 'including some swallowing of pride':

> The design principle of the Logan is a modern car: The dashboard for example, is one injection-molded piece. That minimizes the assembly process and ensures consistency. The dashboard on a midsize Western car is composed of many pieces. We weren't allowed to break up the dashboard with colors or special finishes ... We designed a car for emerging markets from a white sheet of paper.
>
> We kept the production technology as simple as possible. In modern cars, you can apply different decorative films to the interior using a water-bath method, where a film is draped over the false wood or aluminum. It's a high-tech solution. If you go to India, you can't have that kind of technology, so we went for paint.
>
> ('The Making of the Logan', *Bloomberg BusinessWeek*, 4 July 2005)

Design is in tension between a *Sloanian model* that renews and diversifies segmented ranges of cars to reinforce the stylistic identity of the carmaker's brand, and a *design thinking model* that lightens materials and simplifies forms and components in the search for novel architectures to promote new forms of mobility. As will be illustrated in the following section, new players have initiated projects adopting a new approach to mobility. Lotti stresses the tension between the designers' consciousness of environmental responsibilities and the constraints imposed by firms' strategies. The project activity of these professionals depends on the firm's product planning (Lotti, 1998).

Mobility as a product/service, a challenge from new players

Increases in the price of petrol and the exhaustion of fossil fuel reserves, together with government and EU constraints regarding air pollution emissions has encouraged players external to the automotive industry to invest in new experiments. Three case studies will present companies external to the automotive industry that since the year 2000 have proposed original concept designs for new car-share or semi-collective transport projects.

Cybergo and Modulgo Induct projects with Diedre Design

Induct, a company created in 2004, has specialized in the development of embedded equipment. It collaborates with the Institut National de Recherche en Informatique et en Automatique: INRIA (the National Institute for Research in Computer Science and Control) in research on electronics and robotics for driverless cars, obstacle detection, geo-localization and wireless communications. Together with Diedre Design, Induct developed a small electric car, the Modulgo, and an 8-passenger shuttle, the Cybergo.

The Cybergo is equipped with laser sensors, cameras and integral computer solutions developed by Induct that allow totally autonomous, safe travel without any specific infrastructure equipment. The sensors detect the presence of obstacles and the cameras ensure the video surveillance of passengers. The lithium polymer batteries can be replaced or rapidly recharged by induction in 15 minutes at stops when passengers get on and off the shuttle. This ensures a continuous service on a single route without having to return to a charging station. A computer-assisted management and surveillance programme adjusts the number of shuttles in service according to the volume of passengers. This system is more flexible than fixed time schedules, where buses circulate empty or overloaded depending on passenger flow during the day. The Cybergo shuttle can reach a speed of up to 110 km/h, limited to 18 km/h for urban transport use. French legislation prohibits this type of vehicle on normal roads. It can be installed on a predefined route in pedestrian zones or private zones such as airports.

The Modulgo is a small electric car designed for a car-share service. Reservations are made via the internet, a Smartphone or a touch-pad terminal. On returning the vehicle, the user leaves it at the car park entrance. The car park, equipped with sensors, then takes control of the vehicle via a robotized unmanned control system, parks the car, recharges the batteries and manages its allocation to a new user. A computer manages car parking, battery recharging and the circulation of vehicles on the site, thus reducing operational parking costs. The reduced surface area taken up by each vehicle allows for the densification of parking spaces: 100 Modulgos occupy only 30 standard parking places.

One of the areas of design research was to develop a car body using 100% recyclable material injection-moulded using a rotomoulding technique. The aim is to reduce the number of body parts, reduce the number of moulds and simplify assembly. See Figure 3.1 – the design sketches for the Cybergo.

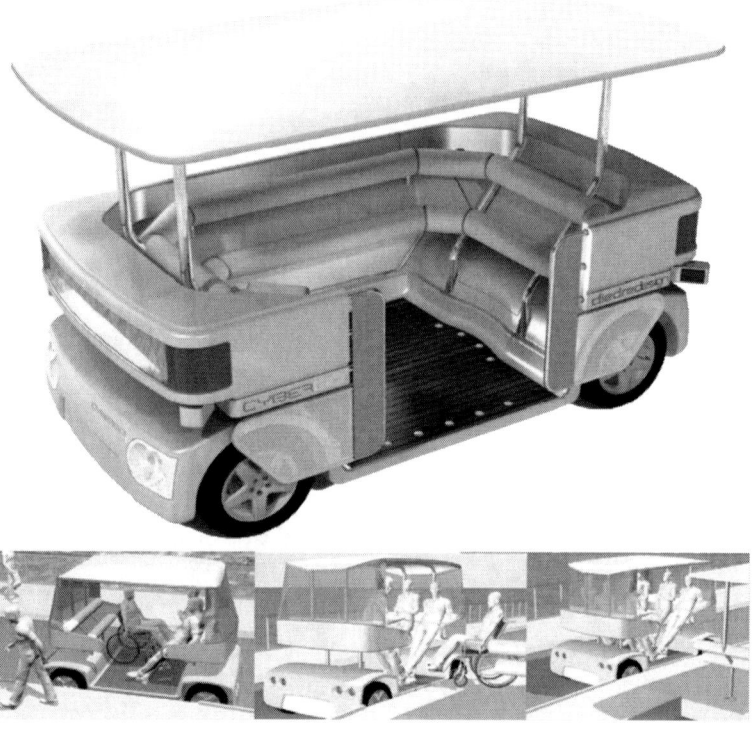

Figure 3.1 Diedre Design sketches for the Cybergo vehicle
Source: ©Diedre Design.

The four Diedre Design sketches for the Cybergo show, in the main sketch, a detailed study of the vehicle; and (left to right) smaller sketches showing a simulation of passengers sitting and pedestrians walking in the street; accessibility for a disabled person in a wheelchair; and the half-sitting position of the Cybergo shuttle passengers.

LOHR and the Cristal project

Cristal is an inter-urban collective vehicle project co-ordinated by the LOHR Industrie group. It will be adaptable for individual use in car-share schemes or coupled to form a convoy for a semi-collective public transport system. In 'convoy' mode, individual vehicles are not coupled mechanically but are equipped with an automated electronic coupling system. LOHR Industrie specializes in the design and development of vehicles and equipment, and innovative public transport solutions (tramways in Padova, Italy; Clermont Ferrand in France; and Tianjin and Shanghai in China) and combined road–rail transport.

The Cristal transport system (see Figure 3.2) is an alternative mobility system, complementary to existing collective and semi-collective public transport systems. LOHR Industrie, in collaboration with GEA, an urban planning consultancy, co-ordinate a multi-disciplinary group of experts: Transitec, a Swiss consultancy specializing in the study of urban travel; and VU Log, a company developing urban mobility information systems and with public research laboratories in Lausanne and Montbéliard. Pioneer users will initially test the Cristal system via virtual scenarios. Pre-series vehicles will then be put into circulation in Strasbourg and Monbéliard. These tests will allow design corrections to be made to improve usage efficiency.

This public transport system can either be individual (car-sharing) or semi-collective (shuttle convoy). The designer, Didier Mandart, has worked on the visual aspects or volume architecture to modify the visual codes applied to individual cars. Passengers enter upright, as in collective public transport. Height and volume have been studied to create the sensation of spaciousness within a limited interior space. The use of transparent Plexiglas opens the vehicle to the urban environment. Wheels are small and differ from car wheels. The framework is composed of cylinders with sufficient thickness to ensure the solidity of the whole structure. The interior is composed of small seats adapted for short travel times. The Cristal transport system rationalizes urban space by limiting the need for parking areas, reconfigures the spatial organization of the street and permits the inclusion of green areas.

Bolloré-Pininfarina, the Bluecar and 'Autolib' in Paris

The City of Paris has entrusted the organization of an electric-car-sharing system to a mixed transport syndicate 'Autolib'. The launch is planned for the end of 2011 with 3,000 vehicles being distributed between 1,100 stations (terminals), of which 700 are in inner Paris. At the end of the

Figure 3.2 The Cristal transport system
Source: ©LOHR Industrie.

consultation period, three vehicle projects remained in competition: Bolloré with the Bluecar; the Avis, RATP, SNCF and Vinci Park consortium with the electric Smart car; and VTLIB (Véolia urban transport) with the Peugeot Ion (based on the Mitsubishi i-MiEV). The criteria taken into consideration by the mixed transport syndicate 'Autolib' included vehicles, service, economic model and risk-sharing conditions.

The Bolloré group proposal was chosen because it was cheaper than its competitors while respecting the criteria outlined above. Several factors account for the lower operational costs. The rental cost of the Bolloré Group's lithium-metal-polymer batteries is lower than that for traditional lithium-ion batteries, but the design process entry no doubt contributed equally to reducing costs.

Compared with the other two competing vehicles (Smart and Peugeot Ion), the Bluecar was from start the only purpose-designed electric car rather than being a modified version of the internal combustion model. The 2011 Bluecar model is the result of an accumulation of research on the architecture, equilibrium, volumes and specific parts that was initiated in 2005. In the current research phase of electric vehicles, an integrated design appears to be more favourable to the reduction of costs than a modular design.

The Bolloré group, through its Batscap division that developed a lithium-metal-polymer (LMP) based automotive battery, initially collaborated with Matra Automobile on an electric car project. Philippe Guédon, CEO of Matra Automobile, worked on the initial bodywork projects for the Bluecar. The first concept car was presented at the 2005 Geneva motor show. After Matra's merger with the Italian Pininfarina group, the partnership was extended.

In 2008, the Bolloré group intervened by injecting capital into Pininfarina, which was encountering financial difficulties.[3] The Bolloré–Pininfarina joint venture on the electric car project ended in March 2011. Bolloré bought back its shares because of Pininfarina debts, which had forced it to find new venture capital alliances.

The different concept cars show the evolution of the design research. The mass-produced vehicle will have a strong visual identity, as a result of the contributions of both Matra and Pininfarina research. The 2005 Matra concept car is a two-door, three-seat vehicle with a half sphere on the side and a cubical rear form, and aims at creating a feeling of security. The Pininfarina–Bolloré Bluecar (2009) is a 5-door, 4-seat car. The Italian style of Pininfarina gave a shape with soft lines, a continuity of lines from the bonnet to the rear of the roof, from the front wheels to the rear lights. Solar cells are embedded in the roof. The Bluecar used for the 'Autolib' project is a small 4-seat, 2-door vehicle. This smaller model keeps the stylistic elements of the larger version with its curved lines which result from research on part assembly (see Figure 3.3). The design and fabrication is the result of transnational intercompany partnerships. Following the financial difficulties encountered by Pininfarina, the Turin body-maker and supplier, Cecomp (Centro Esperienze Costruzione Modelli Prototipi) will manufacture the Bluecar.

The design agency Designers Associés, based in a suburb of Paris, has designed the terminals and charging points, part of the street furniture required to operate an electric car-sharing system. The stations are designed to cater for four to six parked electric vehicles. They are equipped with a

Figure 3.3 The Bluecar for the Autolib project
Source: ©AUTOLIB. All rights reserved.

self-service rental terminal and several charging points equivalent to the number of electric vehicles allocated to a station. Certain charging points will be available for recharging private vehicles (electric cars and electric bicycles). Around a hundred 'Espace Autolib' stations, designed by High Graph Architecture, and composed of a semi-circular transparent structure, are equipped with an information office and a self-service registration terminal designed by the IER Group. It is connected via videoconference to an operations centre accessible 24 hours a day. The IER Group is a subsidiary of the Bolloré Group, specializing in the design, manufacture and marketing of terminals for controlling and reading tickets for air, rail and sea transport networks. Following the acquisition of Automatic Systems, the group gained expertise in security systems and access control equipment.

The self-service rental terminals and charging points have a luminous turquoise summit for easy detection at night. The terminals, charging points and stations were designed taking into account formal codes (volumes, lighting) tested by Patrick Jouin for the 'Velib' system (the Parisian self-service bicycle sharing system). The street furniture for car- and bicycle-sharing systems thus form coherent visual landmarks in the urban space. The stations should permit the creation of around 800 jobs.

The financial conditions applied to the car-sharing system will consist of an annual subscription fee or a 7-day or 24-hour formula to which is added the time of use billed per half hour. According to its 12-year contract, Bolloré will take responsibility for the risk of vandalism and repairs on the basis of €3,000 per vehicle per year. The company will cover any losses in turnover up to €60 million. Three to four years in operation and 200,000 subscribers will be necessary before the 'Autolib' project becomes profitable. Local authorities will finance the stations, each costing €500,000. To create a new market and invent new means of transport, the private sector players benefit from public interventions. 'Autolib' will be demonstrate the Bolloré Group's ability to provide a complete solution in the field of electric cars, and its success will be a commercial showcase for Bascap, its battery manufacturing subsidiary, and the IER Group. This strategy organizes a car design and manufacturing process based on inter-company partnerships that differ from the inter-company relationships around which the automotive sector is traditionally organized.

Conclusion

Production overcapacity or the automotive industry's financial difficulties are only some of the crises currently facing the automotive sector. Since the increase in the price of petrol in 2008, consumers have modified their behaviour – by, for example, taking transport costs into consideration; reducing their fuel consumption; or buying high-fuel-economy vehicles. The current crisis strongly encourages public players and automotive manufacturers to question present forms of mobility and rethink the future of cars by adopting a more social approach to travel.

Experiments with the introduction of electric cars on to the market in the coming years (Renault/Project Better Place; Bolloré/Pininfarina) will be limited to car-share projects and battery rental systems. The methods of introducing electric cars as they have been envisaged will call into question the driver's relationship with the car in terms of use, possession and appropriation. Based on a subscription system, mobility is offered as a service that transforms the perception of a car from being an object of household equipment or property. The age of access evoked by Rifkin (2000), which has already been integrated into the economic model accompanying the development of mobile telephony, is about to transform relationships with means of transport. One must wait for the results of these first experiments to evaluate whether this new economic model will be extended or whether it will remain limited to certain market segments (urban subscription services, or company fleets, for example).

The automobile is experiencing a period of transition. Cars will evolve from fossil fuel motorization to other sources of energy: electricity, hydrogen or biomass. New orientations for future means of transport are in preparation. Answers to short-distance mobility and urban travel are extremely varied, both in terms of the vehicles and services proposed. The projects analysed are, of course, still in their experimental phase. Many projects are still at the concept car or demonstration phase in the first step of development research. Nevertheless, the new actors involved can be identified. Partnerships and inter-company collaborations are learning from their mistakes and developing a better approach to future experiments.

Public interventions stimulate experimentation. By placing orders, government agencies create the market that allows for first-generation vehicles to be tested. The experiments allow companies to capitalize on their trials and learning experience, enabling them to test new economic models, improve the product and mobility service offer, and contribute to its evolution. Will a more sustainable form of mobility contribute to the emergence of new configurations of actors that will profoundly modify automotive design and production systems?

Notes

1. Victor Papanek is a designer of international renown. He has taught at the California Institute of the Arts.
2. The website www.wheelosphere.org/renault-nissan-daimler shows a visual of the Twizy and the Smart Fortwo. The form of the Twizy model doesn't really appear to be the product of a new design process. The external structure basically follows the same profile of the old Smart car.
3. Natahalie Brafman, 'Bolloré vole au secours de Pininfarina pour consolider son projet de véhicule électrique', *Le Monde*, 13 March 2008.

References

Asensio, A. (2008) 'Les usages de la 3D iront bien au-delà de la création d'un produit', *01 informatique*, interview with A. Clapaud, n 1931, 5 February.
Baccaïni, B., Sémécurbe, F. and Thomas, G. (2007) 'Les déplacements domicile-travail amplifies par la périurbanisation', *Insee Première*, (1129) March.
Beaume, R. and Midler, C. (2009) 'From Technology Competition to Reinventing Individual Eco-Mobility: New Design Strategies for Electric Vehicles', *International Journal of Automotive Technology and Management*, 9(2): 174–90.
Bellu, S. (1998) *Histoire mondiale de l'automobile*. Paris: Flammarion.
Bellu, S. (2007) *Art de la carrosserie française*. Boulogne: Etai.
Brown, T. and Katz, B. (2009) *Change by Design: How Design Thinking Can Transform Organizations and Inspire Innovation*. New York: HarperCollins Business.
Brundtland Commission (1987) *Report of the World Commission on Environment and Development: Our Common Future*. Oxford, UK: Oxford University Press.
Calabrese, G. (ed.) (2010) *La filiera dello stile e le politiche industriali per l'automotive in Piemonte e in Europa*. Milan: FrancoAngeli.
Finizio, G. (2005) 'Roberto Giolito: Progettista e designer advanced', *Progettare*, (287): 67–71.
Ford, H. and Crowther, S. (1922) *My Life and Work*. New York: Garden City Publishing.
Francastel, P. (1956) *Art et technique aux XIXe et XXe siècles*. Paris: Éditions de Minuit.
Freyssenet, M. (ed.) (2009) *The Second Automobile Revolution: Trajectories of the World Carmakers in the 21st Century*. Basingstoke: Palgrave Macmillan.
Freyssenet, M., Mair, A., Shimizu, K. and Volpato, G. (2000) *Quel modèle productif? Trajectoires et modèles industriels des constructeurs automobiles mondiaux*. Paris: La Découverte.
Hirt, O. (2004) 'La relation design-ingénierie dans les nouvelles organisations de la conception: la démarche des fondamentaux en design de Renault. Actes de la journée thématique AFM/Audencia marketing et design', [CD-ROM] *Ecole de Commerce Audencia*, Nantes, 30 January, 11.
Kazazian, T. (2003) *Design et développement durable: il y aura l'âge des choses légères*. Paris: Édition Victoires.
Koenig, G. K. (1991) *Il designe'un pipistrello 1/2 topo 1/2 uccello. Storia e teoria del design*. Florence: Gruppo editoriale fiorentino.
Loewy, R. (1951) *Never Leave Well Enough Alone*. New York: Simon & Schuster.
Lorenz, C. (1986) *The Design Dimension: The New Competitive Weapon for Business*. Oxford: Basil Blackwell.
Lotti, G. (1998) *Il progetto possibile: Verso una nuova etica del design*. Monfalcone: Edicom edizioni.
Maldonado, T. (1991) *Disegno industriale: un riesame*. Milan: Feltrinelli.
Manzini, E. (1986) *La materia dell'invenzione*. Milan: Arcadia.
Manzini, E. (1989) 'Il design nella società in transizione: Terreni, livelli e forme di intervento', in E. Mucci (eds), *Design 2000*. Milan: Franco Angeli, pp. 39–48.
Manzini, E. (1990) *Artefatti. Verso una nuova ecologia dell'ambiente artificiale*. Milan: Domus Academy.
Manzini, E. and Vezzoli, C. (1998) *Lo sviluppo di prodotti sostenibili: I requisiti ambientali dei prodotti industriali*. Rimini: Maggioli Editore.
Meadows, D. H., Meadows, D. L., Randers, J. and Behrens, W. W. III (1972) *The Limits of Growth: A Report for the Club of Rome's Project on the Predicament of Mankind*. New York: Universe Books.

Midler, C. (1993) *L'auto qui n'existait pas*. Paris: InterÈditions.
Midler, C. (2007) 'Les challenges de la compétition par l'innovation dans l'industrie automobile', in N. Mottis (ed.), *L'art de l'innovation*. Paris: L'Harmattan, pp. 217–28.
Midler, C., Monnet, J. C. and Neffa, P. (2002) 'Globalizing the Firm through Projects, The Case of Renault', *International Journal of Automotive Technology and Management*, 2(1): 24–45.
Midler, C., Minguet, G. and Vervaeke, M. (2009) *Working on Innovation*. New York: Routledge.
Papanek, V. (1972) *Design for the Real World*. London: Thames & Hudson.
Rifkin, J. (2000) *The Age of Access: The New Culture of Hypercapitalism, Where All of Life Is a Paid-For Experience*. New York: Tarcher.
Rifkin, J. (2003) *The Hydrogen Economy*. New York: Tarcher.
Vervaeke, M. (2010) 'From Watching the Markets to Making Trends: The Role of Industrial Designers in Competitive Strategies', in C. Midler, G. Minguet and M. Vervaeke. (eds), *Working on Innovation*. New York: Routledge, pp. 42–71.

4
Towards New R&D Processes for Sustainable Development in the Automotive Industry: Experiencing Innovative Design[1]

Maria Elmquist and Blanche Segrestin

In this chapter, we shall discuss the design processes of environmentally friendly and profitable competitive products.

In the automotive industry, it is acknowledged that the cars of the future need to have a limited environmental impact and to be profitable. As competition intensifies, car manufacturers find themselves under increasing pressure to find creative solutions and innovative ways of addressing environmental issues and providing enhanced value to their customers. In recent years, they have been working intensively on alternative forms of propulsion and alternative fuels in order to reduce fuel consumption and polluting emissions (Aggeri, 1999). But as these issues have become even more important (Porter and van der Linde, 1995; Porter and Kramer, 2006), the prevailing strategy has reached its limits: indeed, environmentally-friendly technologies are often very costly to develop and customers are not necessarily willing to pay for such a public good that does not directly increase their personal utility (Ottman *et al.*, 2006). In addition, meeting regulatory demands only through improved technical solutions is insufficient in global competition, where hybrid technologies are rapidly becoming a commodity. Instead, manufacturers need to find more creative solutions and innovative ways to address environmental issues, to distinguish themselves from the competition. Thus eco-innovation is no longer limited only to efforts to reduce the environmental footprint of automobiles (Hur *et al.*, 2005; Maxwell and Van der Vorst, 2003); manufacturers must also revisit the value and performance that are included in their offer to build alternatives, and design products that respect the environment and at the same time enhance the perceived value for consumers.

Sustainable development in the automotive industry is thus strongly linked to complementing classical R&D processes with innovative design strategies. Companies are forced to expand on the value, product attributes and solutions of their offers in innovative ways (Le Masson *et al.*, 2010), and this evolution has important implications for the way that companies organize their research and development (R&D) processes. Whereas automotive firms traditionally have organized their R&D processes to improve well-established functional objectives using identified competencies (Le Masson *et al.*, 2010), they instead need to rethink their methods of organization to enable the development of radically different offerings. To succeed in this, new R&D strategies and new managerial techniques are needed.

Based on the design theory framework (Hatchuel, 2001, 2004; Hatchuel *et al.*, 2006; Hatchuel and Weil, 2009; Le Masson *et al.*, 2010), this chapter will discuss how, especially in the automotive industry, innovative eco-design can be organized within the framework of established R&D processes. It draws on the findings from a research study with a European automotive firm that looked for new and innovative ways to expand its environmental contributions. The project was set up as a collaborative research project (Adler *et al.*, 2004; Shani *et al.*, 2007) which aimed at renewing the design alternatives in the company's environmental contributions by introducing a new approach to the existing R&D processes. The project, labelled 'Eco & Eco', used the so-called KCP method, based on the C–K theory of design (see the sub-section below entitled 'The KCP method'), as a way of exploring how the cars of the future could be both environmental and economical. The case provides an example of how companies in the automotive industry could complement the R&D process with an innovative design approach to build a better basis for a more innovative output.

This chapter is organized as follows: the next section introduces the context of the automotive industry and the growing challenge of eco-innovation, presenting the European car manufacturer (hereafter referred to as AutoX) that constitutes the empirical case. It then introduces the design theory framework and the innovative design challenges compared to more conventional R&D processes. The third section describes the research setup, the experiment and the KCP process used to design the experiment and collect the data (based on Hatchuel, 2009). The fourth section provides some insights from the empirical case to relate to the literature and to contrast classical R&D processes with more innovative design regimes.

Sustainable development in the automotive industry: the case of AutoX

The dominant design challenged by environmental issues

In recent years, innovation has become necessary for survival in the automotive industry. Because of structural changes in the marketplace, more intense competition, stricter regulations, growing fragmentation and shorter product life-cycles, manufacturers are required continuously to incorporate new technologies, designs and features in the product development process (Magnusson and Berggren, 2001). The automotive industry has seen tremendous developments in recent decades, and areas such as performance, fuel consumption, safety, comfort and driver information have improved exponentially (Sommerlatte and Karsten, 2001). But, in order to remain competitive, manufacturers must also use, and further develop, new ideas, concepts and technologies (Korth, 2005).

The trend towards sustainable development in the automotive industry is strong, and promises a huge market potential for first-movers (Porter and van der Linde, 1995; Knell, 2001). Regulation and penalty systems are pushing manufacturers to reduce the negative environmental impacts of cars, such as CO_2 emissions. However, there are multiple obstacles to such improvements. First, the dominant design of the all-steel body and the internal combustion engine (ICE) are embedded in an industrial system that is not easy to change (Williander, 2006b). Second, the development of alternative technologies, such as electric cars (Magnusson and Berggren, 2001), is difficult and often contradicts traditional automotive performance: some empirical studies stress the difficulties encountered by proactive management in mobilizing organizational units trapped in existing competences and routines (Williander 2006a, 2006b). There is also an inherent contradiction in the values encompassed by sustainable performance criteria, such as low fuel consumption and traditional performance criteria – illustrated by the high consumption of SUV (sports utility vehicle) models (Luke, 2001). Environmentally friendly technologies are often more expensive but give similar user performance. To differentiate their offers, manufacturers need to develop products that respect the legislation but also offer new and discriminating value to customers: there is a growing strategic need to innovate in the field of 'green' or eco cars.

Complementing the R&D processes: the quest for breakthrough innovation at AutoX

This chapter is based on a study that is part of a broad, long-term collaborative research project (Adler *et al.*, 2004) with AutoX, a European automotive firm. Collaborative research projects are defined co-operatively by researchers and company practitioners to ensure that the knowledge produced jointly is relevant to both academia and practice (Starkey and Madan, 2001) and is also actionable (Argyris, 1993; Argryis *et al.*, 1995; David, 2002).

AutoX was chosen as a representative case of the innovative design challenge in relation to sustainable development. The company had been working on how to develop greener cars for a long time. It had several ongoing projects and research programmes looking at ways to reduce the ecological footprint of the car (new powertrains, green energy sources such as solar, fuel cells, and so on). In 2007, it recognized the strategic need to develop more innovative eco-car offers, in particular within the area of environmentally friendly and inexpensive vehicles (hereafter called Eco & Eco cars),[2] and made an effort to produce innovative concepts.

However, the resulting concepts were all linked to the same strategy of limiting the car's footprint (through its entire life-cycle) on natural resources. Of course, this was already a very large field, with many varying concepts. But the core values were almost identical and did not differentiate the company from what was being offered by competitors. Also, the concepts often proved to be expensive to develop and sometimes had to be sold at a higher cost, though they did not necessarily offer any additional value to the customer. Therefore, AutoX recognized the need to achieve more radical *breakthroughs*.

Basically, two main strategies were applied by AutoX to generate breakthroughs: stimulating the internal ideation phase and sourcing new ideas from the external environment.

- In the literature, the ideation phases, often referred to as the 'fuzzy front end' (FFE) of innovation (Reinertsen and Smith, 1991; Khurana and Rosenthal, 1997, 1998) are often put forward as being essential for companies that want to be innovative. The front end (FE) is defined as 'the period between when an opportunity is first considered and when an idea is judged ready for development' (Kim and Wilemon, 2002). The main objective of these FE activities is to provide strong product concepts that can be refined and developed during the new product development (NPD) process (Khurana and Rosenthal, 1997). However, the literature on FE focuses on how to recognize opportunities (Colarelli O'Connor and Rice, 2001), or on scanning and searching (Tidd *et al.*, 1997). At AutoX, as the internal processes did not generate sufficiently innovative ideas, external sources of innovation appeared to be more promising.
- In their search for external knowledge, AutoX had launched a number of new initiatives: it had participated in and investigated external knowledge sources, such as environmental studies; sociological and ethnographical studies on the future of mobility; technical and marketing benchmarks; prospective scenarios on environmental and markets trends; environmental scanning; and technological research programmes, among others. However, these initiatives of 'open innovation' (Chesbrough, 2003a, 2004) had been disappointing: the potential scope of R&D was infinite, and the diversity and heterogeneity of expertise and ideas made it difficult to evaluate whether the knowledge identified should be incorporated or not. Those responsible for the development of innovative products were also screening the ideas used to predefine performance criteria, but the most innovative ideas were not able to meet the complex and extensive requirements of automotive standards.

Despite the many efforts made by the company on Eco & Eco cars, the results were thus mostly rather conventional offers that did not help AutoX in distinguishing itself from its competitors. In other words, the traditional R&D organization appears inadequate to address innovation processes. More precisely, the case of AutoX shows that the nature of innovation has changed: the R&D processes are no longer as efficient when aiming at breakthroughs.

Introducing the framework of innovative design

As Le Masson and Weil (2008) have shown, R&D organizations were developed at the end of the nineteenth century to structure the development of new products in a planned and controlled way. Most firms organize their R&D processes in a similar way, with stage-gate processes (Cooper, 1990), to generate solutions to well-formulated requirements in a systematic way. Hatchuel, Le Masson and Weil also show that this organization is based on a type of systematic reasoning (Pahl and Beitz, 1988), which is referred to as rule-based design; that is, structured around abstract languages (functional design, conceptual design and so on) (Hatchuel et al., 2005; Le Masson and Weil, 2008). In this design regime, new solutions are produced, but neither new requirements nor new product identities are generated; development activities are structured around known performance parameters; and research activities focus on specific research questions. This organization enables an efficient collaboration between research, engineering and marketing departments and project teams, but also with manufacturing and sales departments. The R&D organization thus enables incremental innovation in well-known performance areas, but has some limitations when it comes to exploration beyond these areas.

At times when there is a need for innovation, and where even the identity of products is questioned, the R&D organization based on the rule-based design regime is no longer necessarily the best. Instead, new performance criteria may emerge, existing competencies may not be sufficient and new expertise may need to develop. It has also been argued that there is a need for adapted management modes depending on the design regimes (Segrestin et al., 2002). Just to highlight the main differences (Le Masson and Weil, 2008), in a rule-based design regime: objectives are known; the interactions between functions are not defined; the knowledge needed is known at the beginning of the process; and evaluation and validation methods are known. In contrast, in an innovative design regime: the objectives are developed or revisited; the interactions between functions are unstable or revisited; the knowledge needed is not identified beforehand; and evaluation and validation methods need to be developed.

In situations where the rule-based design of R&D processes is insufficient, there is a need to develop new methods and ways of reasoning to address the very different situation of innovative design. The purpose of the project at AutoX was to enable the firm to experience an alternative design regime based on the case of Eco & Eco cars.

The KCP method

The experiment was designed on the basis of a method called 'KCP workshops', derived from the C–K theoretical framework and first developed at Mines ParisTech in collaboration with RATP.[3] The method aims at structuring a collaborative exploration of an innovation field. The expected outcomes are a structured set of innovative concepts for further development and the identification of 'missing' resources and competencies needed to enrich ongoing research programmes or external acquisitions (Hatchuel, 2009). It has been used in a number of innovative projects with industrial partners such as Thalès, Sagem and Vallourec, and is being refined systematically by several research teams. The process is not presented in detail here; instead, the description focuses on the main outcomes. For reasons of confidentiality, the content of the experiment has been disguised, but the description provided reflects the actual challenges involved in the design process and covers the different steps in the experiment. The term 'Eco and Eco Innovation' is used to describe a field created for this chapter; it is thus not related to the real analysis. Similarly, the generated values, proposals and research questions are introduced to explain the process, but do not pretend to be either original or robust.

The KCP method is based on the C–K theoretical framework. The C–K (Concepts–Knowledge) theoretical framework models the design process through the interaction between two expandable spaces: *concepts* and *knowledge* (Hatchuel, 2001, 2004; Hatchuel *et al.*, 2006; Le Masson *et al.*, 2010). This separation is made to distinguish between something that does not yet exist (the concept, the starting point of the design process), but which can be formulated, and something that has already been designed and fully determined (knowledge). The K space contains propositions that are considered by the designer to be *true*, while the C space contains propositions that are not part of the designer's knowledge but are interpretable using that knowledge. Design reasoning, then, is modelled as the co-evolution of C and K: the concepts are progressively defined 'through an expansion of their formulation' (Hatchuel *et al.*, 2006, p. 304) and the competencies (knowledge) are extended through learning processes that are driven by the new concepts.

The KCP method involves a series of three workshops: one for *knowledge* sharing (phase K); one for *conceptual* exploration (phase C); and one to structure the *proposal* (phase P) (see Appendix 1 for a more detailed description).

The method aims at structuring the exploration of a set of innovative concepts, here the 'set of innovative offers for Eco & Eco cars'. This set is quite distinct from the set of existing cars, and is mainly unknown, as there is only a small fraction of it that can be formulated or described using existing knowledge. In the process, 'disruptive searchlights' are used to reveal unknown attributes of the innovation field and to identify missing knowledge areas. The main challenge lies in designing these disruptive searchlights. The concepts must be divergent to destabilize the traditional view of the initial concept. In the case of Eco & Eco, the objective was to consider green cars from completely new perspectives (through new lenses). The concepts do not necessarily have to be 'good ones', but they must be expansive. Their value does not lie in their individual business potential, but rather in their ability to allow the designers to revise the identity of the object itself.

Addressing sustainable development through innovative design

A note on research methodology

On the basis of this framework, a process was initiated by a small core team in charge of strategy in the product planning department at AutoX, but involved representatives from several other departments (R&D, advanced engineering, marketing, market intelligence, brand, design and so on) as well as some external experts (designers, urban planners and so on), who mainly joined the process in the first phase (to share their knowledge). In total, about 45 people took part in the three phases (K, C and P) of the process.

Detailed notes were taken in the preparation period and during three days of seminars, facilitated jointly by the researchers and the core team. In the final seminar, key people from different areas of the organization were invited to discuss the proposals put forward.

Following the workshops, interviews were conducted with key people in the organization to validate the results. Key stakeholders acknowledged the originality and robustness of the outcomes. Data collection and analysis were iterative, during which intermediate ideas were discussed with managers in the company in a systematic and abductive analytical approach (Dubois and Gadde, 2002), and several meetings held between the three phases led to the final design of the experiment. We received feedback on the process and its outcomes from 15 people. The method was acknowledged to have supported the renewal of AutoX's strategic capability and provided interesting insights for further innovation strategies.[4] Eight additional interviews were conducted with key stakeholders in the organization in subsequent months to further validate the originality of the output. These interviews confirmed the innovative value of our results. The rest of this section will describe the KCP experiment at AutoX.

The K phase: creating new design spaces

The first phase of the KCP workshop gathered several experts to share their knowledge about the impact of cars on the environment, and the related costs. These included AutoX employees and external speakers from the highways administration, urban planning, regulatory agencies and an eco-magazine, who were invited to discuss future trends in environmental regulation and car usage. It confirmed that existing AutoX projects were exploring a restricted area because of the constraints of the automotive sector. It challenged traditional representations of Eco & Eco cars and made it possible to reopen the notions of 'ecological' and 'economical', whose traditional meanings had seemed stable but perhaps were evolving:

- The notion of 'ecological' in AutoX had referred to clear objectives mainly related to energy consumption and emissions. The experts discussed social trends, which indicated that the term also encompassed other meanings, including well-being, health and social responsibility. Cars could be eco-friendly in terms of their impact on natural resources (energy consumption, emissions, pollution and so on), and in terms of reducing their impact on passengers (for example, obesity, stress) or cities (for example, space, urban landscape, parking, congestion, noise and so on). Moreover, an eco-friendly car could contribute to limiting its own footprint and reducing negative effects on the environment.
- Cost has always been a major concern in the automotive industry. The discussions led the group to reconsider economic aspects. A car could be economical in terms of lower costs and prices (for example, low-cost cars), but also because of lower or no additional expenses. Global offers for cars might include leasing solutions, low-cost maintenance or services, and reduced expenditure on fuel, parking and so on.

The C phase: introducing disruptions

In the second phase, four 'disruptive searchlights' were used to illuminate possible extensions to Eco & Eco cars. The specification of these searchlights was led by the need to find new angles and to motivate an 'innovative search' among the abundance of existing knowledge and ideas. The idea was to address unexplored issues in the field, two examples being a 'low-cost mobility package' – that is, a low-cost package that might include such things as subscriptions to bus networks; and a 'cost-saving green car' – that is, an ecological car that would save on everyday expenses, or an eco-friendly car with 'zero expense' (no maintenance) attached to it.

The searchlights were not designed to suggest concepts to be developed; rather, they were aiming at creating disruptions; that is, at opening up the spectrum of performance parameters in current use in the industry. Based on these 'disruptive searchlights', the groups were set up to investigate attributes unexplored in the existing approaches to Eco & Eco cars.

Interestingly, the groups discussed many proposals, which were more or less disruptive. In the analysis, we can distinguish between two types of proposals:

- Some proposals extended existing strategies by exploring known objectives. For example, there was much discussion about how to reduce the footprint of cars, including the use of more efficient powertrains, green energy sources (solar, wind, waste materials), life-cycle optimization,

green materials and so on. This extension can lead to new proposals. For example, a car could not only minimize CO_2 emissions, but could also destroy the CO_2 particles that are present in the atmosphere.
- More disruptive ideas also emerged, which revealed themselves not to be in agreement with existing strategies. For example, particular attention was paid to concepts such as 'private *and* collective cars'; while car-sharing is a well-known phenomenon, the flexibility of owning a car is lost with car-sharing. The idea of 'hybridizing' private and collective use enables new combinations with public vehicles (taxis, rental cars, trains or buses) and personal transport. But other forms of hybrids are also possible, with functionalities such as extended navigation systems, or systems for sharing batteries. These ideas are valuable to the extent that *they indicate new values or properties*. Most ideas included some integration within a broader transport system: it seemed that Eco & Eco cars would need to be connected to other transport modes, and that communication between users would therefore need to be facilitated. These attributes (connection with other modes and communication between cars) appeared to be critical to developing new types of Eco & Eco cars. They would allow a large area for potential innovations but first require intensive learning (for example, how can private cars be articulated with public transport? What are the potential side-effects? What are the necessary technological means? What would the human machine interface (HMI) look like?)

To summarize, the work on disruptive searchlights generated ideas about new types of attributes, and revealed knowledge gaps in terms of their realization.

The P phase: Proposing a path towards innovative and sustainable strategies

Based on the creative session, a phase of thorough analysis followed, designed to investigate the results of the workshop. A synthesis of these results enabled the identification of new 'strategic spaces' that addressed the issue of ecological vehicles from a novel perspective for the company. This synthesis involved close collaboration between practitioners and researchers. It resulted in proposals for new performance criteria (for example, cars are more efficient when they are articulated within a broader system of transportation) and outlined new research areas and the need for new competencies (for example, to determine how a navigation system could integrate cars with other modes of transport).

Based on this analysis, it was possible to build a collective action plan for AutoX. This included project proposals, research programmes to be addressed by different functions in R&D or by external partners, experiments that could be formulated, and early prototypes that could be developed. The proposals resulting from this phase were all different learning opportunities (to test the concept, to experiment with new technologies, to initiate a new brand positioning). For example, the concept of a 'car that facilitates the use of public transport' could be considered a long-term ambition, and can be refined as:

- Short-term proposals, such as simulations to estimate cost saving for users, or experimentation with a commercial car for particular groups of consumers, or a navigation system linked to public transport; and
- Mid-term proposals, such as research programmes to look at uses and technologies, and to experiment with new systems for particularly demanding fleets and so on.

These proposals are not independent of each other. Together they constitute a consistent 'learning path', where each step generates competencies and contributes to reducing uncertainty in the future. The KCP workshops thus resulted in the formulation of several learning paths that could be used to organize the exploration of new strategic spaces for sustainable and innovative strategies.

Towards new R&D processes in the greening of the automotive industry: learning from an experiment

The experiment made with AutoX is an example of how an expansion from rule-based to innovative design in terms of new environmental values can be organized in the automotive industry. Previous literature on innovation has demonstrated the difficulties involved in expanding rule-based design regime practices (such as those present in the automotive industry with a dominant design and a stable car identity) to an innovative design regime where the identity of the product is deliberately questioned.

How can the R&D organization be expanded to address issues such as sustainable development? We argue that the KCP process could be a prototype of future methods for an 'innovation function' (Hatchuel *et al.*, 2001; Le Masson *et al.*, 2010) that can complement an R&D function that is well able to deal with rule-based design.

Though this single experiment is not enough to allow generalization, the insights from this project show that the KCP process can handle the three tasks that classical R&D processes are not able to support: (i) the identification of missing knowledge; (ii) the regeneration of value criteria; and (iii) the construction of sustainable innovation strategies through the development of new strategies and learning paths.

Producing new knowledge ... but which? A method of identifying 'missing knowledge'

In the ruled-based design regime, the R&D projects at AutoX working on Eco & Eco cars were focusing mainly on the objective of limiting the environmental footprint. These projects had well-identified questions and technological issues, where external expertise could be useful. They had already resulted in new solutions for more efficient engines, and improvements to existing systems, but had not provided a differentiated solution. The complexity was increased by the requirements of R&D projects (regulation, costs, quality and so on) *and* the need to be innovative.

In the literature it is often argued that, to innovate, companies must renew their internal competencies and exploit external ideas and competencies in their internal processes (Cohen and Levinthal, 1990). However, at AutoX, absorbing new knowledge was not enough: in trying to open up the field of Eco & Eco cars, the problem was neither a scarcity nor an abundance of ideas and competencies, but rather that AutoX was *did not know what it was searching for*; it did not know what knowledge and resources were needed. Whereas in dealing with improvements to existing products, the needed knowledge is known, but it is generally unknown for products that do not yet exist. In this respect, the KCP method is a technique *to decide which ideas and competencies should be explored and acquired*. The KCP process thus allows the identification of knowledge gaps which can then be filled through the processes of the rule-based regime, resulting in a development of the firm's core capabilities.

Capturing the value ... but which? A way of introducing new valuation criteria

In the rule-based design regime, AutoX worked on improving existing products. There were clear criteria for evaluating new ideas, expertise and technologies (for example, the NOx emissions filter, energy consumption reduction techniques and so on). But these criteria did not help AutoX to differentiate its innovative solutions.

Previous literature on innovation stresses that companies often do not capture the value of existing ideas or competencies: they need to learn how to select among ideas and projects, and also try to extract value from stopped projects through external exploitation, such as patenting (Chesbrough, 2003b). Yet our empirical case showed that the valuation of ideas and projects is difficult in innovative design processes. In relation to Eco & Eco cars, *the company did not know what was valuable*. There is an abundance of knowledge and new technologies; however, these do not have any value unless they can be applied to the company's products and make a differentiated product. This implies that the performance criteria cannot be constructed *ex ante*, but depend on the concept to which they refer: products are innovative if they address new value spaces and outperform existing products by providing improved performance.

The KCP process enables the addition of attributes to the concept of Eco & Eco cars, describing new value spaces, to define the criteria needed to assess the value of ideas or knowledge, either internally or externally. It is only once the new performance criteria are identified that it becomes possible to organize such valuation: the KCP method helps *to generate new valuation criteria*, which enable the screening of internal and external projects or ideas and the organization of families (lineages) of potential products with similar attributes.

Creating breakthroughs ... but taking limited risks? A way of developing both innovative and sustainable strategies

AutoX made a strategic choice to aim at the Eco & Eco car but, while staying with rule-based design, they did not succeed in developing a product that could differentiate it from its competitors. Each new project was evaluated within the existing model and most of the innovative projects were discontinued because they failed to meet environmental objectives or economic profitability requirements. The literature stresses flexibility to enable projects to be discontinued or outsourced at any stage, thus making investments less irreversible (Chesbrough, 2003b). But long-term innovative ambitions always appear risky when there are missing competencies. On the other hand, an innovative design regime is only sustainable when investment in the exploration of a new value space does not, by making short-term demands, work counter to long-term strategic ambitions.

These perspectives can be combined through appropriate learning paths in which the knowledge produced by short-term initiatives is exploited to revise long-term strategies, and where short-term initiatives contribute to the expansion of the conceptual product family (C) and the competencies (K) (Hatchuel and Weil, 1999; Hatchuel *et al.*, 2001). Le Masson *et al.* (2007) suggest that, in a situation that requires innovative design regimes, companies need to manage new design strategies involving steps where there are multiple outcomes: that is, a root concept (such as 'Eco and Eco Innovation'), a set of projects to be developed (variations of the root concept), the products launched, the new competencies (developed through product development) and the identification of competencies required for the further expansion of the root concept (Le Masson *et al.*, 2007).

Through working on an expansion of concepts (with related product development) and the required competencies (with related questions), organizations can structure learning paths where research and development projects (prototypes, research programmes, small projects) are used as platforms on which to build new competencies, frame design activities, refine innovation strategies and lay down the foundation for long-term developments.

In the KCP process, learning paths were formulated to enable the expansion of concepts and knowledge in a stepwise and controlled way, allowing for the development of a sustainable innovative pathway. For example, in the case of the 'Eco & Eco car that supports the use of public transport', it is impossible to say whether this option will be profitable, sustainable and feasible in the long term. But the ability to introduce alternatives increases the ability to analyse the competitive landscape, to adjust the strategy dynamically, and to identify the knowledge required to prepare the organization for something new. Through the KCP process, AutoX generated a number of potential strategies, not all of which could be pursued, but they enabled AutoX to revise its plans in a stepwise and highly flexible process.

It should be noted, however, that whereas it seems difficult to be innovative within the established rule-based design regime typical of R&D organizations, the KCP process is not separate from the R&D organization; rather, it is a complementary activity that enables the launch of *more innovative* research projects and development projects. The KCP method is thus proposed as a complement to the regular organization of R&D activities.

Fuelling the pipeline of future projects: a way of crafting new potential strategic spaces

The qualitative feedback obtained from the validation interviews carried out revealed that the experiment helped to identifying new value domains and to create a full range of alternative design possibilities (a family of product concepts and services was identified). The objectives of the process were thus fulfilled, in that it generated input to the development of new R&D strategies and identified new knowledge areas. The experiment also contributed to the renewal of the strategic space of Eco & Eco cars, but it did not define a clear environmental strategy for AutoX. Its outcomes were neither clear specifications for future products nor precise advice on product planning. However, the experience of the KCP process helped AutoX to craft a new strategic space through opening up the concept of Eco & Eco cars, thus enabling decisions to be made about more innovative and differentiating strategies. The interviewees therefore perceived that the experience of innovative design through the KCP process offered some new perspectives on ways to restore competitive advantage in the field of Eco & Eco innovation.

Conclusion

In this chapter we have discussed how innovative design activities can be organized around the issue of eco-innovation, using an experience from the automotive industry, which is representative of an environment where conventional R&D methods are no longer sufficient. Based on the design theory framework (Hatchuel, 2001, 2004; Hatchuel *et al.*, 2006; Le Masson *et al.*, 2010), an experiment was set up to renew the design alternatives of environmental offers that are both sustainable and profitable.

In the study, it was shown that the KCP process enables managers to perform three critical tasks in an innovative design regime: identify missing knowledge; design relevant valuation criteria; and developing design strategies in which products and competencies can be expanded in sustainable but innovative ways.

The experiment has fulfilled two main objectives: first, the empirical experiment elaborated on a structured and organized process for expanding both the value space and the knowledge base in a systematic and iterative way. Second, the qualitative feedback from key stakeholders at AutoX confirmed that the experiment supplied ideas that were considered highly innovative, yet both robust and sustainable. Of course, other methods may be equally successful, and this single experimental implementation of the KCP process at AutoX and its relative 'success' cannot be seen as a validation of the method. In future projects, the KCP method needs to be tested in other companies and other sectors, to analyse in more detail the implementation and variables that should be included. But as the discoveries from this experience show, there seem to be some potential in the design theoretical framework as a way of developing new R&D processes that are suitable for the greening of the automotive industry.

Appendix 1: The KCP method (based on Hatchuel, 2007)

1- *Phase K (knowledge phase)*: to build on internal and external knowledge and to be able to produce new concepts collectively, teams of experts and non-experts need to share various types of new knowledge (whether technical or business-related). This first phase is necessarily cross-disciplinary and may involve suppliers, users and other partners; and will eventually indicate potential innovative developments.

2- *Phase C (concept phase)*: this organizes 'oriented creativity' according to precise rules. The innovation team is split into sub-groups, each of which is asked to explore freely concepts called 'disruptive searchlights': surprising and strongly contrasting propositions. They can be seen as searchlights or headlights that extend visibility at night. They need to be formulated according to demanding theoretical principles, ensuring that their selection offers the greatest possible space for innovation. Each sub-group is asked to explore one searchlight concept and to offer appropriate developments and solutions, and is also asked to identify new research areas. During this phase, each sub-group presents the results of its work in order to increase the creative power of the whole group.

3- *Phase P (prototype and proposition phase)*: this phase recombines and further elaborates the propositions of the two previous phases. Hence it formulates a design strategy, which should not be limited to a mere product or new service idea. The goal is to build a 'roadmap' that plans a series of actions, which include immediate solutions, research projects and new prototypes, and the search for new partners. This map should provide a shared vision of the innovation agenda, highlighting alternatives, backups, and radical options, and clarifying the contributions of each actor.

Notes

1. We thank the managers at AutoX who initiated and supported this study. We are also grateful to: the "Réseau Francilien de Recherche sur le Développement Soutenable" (R2D2) as well as the program "R&D innovation et transformation des entreprises" (RITE-ANR) for financial support.

2. This is a recognized field that is being addressed by most car manufacturers. Renault has in fact developed a label called 'eco^2'; however, the case presented here is fictive and does not mirror the actual initiative at AutoX.
3. The French company operating the Paris subway – Régie Autonome des Transports Parisiens (Independent Paris Transport Authority).
4. For reasons of confidentiality we cannot give a detailed account of their judgement on the content, but, for example, one manager stated: 'With this structured method, we have both expanded beyond our current knowledge and found new ideas and areas to explore.'

References

Adler, N., Shani, R. and Styhre, A. (eds) (2004) *Collaborative Research in Organizations, Foundations for Learning, Change and Theoretical Development*. New York: Sage Publications.
Aggeri, F. (1999) 'Environmental Policies and Innovation: A Knowledge-based Perspective on Cooperative Approaches', *Research Policy*, 28(7): 699–717.
Argyris, C. (1993) *Knowledge for Action: A Guide to Overcoming Barriers to Organizational Change*. San Francisco: Jossey-Bass.
Argryis, C., Putnam, R. and McLain Smith, D. (1995) *Action Science – Concepts, Methods and Skills for Research and Intervention*. San Francisco: Jossey Bass.
Chesbrough, H. (2003a) 'The Era of Open Innovation', *MIT Sloan Management Review*, 44(3): 35–41.
Chesbrough, H. (2003b) 'The Logic of Open Innovation: Managing Intellectual Property', *California Management Review*, 45(3): 33–58.
Chesbrough, H. (2004) 'Managing Open Innovation', *Research Technology Management*, 47(1): 23–6.
Cohen, W. M. and Levinthal, D. A. (1990) 'Absorptive Capacity: A New Perspective on Learning and Innovation', *Administrative Science Quarterly*, 35(1): 128–52.
Colarelli O'Connor, G. and Rice, M. P. (2001) 'Opportunity Recognition and Breakthrough Innovation in Large Established Firms', *California Management Review*, 43(2): 95–116.
Cooper, R. G. (1990) 'Stage-gate Systems: A New Tool for Managing New Products', *Business Horizons*, May–June.
David, A. (2002) 'Intervention Methodologies in Management Research', *EURAM Conference*, Stockholm, 9–11 May.
Dubois, A. and Gadde, L.-E. (2002) 'Systematic Combining: An Adductive Approach to Case Research', *Journal of Business Research*, 55(7): 553–60.
Hatchuel, A. (2001) 'Towards Design Theory and Expandable Rationality: The Unfinished Program of Herbert Simon', *Journal of Management and Governance*, 5(3): 260–73.
Hatchuel, A. (2004) 'Organization Theory and Design Regimes: The Impact of Current Evolution in Design Theory', *Academy of Management Proceedings*.
Hatchuel, A. (2007) 'The KCP Method', Working paper, Centre de Gestion Scientifique, Ecole des Mines, Paris.
Hatchuel, A. and Weil, B. (1999) 'Design-oriented Organizations: Toward a Unified Theory of Design Activities', *6th International Product Development Management Conference* (IPDM), EIASM, Cambridge, UK, 5–6 July.
Hatchuel, A. and Weil, B. (2009) 'C-K Design Theory: An Advanced Formulation', *Research in Engineering Design*, 19(4): 181–92.

Hatchuel, A., Le Masson, P. and Weil, B. (2001) 'From R&D to R-I-D: Design Strategies and the Management of "Innovation Fields"', EIASM, 8th International Product Development Management Conference, Enschede, The Netherlands.

Hatchuel, A., Le Masson, P. and Weil, B. (2005) 'Activité de conception, organisation de l'entreprise et innovation', in G. Minguet (ed.) *Travail, entreprise et société. Manuel de sociologie pour ingénieurs et scientifiques*, Paris: PUF.

Hatchuel, A., Le Masson, P. and Weil, B. (2009) 'Design Theory and Collective Creativity: A Theoretical Framework to Evaluate KCP Process'. ICED'09, 9th international Conference on Engineering Design, Stanford, CA, USA, 24–27 August.

Hatchuel, A., Weil, B. and Le Masson, P. (2006) 'Building Innovation Capabilities: The Development of Design-Oriented Organizations', in J. T. Hage and M. Meeus (eds), *Innovation, Science and Industrial Change: The Handbook of Research*. Oxford: Oxford University Press.

Hur, T., Lee, J., Ryu, J. and Kwon, E. (2005) 'Simplified LCA and Matrix Methods in Identifying the Environmental Aspects of a Product System', *Journal of Environmental Management*, 75(3): 229–37.

Khurana, A. and Rosenthal, S. R. (1997) 'Integrating the Fuzzy Front End of New Product Development', *Sloan Management Review*, 38(2): 103–20.

Khurana, A. and Rosenthal, S. R. (1998) 'Towards Holistic "Front Ends" in New Product Development', *Journal of Product Innovation Management*, 15(1): 57–74.

Kim, J. and Wilemon, D. (2002) 'Focusing the Fuzzy Front-End in New Product Development', *R&D Management*, 32(4): 269–79.

Knell, W. (2001) 'Sustainable Development and Mobility – The Challenge for the Motor Industry', in R. Landmann, H. Wolters, W. Bernhart and H. Harsten (eds) *The Future of the Automotive Industry: Challenges and Concepts for the 21st Century*. Society of Automotive Engineers (SAE), Warrendale, PA, pp, 237–49.

Korth, K. (2005) 'The Importance of Innovation and New Product Development', *Automotive Design & Production*, 1: 18–19.

Le Masson, P. and Weil, B. (2008) 'La naissance des bureaux d'études ou la domestication de l'innovation', in A. Hatchuel and B. Weil (eds), *Les nouveaux regimes de la conception*. Paris: Éditions Vuibert.

Le Masson, P., Hatchuel, A. and Weil, B. (2007) 'The Emergence of Innovation Field Management in Companies: From NPD to New Design Strategies', European Academy of Management, Paris, 28–31 August.

Le Masson, P., Weil, B. and Hatchuel, A. (2006) *Les processus d'innovation. Conception innovante et croissance des entreprises*. Paris: Hermès, p. 470.

Le Masson, P., Weil, B. and Hatchuel, A. (2010) *Strategic Management of Innovation and Design*. New York: Cambridge University Press.

Luke, T. (2001) 'SUVs and the Greening of Ford, Reimagining Industrial Ecology as an Environmental Strategy in Action'. *Organization and Environment*, 14(3): 311–35.

Magnusson, T. and Berggren, C. (2001) 'Environmental Innovation in Auto Development, Managing Technological Uncertainty within Strict Time Limits', *International Journal of Vehicle Design*, 26(2/3): 101–15.

Maxwell, D. and Van der Vorst, R. (2003) 'Developing Sustainable Products and Services', *Journal of Cleaner Production*, 11(8): 883–95.

Ottman, J. A., Stafford, E. R. and Hartmar, C. (2006) 'Avoiding Green Marketing Myopia: Ways to Improve Consumer Appeal for Environmentally Preferable Products', *Environment*, 48(5): 22–37.

Pahl, G. and Beitz, W. (1988) 'Engineering Design: A Systematic Approach'. London: Springer-Verlag.

Porter, M. E. and Kramer, M. R. (2006) 'Strategy and Society: The Link Between Competitive Advantage and Corporate Social Responsibility', *Harvard Business Review*, December: 78–93.

Porter, M. E. and van der Linde, C. (1995) 'Green and Competitive, Ending the Stalemate', *Harvard Business Review*, September–October: 120–34.

Reinertsen, D. G. and Smith, P. G. (1991) 'The Strategist's Role in Shortening Product Development', *Journal of Business Strategy*, 12(4): 18–22.

Segrestin, B., Lefebvre, P. and Weil, B. (2002) 'The Role of Design Regimes in the Coordination of Competencies and the Conditions for Inter-Firm Cooperation', *International Journal of Automotive Technology and Management*, 2(1): 63–83.

Shani, R., Mohrman, S., Pasmore, W., Stymne, B. and Adler, N. (eds) (2007) *Handbook of Collaborative Management Research*. Thousand Oaks, CA: Sage Publications.

Sommerlatte, T. and Karsten, H. (2001) 'What Challenges Face the Automotive Industry?', in R. Landmann, H. Wolters, W. Bernhart and H. Harsten (eds), *The Future of the Automotive Industry: Challenges and Concepts for the 21st Century*. Society of Automotive Engineers (SAE), pp. 1–4.

Starkey, K. and Madan, P. (2001) 'Bridging the Relevance Gap: Aligning Stakeholders in the Future of Management Research', *British Journal of Management*, 12(1): 3–26.

Tidd, J., Bessant, J. and Pavitt, K. (1997) *Managing Innovation: Integrating Technological, Market and Organizational Change*. Chichester, UK: John Wiley.

Williander, M. (2006a) 'Absorptive Capacity and Interpretation System's Impact When "Going Green", An Empirical Study of Ford, Volvo Cars and Toyota', *Business Strategy and the Environment*, 16(3): 202–13.

Williander, M. (2006b) 'On Green Innovation Inertia: An Insider Researcher Perspective on the Automotive Industry', Unpublished doctoral dissertation, Chalmers University of Technology, Gothenburg, Sweden.

Part II
Technological Trajectories in Alternative Vehicles

5
Balancing a Strong Strategic Intent and an Experimental Approach to Electric Vehicles

Florence Charue-Duboc and Christophe Midler

Whereas neo-classical thinking argues that price signals (environmental taxes, quotas with tradable permits) provide sufficient incentive for industrial firms to develop green innovations given the potential for profit, it appears that other factors, related to institutional conditions and the distribution of knowledge, help to keep the current technological paradigms in place (Kemp, 1994). To understand how innovation processes lead to a transformation in socio-technical systems, Smith *et al.* (2010) suggest combining an analysis of the new technology design process with a perspective rooted in markets, organizations, regulations and infrastructure. They emphasize the initial deployments in market niches and their role in creating networks of companies pursuing environmental innovation on a broader scale. Meanwhile, the literature on breakthrough innovations and new product development in environments with a high degree of uncertainty (Lynn *et al.*, 1996; Loch *et al.*, 2006) highlights the role of pilot testing as a source of learning (about uses, technical feasibility and so on).

We shall focus on the role and nature of pilot tests in rolling out environmental innovations by combining these two perspectives. In what ways do they generate learning? Do they serve other functions? Given the specific characteristics of the niches in which this pilot testing is carried out, in what way does such testing promote wider deployment? How can lessons learned locally be adapted to a larger scale? And by what means? Our goal is to examine these issues in more detail.

Transportation is responsible for 23 per cent of greenhouse gas emissions by the 27 European member states (European Environment Agency, 2009), of which road transport accounts for 90 per cent,[1] while the OECD reports that cars generate approximately two-thirds of greenhouse gas emissions (International Transport Forum, 2008). Since the mid-1990s, this has resulted in a series of initiatives aimed at reducing vehicle emissions of CO_2, and various forms of technology (reduced engine capacity, stop/start, regenerative braking and so on) are being developed for internal combustion engines. Regulations now set objectives for manufacturers in Europe, and tax incentives are being defined to influence purchasing behaviour.

More radical initiatives are emerging alongside these incremental innovation strategies, such as the hybrid vehicle, Toyota's pioneering programme and the programmes for 100 per cent electric vehicles (EVs) (Aggeri et al., 2009). EVs appear to have a favourable impact on the environment, though this varies with the electric engine's performance, charging methods and the modes of power generation. Several manufacturers are pursuing initiatives, targeted at consumers, that proclaim their commitment to reducing CO_2 emissions, using strategies that are more offensive than defensive. Deployment of these breakthrough strategies, in particular electric vehicles, assumes that the conditions for their use will emerge over time (including charging methods, negotiated electricity rates and a public charging infrastructure). In their attempt to introduce electric vehicles to the market, original equipment manufacturers (OEMs) are being forced to address these issues. We shall concentrate on the approach adopted by Renault in preparing for the commercial launch of its electric vehicle.[2] Alongside its partner, Nissan, Renault is investing €4 billion in an ambitious programme involving four different vehicles. They were unveiled at the Paris motor show in 2010 and will be marketed from late 2011 until 2013. Before analysing the Renault case in more detail, we shall present the literature on these breakthrough projects.

The literature on exploratory projects and environmental innovation

First, we shall describe the research that examined the development process for new products marked by a high level of uncertainty. This work focuses on pilot testing and the lessons learnt from it. Next, we shall discuss the literature that focuses on strategies developed by firms to reduce their environmental impact. This research stream analyses these dynamics either at the company level or in terms of socio-technical systems.

Managing projects with high levels of uncertainty

The literature on project management distinguishes between the development of incremental innovations and 'exploratory' projects; that is, upstream phases in innovation processes, in which both the potential uses and the technical feasibility of these innovations is examined. In such projects, a type of uncertainty called 'unknown unknowns' calls for specific management methods (Loch et al., 2006). To describe this category more precisely, we can distinguish between three types of uncertainty. The first and simplest, risk, refers to a situation in which an activity might culminate in various results based on foreseeable probabilities. In the second type, described as 'Knightian uncertainty' or 'known unknowns', the probability distribution is not known. The third type, 'unknown unknowns', refers to situations where not even the variables that influence the results are known. Traditional risk management methods apply to the first two categories but not the third.

Two specific management methods can be used to address this radical uncertainty. 'Parallelism and selection' consists of performing multiple trials and then choosing the scenario to adopt based on the trial that gives the most satisfactory results (McGrath, 2001; Sobek et al., 1999). 'Learning by trial and error' consists of entering into an iterative learning process, adjusting the design along the way, based on feedback and the greater understanding that develops over time (Lynn et al., 1996; Thomke, 2003). Loch et al. (2006) show that management by iterative learning is best suited to situations with little complexity, while the parallel scenarios approach is best for more complex cases. Other research that looks at specific types of projects tends to emphasize these two methods of management. Lenfle and Midler (2003) analyse exploratory projects and show that they involve innovative offerings aimed at determining the feasibility of a new product or process while exploring new use values. Elaborating on these results, Gastaldi (2007) points to the dynamics of concurrent exploration; that is, the process of building knowledge in various realms (technology and applications) in a simultaneous, collaborative manner.

Le Masson et al. (2006) propose a conceptual framework for innovative design. This framework models the design process as an expansion in two directions at the same time: the concept space and the knowledge space. A tree structure representing various notions being considered, based on one parent concept, helps to generate new models. The derived concepts can be used as a foundation for conducting pilot projects that support learning processes.

Combining two modes of management (learning and parallelism) and producing artefacts (such as prototypes) appear to be key factors in dealing with uncertainty, during the early phases of development as well as subsequently. However, all this research looks at situations in which one company drives the exploration of an innovative offering. But what happens when the innovation results from the combined contributions of a number of organizations as part of an integrated offering? Are the modes of management mentioned above still accessible? Do pilot programmes play any specific role in such a context?

Initiatives related to the environment: environmental innovation

Having analysed and compared the actions taken by multiple companies to reduce their environmental impact, Aggeri et al. (2005) comment that these initiatives stretch well beyond the company's boundaries and involve multiple participants with varying objectives and operating methods, and reciprocal advisory relationships among themselves. They are confronted not with asymmetrical information, but with shared uncertainties (Aggeri and Hatchuel, 1999) that make it impossible for any individual party, working on its own, to define the nature and format of the information that must be produced.

In this context, collective learning and innovative experiments are crucial. Five characteristics of these pilot experiments have been identified: they look beyond the regulatory standard; they set themselves apart from the work of competitors; they have consequences for the various stakeholders; they take a long-term approach; and they cannot be reduced to a cost/benefit analysis.

This research emphasizes the fact that these environmental initiatives involve multiple stakeholders, and highlights the critical need for these stakeholders to accumulate knowledge in order to locate their own interests most effectively and to reduce uncertainty. However, the matter of transitioning from an innovative local experience to a broader application is not analysed.

Working at a macroscopic level, Smith *et al.* (2010) look at the dynamics by which technological innovations help to push socio-technical systems towards more sustainable modes of operation. To understand the connections between company strategy with regard to innovation and sustainable development, and public policy, the authors distinguish three levels: niches; regimes; and landscapes (Geels, 2002).

Niches provide a protected space in which to carry out radically new initiatives. These niches may correspond to forerunner markets, subsidized projects or environments that are culturally favourable to experimentation. They may give rise to changes in the paradigm, but the process of change nourished within those niches must be connected to the larger world over the long term. The determining factors in this process are full-scale networks for testing and experimentation, and the mobilization of participants who either have strong social credibility or are in a position to mobilize other such participants.

Regimes are the modes of operation typically established and instituted to fulfil a given function (such as an aerial mobility regime for communications). They are structures that result from the mutual alignment and accumulation over time of knowledge, investment, objectives, infrastructures, values and standards. Their material and institutional aspects are interdependent, and they evolve incrementally. Breakthroughs tend to emerge in niches.

Landscapes refer to processes such as demographic changes, scientific paradigms and political regimes.

In the overall dynamic, the long-term viability of niches and their influence on regimes are determining factors, defined by the lessons learned within each niche, the more precise and formal definition of (possibly regulatory) requirements, and the involvement of a growing number of stakeholders (including investors and users). As a result, those active in these niches must embark on a significant cognitive, institutional, economic and political undertaking (Smith, 2007).

Given the specific and favourable characteristics of the niches in which the pilot projects are carried out, in what way do they promote a wider dissemination of the innovation? How can lessons learnt at the local level be transposed to a larger scale? The case we analysed can shed light on these questions, which we consider crucial to the large-scale dissemination of electric vehicles.

Case study analysis

Before presenting the case of the Renault Electric Vehicle Initiative, we shall briefly describe how we gathered the data on which our analysis is based.

Our first methodological objective was to identify Renault's strategy with regard to electric vehicles. This approach drew on a variety of sources. We had multiple interactions with a doctoral student who is directly involved in analysing and monitoring the progress of Renault's electric vehicle programme (Midler and Beaume, 2010). We attended several conferences at which company executives presented their strategy for electric vehicles.[3] In addition, we monitored the firm's communications in this area on its website and in the media.

Our second aim was to understand the dynamic behind local pilot testing and the network involved. We focused on a specific department – EV business development – which was created in July 2009 with the intention of forming partnerships and developing the infrastructure and ecosystem required to market electric vehicles on a large scale. We based our understanding of these tasks on the work of students who completed their Master's theses in the department, one working from September 2009 to August 2010, and a pair of students working from September 2010 to August 2011. By holding regular meetings with the line staff, we were able to develop a broad perspective on the entity's areas of activity and a more detailed understanding of several projects, which we shall discuss in this chapter.

Characteristics of the overall strategy: the zero emissions vehicle (ZEV) initiative

First, we shall highlight the various aspects of Renault's EV strategy that lead us to consider it a 'breakthrough'.

The first component of this breakthrough involves the battery. Battery technology has progressed substantially alongside tools for mobility (computers, telephones, consoles and so on) with the development of industrial sectors. These technological advances have led to the development of lithium-ion batteries for automobiles. This technology, never before used in vehicles, combines greater vehicle autonomy with the necessary power at a reasonable cost.

The second unique characteristic of the ZEV initiative involves the design of vehicles that have been optimized for electric engines, rather than electrifying vehicles that have been designed to use internal combustion engines. A line of four models (commercial, sedan, compact and an innovative urban vehicle) will be marketed between 2011 and 2013. The models Renault will market in 2011 will also be offered in ICE versions, while the two other models in the programme will be available only with an electric engine. The scale and ambition of the current initiative and its stated objective (10 per cent market penetration for electric vehicles by 2020, as part of an investment totalling €4 billion) are unique.

The third aspect of this breakthrough is the business model. The economic benefits for customers come with use, since power consumption is far less expensive than fuel consumption for a corresponding distance travelled. To realize these benefits, however, customers must invest in a vehicle, and specifically a battery, that is much more expensive than a fuel tank. The company's strategy calls for selling the vehicle without a battery and offering battery rental and, optionally, electricity for charging the battery at a flat rate. This is a radically new business model.

The fourth aspect is the charging infrastructure. A range of charging systems has been considered: equipment that customers may have at home; quick-charge stations where vehicles can be charged in 10 or 20 minutes; and, finally, a battery exchange system to ensure maximum battery life in just a few minutes. This latter option is completely novel and was initially devised by BetterPlace, a company that is using the system in Israel. It entails severe restrictions on the vehicle's design, since the battery must be very easy to remove, and requires a business model in which the driver of the EV does not own the battery but instead is guaranteed the use of a reliable, charged battery and billed at a flat rate based on mileage.

The final area in which breakthroughs are anticipated is the offering of innovative transportation services. As use rather than ownership becomes an option for various types of goods (house, car, boat and so on), we are seeing the emergence of car-sharing solutions and an expanded reliance on carpooling. These changes are tied to economic and environmental conditions, and are likely to have synergies with the characteristics of electric vehicles.

Conducting parallel pilot testing to support the formation of the ecosystem

We focused on the aspects of the Renault breakthrough relating to the business model and charging systems. To offer a complete package for battery use and recharging, the company must work with entirely new partners (notably energy companies) to define responsibilities, formulate reciprocal agreements and decide on service rates based on the roles of each party involved. Renault's BDVE (Electric Vehicle Business Development) entity in charge of this development combined two approaches. First, it conducted studies to refine the forecasts based on the various information that could be collected to provide a frame of reference. Second, it designed detailed scenarios for offerings in a set of clearly defined geographical areas. The design was focused on a pilot experiment in a specific city with prototype vehicles. These scenarios were developed in co-operation with other interested partners, taking into account each party's objectives and any specific local characteristics, and drawing on the skills of those involved.

Finding partners in the difficult economy following the 2008 financial crisis was no easy task, in so far as the projected volumes involved were still no more than estimates. To enlist a range of partners, the decision was made to conduct several pilot tests. This approach was developed alongside other negotiations aimed at designing larger-scale offerings with an eye to a commercial launch.

The role of pilot testing in Renault's EV strategy

The pilot testing, conducted within a city or a limited geographical area, involves providing access to vehicles and a recharging infrastructure, coupled with a sales proposal regarding the use of the vehicle, battery and equipment, along with related services and guarantees. One objective of these pilot projects is to measure CO_2 emissions based on the data of actual distances travelled and electricity consumption, as recorded each time the battery is recharged. At the time of writing, several pilots are being conducted simultaneously. Each is unique with respect to issues that might affect one region more than another, the energy companies involved, other partners responsible for charging stations or public infrastructure, and the users of the prototype vehicles under consideration. The pilot projects also have several characteristics in common.

The pilot designed for the island of Réunion (see Appendix 1) is fairly typical and addresses the issue of countries where energy production has a large carbon footprint. We examine this case in detail in order to illustrate the types of learning and negotiations that took place.

Pilot projects as a tool for building a micro-local ecosystem

This example highlights the issues that must be addressed jointly in designing solutions for recharging EVs. It demonstrates the need to attract partners in creating solutions with a positive impact on CO_2 emissions, and the importance of adapting solutions to local conditions. In addition, the pilot project has fostered co-operation among very different stakeholders, who previously had no reason to co-operate.

Local politicians have led the way in uniting various stakeholders around the project. The financing opportunities tied to these political initiatives and the anticipated media coverage have encouraged these partners to join forces in preparing the pilot project and completing it on time. While each of the protagonists had very different ideas about how soon electric vehicles will be arriving on the market and how widely they will be adopted, the pilot project has helped to synchronize their expectations. Because the project is geographically limited in scope, so is each stakeholder's involvement. Each retains the ability to modify its strategy as opportunities or difficulties arise during the project. For all these reasons, we believe that the pilot projects support the building and leveraging of a micro-local ecosystem.

Pilot projects as an opportunity for collective learning

Once parties commit to participating in a pilot project, the offering can be defined in more detail, as the participants examine issues of key interest to each partner as well as any opportunities that can be seized and that may emerge only once each partner's skills have been integrated.

The opportunity to learn is also a motivation for various participants to commit to the pilot. Other pilot projects are now at a more advanced stage than the one on Réunion, with prototype vehicles already on the road. In these cases, the various partners are seeking data regarding use of the vehicles: distances travelled, time and frequency of charges, recharging stations used, consumption, maintenance requirements, breakdowns and so on. In this sense, experimental projects educate the manufacturer about how the vehicles are being used and what maintenance issues might arise. In addition, they also provide information to other partners in the project.

During the pilot, one goal is to address any problems that emerge, whether they involve the reliability of the terminals, batteries or payment methods, and to adapt to the sometimes surprising ways that the vehicles are used. This draws the focus of the researchers towards issues that appear critical. As such, a pilot project can reduce the apparent complexity and serve to define priorities.

Pilot projects as a means of developing assets and generating support

In addition to these factors, pilot projects trigger a process of asset development by providing a local recharging infrastructure that can be used in the future. The partner then takes a greater interest in operating and turning a profit from these assets over time, and as a result this asset development helps to promote the broader dissemination of EVs.

Another key aspect of pilot projects is the potential for communication. The pilot project on Réunion, for example, received support from the French president, who flew in to give official recognition to the project and enjoy a ride in an electric vehicle. This visit set off an impressive media buzz and encouraged the parties involved to redouble their efforts in preparation for the president's arrival.

Discussion

The case we have examined spotlights the role of pilot projects as a vehicle for collective learning that extends work on projects with radical levels of uncertainty and on environmental initiatives. It also underscores the fact that pilot projects provide leverage for building a (micro-local) ecosystem, a factor that is emphasized in the literature on environmental initiatives but not often mentioned in the project literature. The third aspect we would like to discuss is the connection between 'micro-local' experimental projects and the broader ecosystem.

The literature on new product development stresses the role of prototypes as a tool for validation and learning. Any steps taken in the wake of prototype testing incorporate the feedback that was obtained and offer an opportunity to redefine the product or production process. These steps are supervised internally by the firm's development teams (Thomke, 2003; Loch et al., 2006). In the EV case, however, more complex mechanisms are involved in the transition from a pilot project to the later phases of deployment. There is even a contradiction between the pilot project approach and the logic involved in a large-scale global operation.

Our analysis of the pilot designed for the island of Réunion (see Appendix 1) prompts us to conclude that pilot projects are based on specific local features. They work by attracting the interest of those who have a strong local position and influence. This may involve offering tax incentives and grants at the local level, or considering the routes travelled by users when making decisions such as where to place recharging stations. Pricing assumptions and incentives must be adapted to the characteristics of the local energy mix and its hourly and seasonal variations. Furthermore, to some extent, pilot locations are selected because they offer specific configurations, different in each case, that are favourable to the development of EVs. The pilot approach consists of using this favourable environment to co-ordinate the involvement of the participants and help them converge on a proposal that works for all of them. This result is consistent with the work of Smith et al. (2010), who point out that innovative initiatives develop in, and take advantage of, protected niches.

Conversely, because of the scale involved, the logic governing broader deployment entails a certain level of standardization. In our case, this means simplifying the retail offering available through the manufacturer's sales network with regard to flat-rate battery rental, recharging equipment and the options for using public charging stations. Similarly, manufacturers will seek to sign relatively long-term agreements with their various partners to ensure that their sales offering remains stable over time.

How can pilot projects with a strong local component contribute to a learning process when their subsequent evolution is based on large-scale deployment and a certain level of standardization? Several mechanisms that help to resolve this tension can be outlined.

The first is that, while pilot projects are conducted within a very specific geographical area and are rooted in equally specific partnerships, they generally involve national or even international partners (such as EDF, Total,

Schneider Electric or even supermarket chains). As a result, local negotiations over a pilot project can set a precedent by providing experience on which each party can build.

Moreover, among the participants in these pilot projects are agencies that are linked closely with public policies on sustainable development, such as ADEME, the French environment and energy management agency. They can help to call attention to the reduced CO_2 footprint of electric vehicles, or contribute to the development of standardized methods for measuring that impact. In this way, they too exert an influence that goes well beyond the local pilot project.

The second mechanism relies on conducting several pilot projects in parallel. Simultaneous projects make it possible to form agreements with a range of energy producers and electrical equipment manufacturers, and to work within a variety of tax incentives and regulatory constraints. By conducting several different pilot projects at the same time, manufacturers can address an array of possibilities and identify different types of constraints. Various scenarios can be constructed: on Réunion, for example, some company parking areas have been equipped with solar panels. In France, solutions will be focused more on night-time recharging, when the energy mix has a very low CO_2 impact. These various projects will provide information on a whole range of possibilities; and large-scale deployment in each country may rely on some innovative combination of pilot features.

The third mechanism is to negotiate long-term framework agreements, concomitant with the pilot testing, to support the commercial launch of EVs with various partners who help to provide an infrastructure for recharging electric vehicles: energy companies, equipment and charging station manufacturers and so on. Contrary to the pilot strategy, which tended to crystallize concrete projects in favourable locations, the method here is systematic and proceeds country by country, even if only some potential partners are ultimately approached. The preparations for large-scale deployment are made concurrently and by the same teams responsible for setting up and monitoring the pilot projects.

Given the diverse range of partners approached and pilot projects conducted, the deployment process evolves not by seeking to generalize the model that produced the most satisfactory results among the various projects. Instead, it involves studying these pilot projects and consulting systematically with partners to identify differentiating criteria and a typology of situations, to develop a range of offerings derived from the lessons gained through this process that can best address this diversity. Contrary to assertions in the literature about parallelism in development processes (Sobek *et al.*, 1999; Loch *et al.*, 2006), the aim is not to select one offering for widespread deployment from among the various models tested on different pilot sites.

The fourth mechanism we would like to emphasize is the 'product support' behind the recharging services and the structured EV programme within the company. In this chapter we have mainly emphasized the need to prepare the environment for electric vehicles. The vehicles provided for use in these pilots are prototypes and pre-production vehicles, which create their own dynamic in terms of moving towards large-scale production. To some extent, the investment made in the programme demands that the sales network and business development teams prepare for global deployment.

Conclusions

The EV represents an engineering breakthrough via its innovative battery technology and the changes in the vehicle architecture. It is also a breakthrough with regard to its ecosystem, which involves different players in the recharging system. By grasping this ecosystem in its entirety, we can assess the impact of electric vehicles on the environment as well as their economic benefit. Renault's strategy for rolling out electric vehicles combines a proactive product development strategy with a more exploratory, emerging strategy for infrastructures, recharging plans and so on. We have highlighted the role of pilot testing in this second strategic approach as a means of building 'micro-local' ecosystems that prove educational for the various participants involved, in so far as they provide guidance for deployment on a broader scale and help to create long-term assets.

In particular, we note that developing an environmental innovation entails a joint design/exploration process involving various stakeholders in which pilot projects play a pivotal role.

While the two strategic approaches – deliberate and emerging – may at first glance appear to be contradictory, we have emphasized that they actually reinforce one another and are interdependent in numerous respects. The pilot projects use prototype vehicles from the product development programme. Negotiations for commercial rollout in each country are conducted alongside the pilot testing, often with the same partners who were critical to development of the pilot project, and some of the feedback may prove useful. Conversely, pilot project sites may be selected on the basis of sales priorities rather than favourable circumstances and opportunities.

The approach analysed here includes specific features we have attempted to characterize. Our results appear to complement other work in this area. By observing this dynamic in real time, we were able to identify more clearly the difficulties, uncertainties and possible options as they arose. It is still too early, however, to evaluate the consequences of this process for the dissemination of electric vehicles. The regulatory environment and tax incentives also merit analysis, since they will probably have an impact on the adoption of the vehicles, but these are still being defined. This restricts the scope of our contribution while offering an invitation to pursue further research in these areas.

Appendix 1: Pilot project on the island of Réunion: VERT (Electric Vehicle Technology for Réunion)

The geographical characteristics of the island of Réunion make it a particularly appropriate site for electric vehicles. It is less than 100 km wide and 200 km in circumference, and as a result drivers can cover the whole island on a single battery charge. The island has strong political motives to invest in innovative technological development, which is why in 2007 France's Minister for the Environment, Jean-Louis Borloo, launched the GERRI project (Grenelle de l'Environnement à la Réunion – Réussir l'Innovation [Green Energy Revolution – Réunion Island]). Finally, environmental issues pose a major challenge on the island, partly because tourism represents such a large share of Réunion's economy and partly because its electricity is generated largely from coal (50 per cent). These circumstances gave birth to the pilot project known as VERT (Electric Vehicle Technology for Réunion), with strong support from local politicians, who played a decisive role in encouraging various parties to sign on to the project. These participants included not only energy companies and automobile manufacturers, but also an oil company with expertise in operating petrol stations, a financial company with a large stake in automotive financing, a retail manufacturer, and an installer of photovoltaic systems.

However, demonstrating the environmental benefit of electric vehicles in an area where most electricity comes from coal is a delicate process and much more difficult than it would be in mainland France, where nuclear production accounts for a large proportion of the electricity generated. Indeed, this challenge was a key component of this pilot programme.

Emissions from ICE vehicles are usually measured around the perimeter of the vehicle. A stable measurement method and standardized New European Driving Cycle (NEDC) have been established. If this standard is applied, electric vehicles are 'zero emissions' vehicles. There are also measurements used for power generation, expressed in gCO_2/kWh. In order to determine the overall well-to-wheel environmental impact of a particular mode of transportation, we must define a new measurement tool that addresses energy production, transport and use in the vehicle. Currently, there is no such shared, standardized indicator; the relevant parties have not agreed on any calculation conventions for ensuring that the indices calculated by different entities are actually comparable.

An initial approximation suggests that the overall impact of EVs on Réunion is comparable to that of optimized ICE vehicles. A more detailed analysis of the carbon impact from power generation was performed to examine potential patterns of EV use that would reduce the vehicles' carbon footprint. Extreme variations in the carbon impact from power generation have been updated (the carbon footprint in winter is approximately half that of summer, and half as much during the day as at night. These effects can be attributed to the use of intermittent sources to generate electricity, namely bagasse – a biofuel made from sugar cane stem residue – in the winter and photovoltaic power during the day). A co-operative partnership with the local energy provider, which recognized the value of expanded EV use, provided insight into these wide variations as well as access to specific data.

Based on this understanding, an effort was made to develop offerings that encouraged customers to recharge their vehicles during the day. Two types of incentives for daytime recharging were considered. The first was rate-based, and consisted of setting higher rates at night. The second was physical: charging stations were installed in locations that see higher traffic during the day, such as office and supermarket parking lots. Here too the energy company played a crucial role in defining the rates, and mass retailers were also enlisted so they could discover the costs and value associated with installing charging stations in their parking areas. Daytime recharging was also brought into line with the target market segments, and a test project was envisioned using fleet vehicles.

A second option was also explored, in which solar energy was used to produce the electricity needed to recharge the battery. The idea of equipping customers with both an EV and a photovoltaic installation to supply power to a charging station was considered. Analysis of this option was made easier by the fact that an energy provider and an installer of photovoltaic systems were among the partners on the project taskforce. The problems that must be overcome in order to offer this type of solution have been identified. One is related to the necessary investment, and the other to the stability issues that might arise when these installations are connected to the general electrical grid. Companies in the photovoltaic sector are currently working on projects to address these issues, and the manufacturer is monitoring their results. At the time of writing, the vehicles have not yet been made available; only the preparation and planning phases have been completed. Based on these adapted recharging methods, the estimated overall well-to-wheels CO_2 emissions index was reduced by roughly half in comparison with the estimate made at the outset of the initiative, which makes EVs extremely appealing as an alternative to ICE vehicles in terms of their CO_2 footprint.

Notes

1. Annual European Community greenhouse gas inventory 1990–2005 and inventory report 2007, Technical Report No.7/2007. European Environment Agency, 2007: data for EU-15 (2005): total emissions of 4,192 Gg of CO_2 equivalent; transportation sector emissions: 879.7 Gg of CO_2 equivalent; road transport emissions: 793.9 Gg of CO_2 equivalent.
2. We would like to thank Renault, the Sustainable Mobility Institute, Paris, and the BDVE (Electric Vehicle Business Development) division; this research would not have been possible without their support.
3. Ghosn, 9 March 2010, Paris-Dauphine University; Pelata, 2 February 2 2010, École de Paris; Perrin and Tennenbaum, 29 September 2009; École Polytechnique Ecomobility Conference.

References

Aggeri, F. and Hatchuel, A. (1999) 'A Dynamic Model of Environmental Policy: The Case Of Innovation Oriented Voluntary Agreement', in C. Carraro and F. Leveque (eds), *Voluntary Approaches in Environmental Policy*. Dordrecht: Kluwer Academic Publishers.

Aggeri, F., Elmquist M. and Pohl, H. (2009) 'Managing Learning in the Automotive Industry – The Innovation Race for Electric Vehicles', *International Journal of Automotive Technology and Management*, 9(2): 123–47.

Aggeri, F., Pezet, E., Abrassart, C. and Acquier, A. (2005) *Organiser le développement durable*. Paris: Éditions Vuibert.

BenMahmoud-Jouini, S. and Charue-Duboc, F. (2008) 'Concept Generation Processes: Customer Involvement for Radical Innovation', Best Paper Proceedings, Academy of Management, Technology and Innovation Management Division, Anaheim, CA, 8–13 August.

European Environment Agency (2009) 'Annual European Community Greenhouse Gas Inventory 1990–2007 and Inventory Report 2009', Submission to the UNFCCC Secretariat, EEA Technical, 4/2009.

Fréry, F. (2000) 'Les produits éternellement émergents: le cas de la voiture électrique,' in D. Manceau and A. Bloch (eds), *De l'idée au marché. Innovation et lancement de produits*. Paris: Éditions Vuibert, pp. 234–64.

Gastaldi, L. (2007) 'Stratégies d'innovation intensive et management de la recherche en entreprise. Vers un nouveau modèle de recherche concourante', Doctoral thesis in Management Science, Université de Marne-la-Vallée and École Polytechnique, Paris.

Geels, F. (2002) 'Technological Transitions as Evolutionary Reconfiguration Processes: A Multi-Level Perspective and a Case Study', *Research Policy*, 31(8–9): 1257–74.

International Transport Forum (2008) 'Highlights of the International Transport Forum 2008 – Transport and Energy: The Challenge of Climate Change', OECD.

Kemp, R. (1994) 'Technology and Environmental Sustainability: The Problem of Technological Regime Shifts', *Futures*, 26(10): 1023–46.

Lenfle, S. and Midler, C. (2003) 'Innovation in Automotive Telematics Services: Characteristics of the Field and Management Principles', *International Journal of Automotive Technology and Management*, 3(1/2): 144–59.

Le Masson, P., Weil, B. and Hatchuel, A. (2006) *Les processus d'innovation: conception innovante et croissance des entreprises*. Paris: Hermès.

Loch, C. H., De Meyer, A. and Pich, M. T. (2006) *Managing the Unknown: A New Approach to Managing High Uncertainty in Projects*. New York: John Wiley.

Lynn, G. S., Morone, J. G. and Paulson, A. S. (1996) 'Marketing and Discontinuous Innovation: The Probe and Learn Process', *California Management Review*, 38(3): 8–37.

McGrath, R. G. (2001) 'Exploratory Learning, Innovative Capacity, and Managerial Oversight', *Academy of Management Journal*, 44(1): 118–31.

Midler, C. and Beaume, R. (2010) 'Project-based Learning Patterns for Dominant Design Renewal: The Case of Electric Vehicles', *International Journal of Project Management*, 28(2): 142–50.

Smith, A. (2007) 'Translating Sustainabilities Between Green Niches and Socio-technical Regimes', *Technology Analysis and Strategic Management*, 19(4): 427–50.

Smith, A., Voss, J.-P. and Grin, J. (2010) 'Innovation Studies and Sustainability Transitions: The Allure of the Multilevel Perspective and Its Challenges', *Research Policy*, 39(4): 435–48.

Sobek, D. K. II, Ward, A. C. and Liker, J. K. (1999) 'Toyota's Principles of Set-based Concurrent Engineering', *Sloan Management Review*, 40(2): 67–83.

Thomke, S. (2003) *Experimentation Matters*. Boston, MA: Harvard Business School Press.

6
'Sailing Ship Effects' in the Global Automotive Industry? Competition Between 'New' and 'Old' Technologies in the Race for Sustainable Solutions

Dedy Sushandoyo, Thomas Magnusson and Christian Berggren

Facing the threat of rising fuel prices, a looming shortage of fossil fuels and increasingly stringent regulation to curb greenhouse gas (GHG) emissions, vehicle manufacturers are intensifying their R&D, in particular powertrain technologies. These efforts involve both incremental improvements of the internal combustion engine, and a search for entirely, or partially, new powertrain configurations. New trajectories are emerging in a once-mature industry, and dramatic competition is unfolding. After the oil shocks of the 1970s, carmakers struggled with issues of cost and appropriate technology to combat noxious emissions from petrol engines. This created a short period of uncertainty, but now these manufacturers face the daunting challenge of developing an effective and cost-competitive GHG-reduction trajectory to be sustained for a much longer time.

This challenge has ushered the industry into an 'era of ferment' (Anderson and Tushman, 1990), characterized by increased variation, experimentation and uncertainty. There is as yet no dominant design for the powertrain of the future, and strategic decisions today will have considerable effects on future competitive positions. These decisions involve which 'old' technologies to develop or to drop, which new ones to invest in, which partnerships to engage in and so on. Many studies have been devoted to the potential performance of 'alternative' technologies, such as fuel cells and hydrogen-powered cars. In this chapter we take a less futuristic approach by looking at the competition being played out in the marketplace today and over the next 8–10 years. This is the contest between the established and still vital diesel technology, and well-publicized new contenders, the petrol–electric hybrid vehicles in various configurations. The former is the technology of choice of European manufacturers; and the second the alternative favoured by Japanese producers, with US firms playing a catch-up game with the

Japanese. Purely electric vehicles are being developed in Japan, Europe and the USA, but for the next decade they are expected to target mainly narrow niche markets, such as small city cars. The diesel–hybrid contest could be perceived as being a competition between 'old' and 'new' technologies (though the hybrid powertrain includes several 'old' key components, including the internal combustion engine), but it is also a competition involving different types of innovation and different corporate actors and destinies.

This chapter starts with a brief overview of studies on the new powertrains, followed by a section on competition between 'old' and 'new' technologies. Then we present data on the development, market share and sales projections of so-called clean diesels versus hybrid electric cars in Europe and the USA. This is followed by an analysis of the different innovations involved, architectural versus modular, how they relate to the integration capabilities needed in the future, and to the sourcing and supply strategies used by the main protagonists of this competition. The chapter deals with a limited timeframe, corresponding to two auto development cycles. In the final section, we tentatively discuss the most promising technology choice for greenhouse gas abatement in the mid-to-long term (10–20 years) and how the strategies of manufacturers today are positioning them for this evolution.

The chapter analyses the competition between technological trajectories, and the major firms involved. The Japanese companies, Honda and Toyota, are perceived to be the chief drivers of the hybrid trajectory, with a substantial commercial presence and strong patent portfolios. To further simplify a complex pattern, we single out the European mass producers, above all the VW – and PSA Peugeot Citroën – groups as the key drivers of clean and efficient diesel engines. Component specialists such as Bosch, Valeo, Continental and so on also play an important part in shaping the technological future, particularly in the European automotive industry, but it has not been possible to analyse their contribution within the limits of this chapter.

Measures of car manufacturers' innovativeness

In this study, we attempt to capture three different components of the innovativeness of carmakers: (i) technological (inventive) competence; (ii) product development competence; and (iii) industrialization and marketing competence (or capability, as we use these terms interchangeably; see Prencipe, 2000). The first component will be measured by patenting; the second by the launch of new products (to be distinguished from concept cars); and the third by actual sales. This is, of course, a simplification. Successful sales are a result of all three capabilities combined, but in this context it is important to analyse them separately, since inventive capabilities (patents) are not necessarily accompanied by innovative performance.

In the development of hybrid powertrains there is also a fourth type of capability, which may be labelled as integrative capability seems to warrant attention for several reasons. First, developing hybrids requires an ability to integrate conventional technological specialties – for example, mechanical engineering – with new fields such as electrical engineering, high-voltage systems and power electronics. Second, as a new technological system, hybrid power trains are immature in many respects, both with regard to key components (batteries, control systems and so on) and the overall system configuration or architecture. Third, automotive markets are highly cost-sensitive. Current hybrid cars are uncompetitive in mass markets, where their fuel efficiencies do not justify their high price premiums compared to the evolving 'old' technologies. To cut costs, hybrid makers need both to reduce component costs and to integrate them into optimized systems, which are very different from conventional powertrains. So there are good reasons to study this integrative capability. Unfortunately, it has not been possible to find independent indicators of this capability; here we are dependent on interviews with R&D managers.

Measuring technological competence

The advantages and disadvantages of patent data for measuring R&D activity or the technological competence of firms have been discussed extensively. Patel and Pavitt (1991) argue that patent data can provide details about firms and their technological activities in both R&D departments and other departments (for example, production engineering) over long time periods (Patel and Pavitt, 1991). Holmén and Jacobsson (2002, p. 336) suggest that patents, in most cases, can be considered as 'the best available indicator of technological development activities, excluding software'. As argued by Hagedoorn and Cloodt (2003), patents are frequently used to measure *inventive performance* of firms. In their analysis, inventive performance is defined as 'the achievements of firms in terms of ideas, sketches, models of new devices, products, processes and systems' (p. 1366). By contrast, these authors suggest that innovative performance refers to the degree to which firms actually introduce inventions into the market in terms of new products, new process systems or new devices. However, this does not necessarily mean that the market launch of new products will result in economic success (Hagedoorn and Cloodt, 2003).

In this study, we used two patent databases: the US Patent and Trademark Office (USPTO) and the European Patent Office (EPO), since we are principally interested in the European and US markets. Japanese automakers active in these markets also patent here, whereas European and US firms seldom use Japanese patent protection (Patel and Pavitt, 1991). There are some practical problems involved in using patents as a method to assess the competence base of firms: (i) the US patent classification system is not structured around specific knowledge area; and (ii) patents reflecting the knowledge to develop a certain technology may be found in many classes (Carlsson *et al.*, 2002, p. 241). To identify patents related to hybrid vehicles and diesel engines, and the related patenting activities of major vehicle manufacturers, we implemented the following steps:

- For the USPTO data, we searched patents that had the word 'hybrid' in their abstracts and 'vehicle' in the content of document. We searched the patents that were granted to vehicle manufacturers between 1990 and 2007. In a slightly different way, for the EPO data, we used 'hybrid vehicle' as a search term. By so doing, the search engine showed all patent documents containing 'hybrid' and 'vehicle' in their titles and abstracts. Here we also selected the 1990–2007 period.
- In our second step we examined the abstracts of the patents to eliminate irrelevant patents. We also read the specifics of the documents, particularly the 'Field of the Invention' ('Technical Field' in the EPO data) to make sure of the relevance of the patents to our study.

These steps allowed us to analyse the publicly documented inventive performance of firms related to hybrid technology. We applied similar steps to the study of patents relating to diesel engines, especially R&D activities related to NOx reduction. Employing patent data also enabled us to assess to some extent the accumulation of technological knowledge (stock of knowledge) of firms, as indicated by the aggregate number of patents possessed by individual firms. Patent holding does not automatically reflect a deep knowledge in a particular knowledge area, however (Carlson et al., 2002, p. 241). There are also types of technological capability possessed by firms that cannot be articulated in patents, such as the tacit knowledge embedded in employee skills and production systems (Prencipe, 1997). Further, there is a huge variation in propensity to patent inventions across firms and continents, which detract from the general comparative value of this analysis.

Product development, industrialization and marketing competence

To identify the *product development* and *industrialization and marketing* competence of firms, we searched for data on new product releases and sales regarding hybrid and diesel cars. Concerning product releases, we mainly used trade journals such as *Automotive News* and internet sources such as Hybridcar complemented by business news sources, such as the *Financial Times*. Regarding sales, we compiled data on hybrid and diesel cars sold on the US and European markets, using mainly online news as referred to in the text, complemented by special reports – for example, the UBS report on diesel vehicles in the USA (UBS, 2007). This strategy enabled us to identify the dominant players in the market.

The origin of the 'sailing ship effect'

Since the mid-2000s there has been a rapidly growing literature on new and promising powertrains and vehicle types. Characteristic titles of publications include 'Which energy source for road transport in the future? A comparison of battery, hybrid and fuel cell vehicles' (Van Mierlo et al., 2006); 'Roles of plug-in hybrid electric vehicles in the transition to the hydrogen economy' (Suppes, 2006); 'Hybrid cars now, fuel cell cars later' (Demirdöven and Deutch, 2004); and 'Hybrid vehicles: a temporary step' (Chanaron and Teske, 2007). Many of these studies tend to reach a similar conclusion – that fuel cells are the technology of the future. Most of them limit their comparisons to fuel use on the road, and do not include assessments of energy used in the fuel and vehicle production life-cycles. And, with the exception of Chanaron and Teske (2007), they do not examine the real-world performance and costs of the studied technologies compared to the currently dominating powertrains: the ICE that are used in more than 98 per cent of new cars sold and – which is often neglected – continue to evolve incrementally every year.

Students of the history of technology have been inspired by the so-called 'sailing ship effect' first discussed by Gilfillan (1935) who showed how the 'old' type of sailing ship was dramatically improved when steamships emerged during the nineteenth century. These improvements concerned nearly all of the components and materials of the sailing ship, which was transformed from a wooden to a metallic structure with a massively improved carrying capability and speed performance. Studying more recent cases, Cooper and Schendel (1988) considered seven different instances of competition between 'old' and a 'new' technologies: vacuum tubes versus transistors, steam locomotives versus diesel-electric, fossil fuel boilers versus nuclear power plants, safety razors versus electric razors, and propeller-driven aircraft versus jet engines. They concluded: 'In every industry studied, the old technology continued to be improved and reached its highest stage of technical development after the new technology was introduced. For instance, the smallest and most reliable vacuum tubes ever produced were developed after the introduction of the transistor' (Cooper and Schendel, 1988, p. 255). These improvements made the old technology survive longer, and reach much higher performance levels than had been forecast when the competition began. Nevertheless, ultimately, the 'new' technology carried the day. 'Initially crude and expensive, it expands through successive submarkets, with overall growth following an S-shaped curve. Sales of the old technology may continue to expand for a few years, but then usually decline, the new technology passing the old in sales within five to fourteen years of its introduction' (Cooper and Schendel, 1988, p. 249). The authors found two significant exceptions, however, where the new technology failed to create any new market and defeat the incumbent technologies. However, these interesting exceptions were not further analysed.

Technology historians such as Lindqvist (1994) and Edgerton (2007) have warned that conventional histories of technological evolution are biased towards the 'new' and its emerging S-curve, which 'exaggerates the importance of the initial stage of growth' and neglects the importance of 'the prevailing technological volume, the sheer amount of existing technologies at any given time ... Nor have we acknowledged the fact that a number of competing technological systems co-exist in the technological landscape at any given time, superimposed upon one another and in a state of competition' (Lindqvist, 1994, pp. 271-2).

A general problem in the studies of technology competition is a tendency to reach foregone conclusions: the new technology ultimately wins, even if it might take a long time. In a discussion of the problems of projecting the future cost/performance of potential low-carbon technologies, the Stern Report on the economics of global climate change draws attention to the 'selection bias' in studies of technological learning curves: 'technologies that fail to experience cost reduction drop out of the market and are then not included in the studies' (Stern, 2007, p. 254). To compensate for this in-built bias towards 'success of the new', it is important to distinguish between different types of industries. In system-dependent and capital-intensive technologies, for example, the attacker seldom has the advantage assumed by Foster (1986) and others.

Another problem is how to analyse complex technologies, where key components may be added or radically changed, while other 'old' components or subsystems are modified only incrementally. Is the resulting new complex to be seen as a 'new' technology, or a continuation of an 'old' one? This is related to the perception of the innovation. Is it perceived as a modular innovation with a focus on changes in the individual component, or as an architectural innovation, where the integrated product system is in focus (Henderson and Clark, 1990)? As we shall demonstrate below, this is not only an analytical issue but also related to firm strategies of innovation, competence building and sourcing strategies.

Hybrid electric/petrol cars versus clean diesels: technologies and market performance

Electric/petrol hybrids: Toyota takes the initiative

The modern hybrid electric/petrol car was developed by Toyota and marketed as a new and unique model, the Prius, in 1997. To solve tricky problems such as the function of the electronic control system, the performance of the battery system, and to ensure a smooth ride despite a constantly changing power source, the R&D engineers had to work in a very different way from ordinary Toyota projects (Magnusson and Berggren, 2001). Since its debut, its engineers have been working hard to reduce the cost and weight of the hybrid components, in particular the nickel-metal hydride (NiMH) batteries, and to improve torque and power, but have retained the complex architecture of 1997. Supported by government incentives and positive press reports, the Prius was well received on the Japanese market and has become the country's best-selling car. In the USA, Prius sales were slow

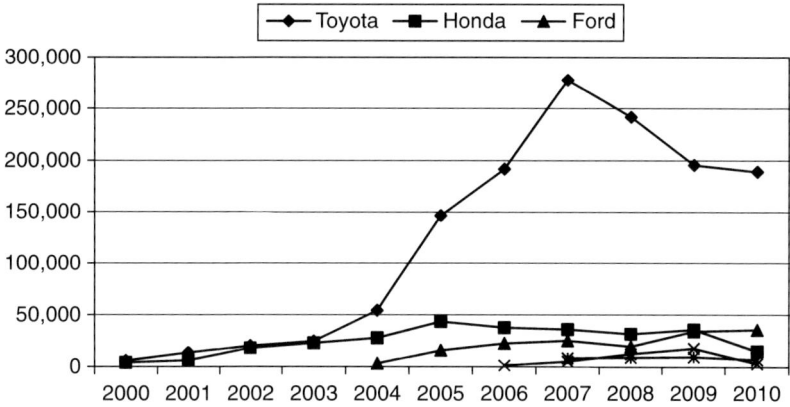

Figure 6.1 Hybrid car sales, USA, 2000–10
Source: Data compiled from Hybridcar.com (2010).

in the early years. But when fuel prices and environmental concerns began to take off, the Prius became a pioneering model of a modern green car, and an economy car for high-end consumers (UBS, 2007, p. 25). In 2004, 88,000 hybrid cars were sold on the US market, and two years later sales had trebled to 256,000. This was a remarkable development, but still only 1.6 per cent of total vehicle sales (Baum, 2007). Honda presented its less expensive hybrid model, the 'Insight', shortly after the Prius was introduced. The Insight was launched in the USA before the Prius, but as it was less convenient and distinctive it largely failed in the market. See Figure 6.1.

American producers initially failed to respond to Toyota's hybrid offensive. However, when gasoline prices began to rise in 2004–5, sales of SUVs levelled off. A milestone was reached in 2007, when the Toyota Prius surpassed the sales of the Ford Explorer, America's top-selling SUV for more than a decade (FT, 2008a). The new fuel-economy standards decided by Congress in 2007 requiring a 40 per cent improvement in gasoline mileage by 2020 increased this pressure. Since 2007, GM and Ford have ramped up their technological activities in the hybrid field (see the section below entitled 'Technological activities – weak patent performance of European producers'). Both companies have now launched their own hybrid cars, but with little commercial success. A more daring initiative is the effort at GM to overtake Toyota by developing a plug-in hybrid car, the Chevrolet Volt, launched in 2011 in the USA.

The evolution of diesel technologies: a European success story

The first patent in diesel technology, where high compression is used to ignite an air/fuel mixture, was granted in 1893, some time after the petrol engine was developed by French and German engineers. Step by step, diesel engines were improved to become the powertrain of choice for heavy-duty vehicles, first in Europe and later in the USA, based on the virtues of its superior fuel efficiency and torque. In the car market, however, diesels remained a hard sell, because of their unpleasant smell, noxious emissions, and high level of noise and vibration. Whereas the introduction of the catalytic converter basically solved the problems of emissions from petrol engines in the 1980s (Bauner, 2007), diesel emissions, especially NOx and particulate matters, were long perceived to be almost impossible to eliminate. In the 1990s, however, a hundred years after its birth, European diesel technology embarked on an impressive improvement trajectory, stimulated by high fuel taxation, advances in component technology and a stepwise tightening of emissions legislation (Euro I in 1992; Euro II in 1996; Euro III in 2000; Euro IV in 2005; and Euro V in 2008) (Dieselnet, 2009).

By incorporating turbochargers, modern control systems, new injection devices (common rail injection), and composite after-treatment systems, diesel engines were able to improve performance and convenience, and reduce emissions markedly. As a result, sales increased year on year. In 2010, more than 50 per cent of new cars in Western Europe were powered by diesel engines (see Figure 6.2). European producers such as Peugeot PSA, Volkswagen and Renault dominate this market, and the presence of Japanese makers is insignificant. The success of diesel powered cars in Europe contrasts to their meagre sales in the USA, where diesel vehicles make up only around 2–3 per cent of the total vehicle market (see Figure 6.2).

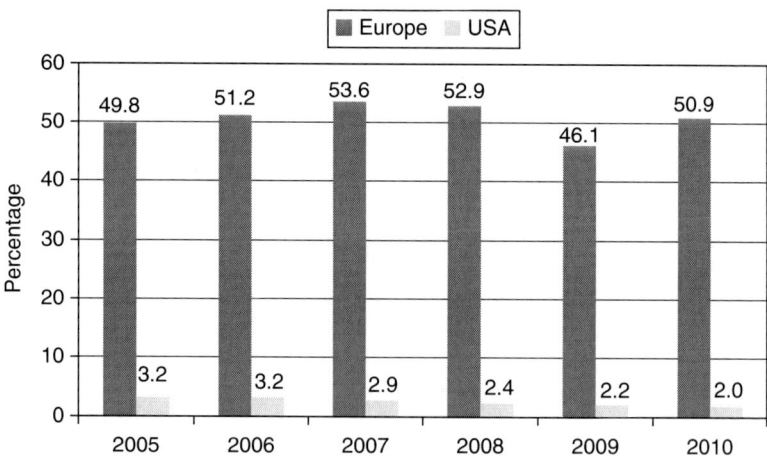

Figure 6.2 Diesel car sales, Europe and USA, 2005–10
Source: Data compiled from Power Information Network (*Automotive News*, 2008a) for US sales 2005–7; Polk (2011) for US sales 2008–10 (covers only Sales Quarter 1: January–March); ACEA (2010) for Europe car sales (15 countries; 2010 covers only January–August).

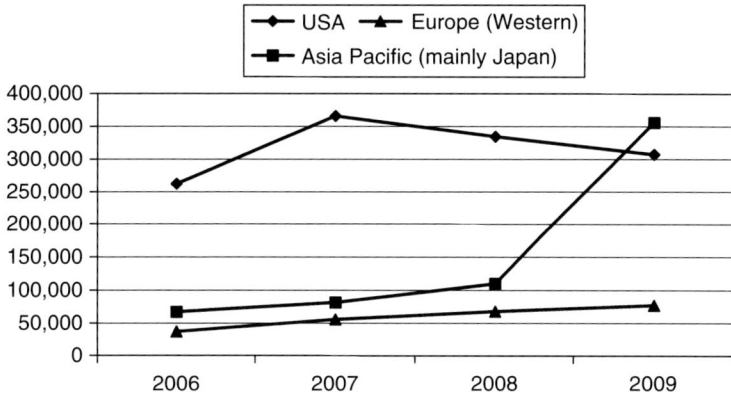

Figure 6.3 Hybrid car sales in USA, Asia Pacific (mainly Japan) and Europe, 2006–9
Source: Adapted from Polk (2010).

European hybrid sales are slowly increasing but, with fewer than 100,000 vehicles sold, did not even reach 1 per cent of the market in 2009. The contrast with the success of hybrids in the USA, and in particular in Japan, is striking (see Figure 6.3).

How can this huge difference be explained?

One reason for the failure of diesels in the USA is the huge difference in general fuel prices, with European pump prices regularly up to three times higher than in the USA. As noted above, the two oil-shocks of the 1970s forced Americans to search for more fuel-efficient cars, but only for a short period. When oil prices stabilized at low levels in the 1990s, consumers returned to treating fuel costs as a non-issue. Moreover, cheaply produced and malfunctioning diesels developed in Detroit in the 1970s created a reputation for compression-ignition machines being unreliable, dirty and inconvenient. In particular, GM 'helped to sully diesel's name during the last oil crisis in 1990 by converting the petrol engine of an Oldsmobile model to diesel with disastrous results' (FT, 2007b).

In contrast, high European taxes stabilized a general interest among consumers regarding fuel-efficient cars, which provided a potential market for diesels. Thus car buyers were ready to appreciate the improvements in the diesel generations coming off the line in the 1980s and 1990s. This strengthened a virtuous circle of more investment in diesel R&D among all the major manufacturers. When French manufacturers, led by the PSA group, introduced an effective particulate filter as standard equipment in the early 2000s, the resistance in such 'diesel-negative' markets as in Sweden began to erode. In 2007, when the EU Commission suggested legislation requiring a cap on GHG emissions to 120 g/km by 2012, several manufacturers were able to quickly launch diesel products complying with this limit.

In France, in particular, the trend towards efficient diesel cars is being further reinforced by a system of incentives and taxes (the 'bonus-malus system') introduced in 2008 to stimulate sales of cars with low emissions and penalize high-emission vehicles (Delmas, 2008). For most driving conditions, apart from city driving, diesel cars are viewed as being more fuel-efficient than hybrids in Europe. Moreover, they are less expensive than hybrids, both because their technology is less complex, and because of the advantage of large-scale production. As a result, petrol hybrid vehicles have much less appeal than in the USA, and have remained confined to small niches. In conjunction with the Geneva salon in March 2008, the chief Europe correspondent of *Automotive News* summarized this sentiment with the question (*Automotive News*, 2008b) 'Who needs a hybrid?', referring to 'the more affordable and practical five-seat green cars' announced by VW, Fiat, and Ford Europe, which emitted less than 100 g/km of CO_2.

Thus 14 years after the first launch of modern hybrids in Japan, the pattern of strong diesel investment and sales in Europe contrasted starkly with the US situation, where hybrid electric cars by Toyota, and less so by Honda, had emerged as an alternative to conventional petrol engines. However, on both sides of the Atlantic, markets and technologies were in an 'era of ferment' with increasing rates of change and intensified technological competition. In the next section we turn to one important indicator of R&D efforts: patenting.

Technological activities: the weak patent performance of European producers

Both petrol hybrid and diesel technologies are in a process of continuous development, demonstrated by active patenting by major manufacturers. Petrol engines are also being refined continuously but with much less patenting activity (see Figure 6.4).

In the case of hybrid technologies, the major challenges for existing platforms are to reduce cost and weight, and to find simpler solutions. For the next hybrid step – plug-in hybrids that can be recharged from external sources to make purely electric rides of 40–50 km possible – there is a demand for more effective energy storage systems (batteries and battery control systems). These challenges stimulate inventive activities in component technologies, in control systems and in integrative technologies. Toyota pioneered patents in hybrids (Yarime and Kuroki, 2006) and was the first, together with Ford, to patent these in the USA (see Figure 6.5). In the early 2000s, Honda's US patenting outperformed Toyota's, but its performance is uneven and has been falling. GM started late but emerged as a leader in the last period of 2006–7, an indicator of its accelerated efforts

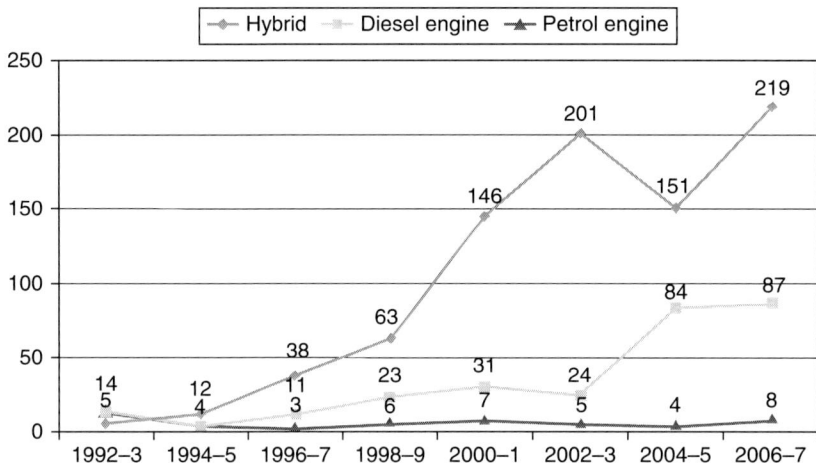

Figure 6.4 Total number of hybrid, diesel and petrol engine patents, 1992–2007
Source: Data from USPTO and EPO; patents submitted by Toyota, Honda, GM, Ford, Volkswagen, BMW, Nissan–Renault, Peugeot–Citroën, Fiat and Daimler.

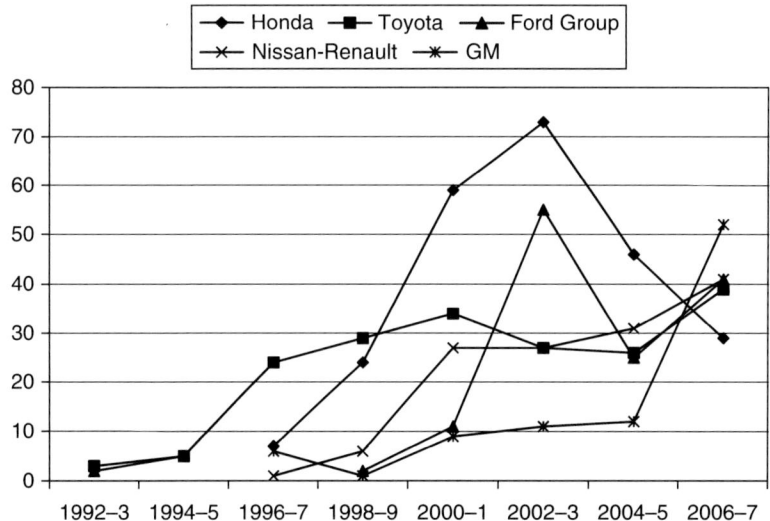

Figure 6.5 Patents by carmakers related to hybrids, 1992–2007
Source: Data compiled from USPTO and EPO.

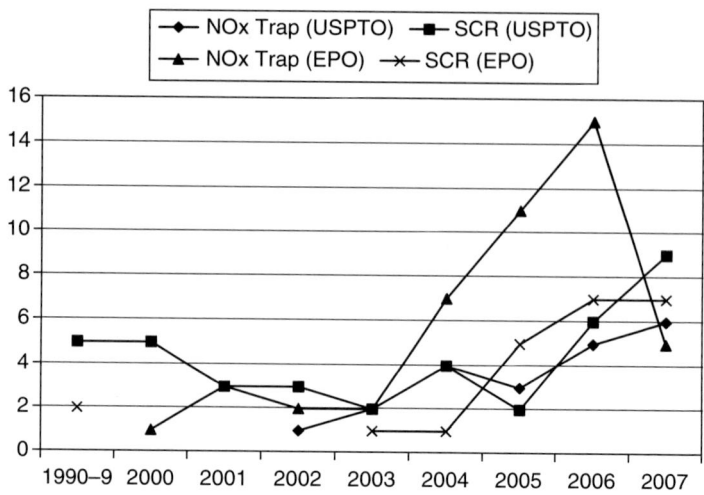

Figure 6.6 Diesel patents related to emissions reduction, 1990–2007
Note: SCR = selective catalytic reduction.
Source: Data compiled from USPTO and EPO.

to develop technologies able to compete with Toyota in the next round of hybrid evolution.

An increasing amount of patenting activity could also be found among the large component firms. The Japanese seem to be leading here too, but there is no clear and stable pattern. In diesel technologies too there has been an increased level of patenting activities, as indicated by Figure 6.6. Considering the established nature of the diesel engine, this is an interestingly high figure. Most of the patents are related to emissions-reduction devices: new control technologies, new fuel injection features, and improved injection/combustion methods.

Of the manufacturers, Japanese and US firms are generally the most active in diesel patenting, when considering both US and European patents (see Figure 6.7). In contrast, the European companies who are the actual market leaders are applying for relatively few patents. In the Nissan–Renault case, the European partner has contributed around 10 per cent – 9 out of 88 patents.

As can be seen from Figure 6.8, the patent activities of European vehicle manufacturers related to diesel technologies have been at a low and uneven level over the years. Peugeot–Citroën is the most active European patent assignee, but its patenting activity began only recently. This may be interpreted as a late awakening to a strategy of protecting its inventions more aggressively. It is well known in the patent data literature that companies

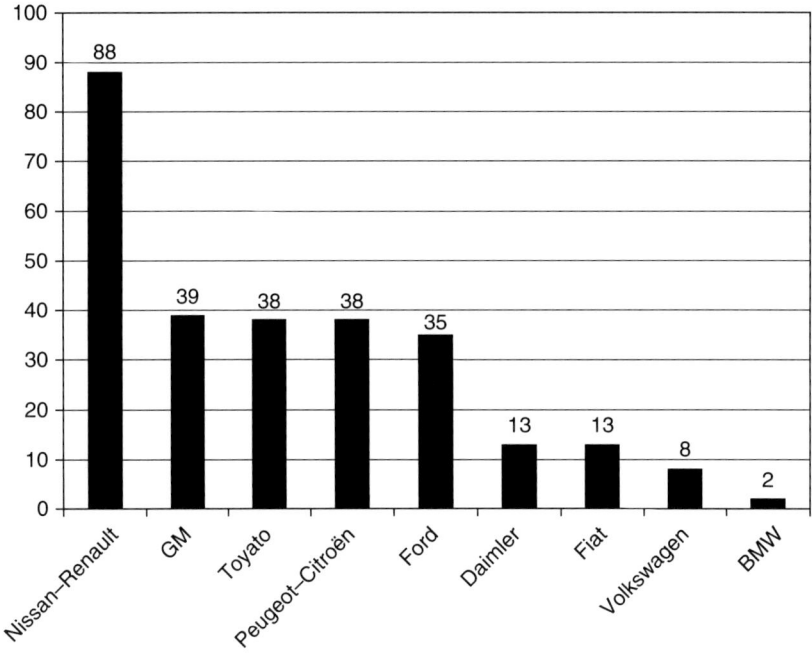

Figure 6.7 Diesel engine patents by major carmakers, 1990–2007
Source: Data compiled from the USPTO and EPO.

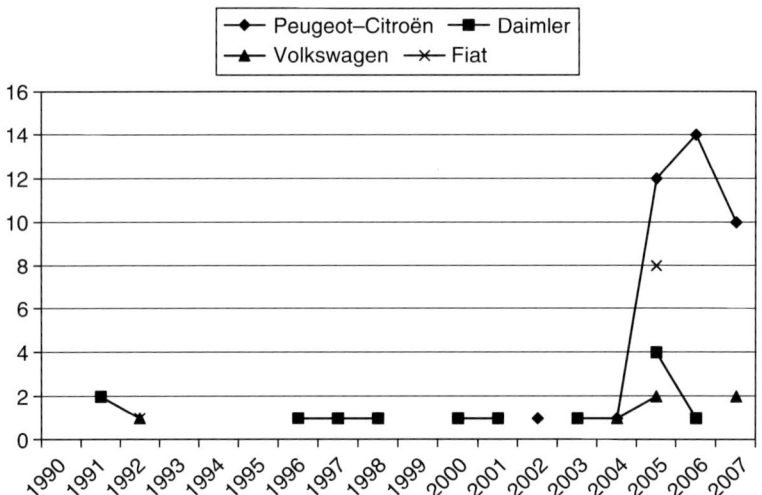

Figure 6.8 Diesel patents by European carmakers, 1990–2007
Source: Data compiled from the USPTO and EPO.

exhibit big differences in the propensity to patent across countries (Patel and Pavitt, 1991). The low level of European patenting in the USA is probably also related to the fact that the car sales of European diesel manufacturers have been confined to Europe. In 2008 this began to change, however, with the espoused strategies of German firms to launch clean diesels in the USA (see the next section). This may have an influence on their future patenting. The French carmakers, however, had no presence in North America, so their patenting behaviour would be unlikely to be influenced by new marketing strategies.

The US market: a battlefield between diesel and petrol hybrids

The technological activities in the petrol hybrid and diesel technology fields are underpinning aggressive product strategies of both Japanese and European firms (mainly German) regarding the US market. As for the hybrids, both Toyota and Honda plan rapid expansion. Toyota is developing its hybrid system into a generic technology in the company. Honda sold only 55,000 hybrids globally in 2007, but has set a target to sell 400,000 annually in 2012, the majority in the USA (*Automotive News*, 2008c). A major hurdle for both companies is the cost of the hybrid powertrain compared to the simplicity of the petrol engine. A report published in the late 2000s (UBS, 2007, p. 28) estimated the cost penalty of a large hybrid such as the Lexus LX400h to be at least US$7,000–8,000. Toyota is pursuing an aggressive cost-cutting strategy, but experts estimate a considerable cost penalty to remain in the mid-term perspective. As for diesel engines, European makers, led by Volkswagen–Audi, began to roll out several new vehicles on the American market in 2008. The Volkswagen Jetta was launched as the first diesel to comply with the emissions standards adopted by the most demanding states around California (the so-called Tier 2, Bin 5 standard). Mercedes-Benz also announced its intention to launch diesel vehicles complying with the same standards (FT, 2007b; *Automotive News*, 2008d). Complying with increasingly stringent emissions legislation raises the cost of diesels, however. In 2007, clean diesels were estimated to suffer from a cost penalty of US$3,000–4,000 compared to petrol engines, but to have a cost advantage of US$2,500–5,000 compared to hybrids (UBS, 2007, p. 8). Since this prediction was published, however, Toyota has reduced the price of the hybrid versions in its main programme (see the next section), which has removed much of the cost advantage of clean diesels.

A major hurdle for expanding sales of diesels on the US market seems to be their negative reputation; and another is the lack of availability of low-sulphur diesel fuel in the USA. A further disadvantage has been the many local and state incentives offered to hybrid owners, from tax rebates to parking and driving privileges. Some of these incentives are, at the time of writing, in the process of being dismantled, but even more generous benefits are being awarded to new, plug-in hybrids and purely electrical cars. In its special report, UBS estimates that total US sales of hybrids and diesels will grow from 800,000 vehicles in 2007 to 2.7 million in 2012, with diesels being responsible for a major part of the growth (UBS, 2007, p. 53). The many subsidies for advanced HEVs and EVs make it difficult to forecast the trend, however.

Sourcing strategies: in-house sourcing in Japan, specialization in Europe

For a long time, the European mass producers, such as the PSA and VW groups, have continued to refine their combustion engines and drive-train components, with new transmission systems, for example. But the relationship between technologies and manufacturers is now becoming less clear-cut. The Europeans have started to add micro hybridization as a way to reduce emissions without having to invest in entirely new power train configurations. VW, for example, offered a start–stop feature promising a fuel-efficiency of 3.8 litres/100 km (62 mpg) in the BlueMotion model of the sixth-generation Golf platform launched in 2008. At the same time Peugeot–Citroën announced their plans to make start–stop standard in all its small and medium cars by 2010 (*Automotive News*, 2008e). Other manufacturers are doing the same, including BMW with its 'EfficientDynamics' concept, which also features an increased use of electrical rather than belt-driven support systems (Chanaron and Teske, 2007, p. 281). Several European firms have also started to offer hybrid models, albeit in low production numbers for premium markets. The major reason among the previous reluctant Europeans for investing in hybrid systems seems to be a perceived necessity of offering this technology on the US market, and a need to develop broader capabilities. There are also huge differences in the innovation strategies related to hybrids pursued by the Japanese market leader, Toyota, and the major European makers.

Modular versus architectural innovation

A way to reflect on this difference is to use the innovation typology suggested by Henderson and Clark (1990). The authors argue that, if a product consists of interrelated components, technological discontinuities may either refer to new components or to new linkages between components. Thus different categories of innovation arise: *modular innovation* involves new design concepts of individual components but leaving the established linkages between components untouched; and *architectural innovation* changes the product configuration and combines components in new ways without introducing fundamentally new component technologies. The evolution of the diesel engine can be seen as a case of modular innovation, involving significant redesign and the addition of components, but no change in the technology's basic architecture. Hybrid systems can be seen as architectural innovations (Magnusson et al., 2003, p. 13), but the evolution of the Toyota Hybrid System (THS) is also a radical innovation, involving new interfaces between components as well as substantial innovation in batteries, battery control systems, transmission and so on, as witnessed by the many patents relating to these sub-fields, and the importance of new competencies in power electronics, electrical engineering and so on.

There is an important point in analysing hybrid power trains as architectural innovation, since this draws attention to the critical role played by a company's integration competence. This is highlighted in an interview with Toyota Europe's vice president of R&D:

> We don't believe in black box design, where suppliers have all design responsibility and design insight. We want to know all the details of new important technologies. With black box design the OEM cannot do anything if there is a problem. To develop hybrid systems is about integration. Then you need to have detailed knowledge of key components. After we have acquired this detailed knowledge, we may subcontract production. (Personal interview, November 2006)

Sourcing strategies

This insistence on detailed component knowledge as a basis for successful hybrid development is in line with Toyota's sourcing strategy. Unlike other manufacturers, Toyota is building strong internal capabilities, both in new R&D fields such as power electronics and in critical component production – for example, electric motors – where Toyota has become a significant mass producer (Alakula, 2008). For other critical components, batteries and battery systems, Toyota has acquired a controlling interest in its joint venture with Panasonic. With its comprehensive R&D and industrial base, Toyota is building both technical integration competence and industrial capabilities for mass production and cost reduction. As noted above, European firms have invested much more cautiously in hybrid powertrains and the importance of specific integration and industrialization capabilities are played down (interview with R&D vice president, VW, 4 July 2008). Instead, European makers rely on the capabilities of component makers, such as Bosch, ZF, Valeo and Siemens VDO, to develop integrated solutions. In strategies that do not plan for immediate mass production, this reliance on independent component specialists is justified by simple economics. At the same time, the European carmakers are forgoing opportunities to build capabilities for in-house product optimization, and industrial and commercial testing, on a large scale. If hybrid powertrains continue to use both modular and architectural innovation this might constitute a considerable risk (Magnusson and Berggren, 2011).

To complicate the picture, Honda has adopted a position somewhere between Toyota's integration strategy and the European pattern of specialization. On the one hand, Honda is building broad market experience and mass production capabilities for hybrids, but on the other they have not invested the same massive resources in in-house component resources in the way that Toyota has. Thus there are several distinct innovation and sourcing strategies. The key question is what these will mean for the future competitiveness of the companies involved in an era of increased technological competition.

Plug-in hybrids: the next step?

To compensate for its late start in the hybrid race, GM has, as noted above, invested heavily in developing plug-in hybrids. Presented in late 2010, the new Chevrolet Volt sold 2,700 units in the first half of 2011, beating Toyota to the market in this new competition (*Automotive News*, 2011a). Plug-in cars offer emissions-free rides for the ordinary home–work commuter and city driver, but demand considerably a higher electrical storage capacity, in the range of 16 kWh for a mid-sized car, 10 times higher than the battery capacity of a standard Prius car. This, plus the more powerful electrical motor, adds a substantial cost penalty, and thus plug-in hybrids are dependent on generous government subsidies.

Toyota has worked hard to reduce costs in hybrid powertrains, and in the Camry programme introduced in the USA in 2011, the cost penalty of the hybrid version was reduced to US$3,400 (*Automotive News*, 2011b). As for plug-in hybrids, there is also the key issue of the fuel mix used in generating the electricity feeding these vehicles. If plug-in cars operate on power from coal-burning plants, the real progress in GHG abatement will be largely illusory. A related advance is the new development of mass-produced, purely electric vehicles. These efforts have their strongest proponents in Mitsubishi, collaborating with Peugeot in Europe, and Nissan–Renault. Launched at the same time as the Chevrolet Volt, Nissan's electric Leaf car was selling at double the rate of GM's plug-in car in the USA (*Automotive News*, 2011a). With a simpler technological configuration and a narrow target ('dedicated city cars'), EVs may be more successful commercially than the plug-ins. Their impact on the mass market will probably be small in the foreseeable future, but for involved automobile companies, they will build new capabilities in electrical engineering and components.

Conclusions and discussion

In a short-to-mid-term time-frame, there are two major powertrain contenders in the global automobile market: the evolving petrol engine, which at the time of writing accounts for more than 95 per cent of cars sold in the

USA and Japan, and the 'clean diesel' that is the European market leader. The third, and much smaller, contender is the petrol/electric hybrid. Increasing stringency in emissions legislation, such as the Euro VI (mandatory from 2013/14) (Dieselnet, 2009) will change relative costs to the disadvantage of diesel vehicles, but diesels will probably retain an advantage, and a much stronger market share, than petrol hybrids in Europe. The technological competition is in a dynamic phase, however. Petrol hybrids are challenged by more advanced electrification, plug-in hybrids and purely electrical vehicles. Diesel vehicles are adding micro hybridization, and full-blown diesel hybrids are been introduced, such as the Peugeot version presented at the Frankfurt motor show in September 2011. Looked at from a 5–10-year perspective (depending on fuel prices and future GHG regulations), diesel hybrids stand a good chance of offering a cost-competitive and fuel-efficient vehicle alternative, if emissions from the entire life cycle are included (Schäfer et al., 2006). Thus the contrast between 'old' and 'new' technology discussed in the introductory section could be dissolved in a new synthesis. The 'old' diesel technology will continue to evolve, partly as a result of being challenged by other technology options.

The assessment of various technologies must be separated from the analysis of the firms involved in the different trajectories. To be successful in diesel hybrids, a deep knowledge of their specific characteristics will be an important aspect for carmakers, but how important in relation to the mastery of electrical components and systems (batteries, power electronics, electric motors and transmission devices) depends on the evolution of future powertrain systems. If they are going to evolve into an ensemble of discrete modular innovations, a knowledge of system integration will suffice. But if hybrids and other powertrains continue to employ a combination of architectural and modular innovations this will not be enough. For the near future, there seems to be a case for European manufacturers to retain their focus on clean diesels, projecting this strength on to the US market and adding elements of micro hybridization supplied by external specialists. At the same time, Toyota is building a comprehensive R&D and industrial system around hybrids, with strong links to leading battery suppliers. The European model of reliance on a strong supplier industry means that carmakers can take advantage of independent innovation in key components, but a major problem is the lack of strong suppliers of advanced battery systems in Europe.

A key issue will be the evolution of new vehicle systems as a whole: will the main trend be component innovation and standardization of modular interfaces, or will product performance and cost remain closely allied to the combination of component knowledge and system optimization skills of the car assembler? In an analysis of similar challenges in the disk drive industry, Christensen et al. (2002) found that, if the first trend dominates, the specialization model ('the European pattern') could be most beneficial. But if the second trend prevails, the integrated model ('the Toyota way') will be superior. The question of whether specialization or integration strategies will be the most appropriate in relation to new powertrain development remains a strategic issue for the future.

References

Anderson, P. and Tushman, M. L. (1990) 'Technological Discontinuities and Dominant Designs: A Cyclical Model of Technological Change', *Administrative Science Quarterly*, 35(4): 604–33.

Baum, A. (2007) 'Market Penetration of Hybrid and Diesel Vehicles in the U.S. Market, 2004–2015', A presentation to the Fuel Economy Technology Trends and Policy Options Forum. Washington, DC, 1 October 2007.

Bauner, D. (2007) 'Towards a Sustainable Automotive Industry: Experiences from the Development of Emission Control Systems', Doctoral dissertation., KTH Royal Institute of Technology, Stockholm.

Carlsson, B., Jacobsson, S., Holmén, M. and Rickne, A. (2002) 'Innovation Systems: Analytical and Methodological Issues', *Research Policy*, 31(2): 233–45.

Chanaron, J.-J. and Teske, J. (2007) 'Hybrid Vehicles: A Temporary Step', *International Journal of Automotive Technology and Management*, 7(4): 268–88.

Christensen, C. M., Verlinden, M. and Westerman, G. (2002) 'Disruption, Disintegration and the Dissipation of Differentiability', *Industrial and Corporate Change*, 11(5): 955–93.

Cooper, A. C. and Schendel, D. (1988) 'Strategic Responses to Technological Threats', in M. L. Tushman and W. L. Moore (eds), *Readings in the Management of Innovation*. London: HarperCollins, pp. 249–58.

Delmas, S. (2008) 'Sustainable Development Requirements and New Product Policies', GERPISA XVI symposium, 'The Automobile Industry and Sustainable Development', Moncalieri, 18–20 June.

Demirdöven, N. and Deutch, J. (2004) 'Hybrid Cars Now, Fuel Cell Cars Later', *Science*, 305(5686): 975–6.

Dieselnet (2009) 'Emission Standards (European Union): Heavy-Duty Diesel Truck and Bus Engines'. Available at: http://www.dieselnet.com/standards/eu/hd.php.

Edgerton, D. (2007) *The Shock of the Old. Technology and global history since 1900*. London: Profile Books.

Foster, R. (1986) *Innovation: The Attacker's Advantage*. New York: Summit.

Gilfillan, S. C. (1935) *Inventing the Ship*. Chicago: Follett.

Hagedoorn, J. and Cloodt, M. (2003) 'Measuring Innovative Performance: Is There an Advantage in Using Multiple Indicators?', *Research Policy*, 32(8): 1365–79.

Henderson, R. and Clark, K. B. (1990) 'Architectural Innovation: The Reconfiguration of Existing Product Technologies and the Failure of Established Firms', *Administrative Science Quarterly*, 35(1): 9–30.

Holmén, M. and Jacobsson, S. (2000) 'A Method for Identifying Actors in a Knowledge Based Cluster', *Economics of Innovation and New Technology*, 9(4): 331–51.

Lindqvist, S. (1994) 'Changes in the Technological Landscape: The Temporal Dimension in the Growth and Decline of Large Technological Systems', in O. Granstrand (ed.), *Economics of Technology*. Amsterdam: Elsevier Science.

Magnusson, T. and Berggren, C. (2001) 'Environmental Innovation in Auto Development: Managing Technological Uncertainty within Strict Time Limits', *International Journal of Vehicle Design*, 26(2/3): 101–15.

Magnusson, T. and Berggren, C. (2011) 'Entering an Era of Ferment – Radical vs. Incrementalist Strategies in Automotive Power Train Development', *Technology Analysis and Strategic Management*, 23(3): 313–30.

Magnusson, T., Lindström, G. and Berggren, C. (2003) 'Architectural or Modular Innovation? Managing Discontinuous Product Development in Response to Challenging Environmental Performance Targets', *International Journal of Innovation Management*, 7(1): 1–26.

Patel, P. and Pavitt, K. (1991) 'Large Firms in the Production of the World's Technology: An Important Case of "Non-Globalisation"', *Journal of International Business Studies*, 22(1): 1–21.

Prencipe, A. (1997) 'Technological Competencies and Product's[sic] Evolutionary Dynamics: A Case Study from the Aero-Engine Industry', *Research Policy*, 25(8): 1261–76.

Prencipe, A. (2000) 'Breadth and Depth of Technological Capabilities in CoPS: The Case of the Aircraft Engine Control System', *Research Policy*, 29(7–8): 895–911.

Schäfer, A., Heywood, J. B. and Weiss, M. A. (2006) 'Future Fuel Cell and Internal Combustion Engine Automobile Technologies: A 25-year Life Cycle and Fleet Impact Assessment', *Energy*, 31(12): 2064–87.

Stern, N. (2007) *The Economics of Climate Change*. Cambridge, UK: Cambridge University Press.

Suppes, G. J. (2006) 'Roles of Plug-in Hybrid Electric Vehicles in the Transition to the Hydrogen Economy', *International Journal of Hydrogen Energy*, 31(3): 353–60.

Van Mierlo, J., Maggetto, G. and Lataire, P. (2006) 'Which Energy Source for Road Transport in the Future? A Comparison of Battery, Hybrid and Fuel Cell Vehicles', *Energy Conversion and Management*, 47(17): 2748–60.

Yarime, M. and Kuroki, Y. (2006) 'The Strategies of the Japanese Auto Industry in Developing Hybrid and Fuel Cell Vehicles', Working paper, University of Tokyo.

References to trade journals, internet sources and special reports

ACEA (European Automobile Manufacturers' Association) (2010) 'New Passenger Car Registrations – Breakdown by Specification Share of Diesel'.

Automotive News (2008a) 'Final Tally: Hybrids Higher, Diesel Dip', 7 April.

Automotive News (2008b) 'Who Needs a Hybrid?', 3 March.

Automotive News (2008c) 'Honda Sets Lofty Hybrid Goal with Civic's Help', 3 March.

Automotive News (2008d) 'M-B Will Hit the Road with 50-State Diesels', 3 March.

Automotive News (2008e) 'PSA: Standard Stop-Start by 2010', 5 March.

Automotive News (2011a) 'How Goes the EV Race?', 8 August.

Automotive News (2011b) 'Toyota Proceeds Cautiously with New Camry', 29 August.

FT (*Financial Times*) (2008a) 'Toyota Prius Sales Pass Ford Explorer in US', 10 January.

FT (*Financial Times*) (2007b) 'VW Clears Up Diesel's Grimy US Image', 4 May.

FT (*Financial Times*) (2008c) '"Diesel" No Longer a Dirty Word in the US', 14 January.

FT (*Financial Times*) (2008d) 'Toyota Move Puts GM Under Pressure', 14 January.

Hybridcar (2010) 'Market-dashboard'.

NYT (*New York Times*) (2008) 'Interest Fades in the Once-Mighty V-8', 16 January.

Polk (2010) 'Asia Pacific Region Propels Growth of Hybrid Market'.

UBS (2007) 'Will Diesel Boom in the US?' Special report.

Personal interviews

Alakula, M., Professor of Electrical Engineering and hybrid powertrain expert, University of Lund, 11 February 2008.
Volvo, Product portfolio officers and technology strategists, 23 February 2007.
Toyota vice president of R&D, Europe, 7 November 2006.
VW vice president of R&D, 4 July 2008.

7
Firm Perspectives on Hydrogen

Marc Dijk and Carlos Montalvo

A glance at the product range in today's car showrooms shows a strong commitment of all mainstream vehicle manufacturers to conventional petrol and diesel internal combustion engine (ICE) technology. These types of engines have been built and refined for more than 100 years (Cowan and Hulten, 1996). Since 1990, alternative ways of propelling vehicles have been presented by various automobile manufacturers: full electric and hydrogen driven, as well as hybrid-electric models. In general, these three alternatives have received much less R&D attention than has the dominant IC engine. Hybrid-electric engines have by now been incorporated in most car manufacturers' research efforts, at least to some extent, and most firms consider them important in the short or medium term (Chanaron and Teske, 2007). From an environmental perspective, they should be considered to be neither a minor nor a major innovation, decreasing harmful emissions (from driving) by 10– 20 per cent (Lave and MacLean, 2002). Alongside hybrid-electric systems, more radical alternatives have also been presented. There are full electric vehicles, (re)introduced in the 1990s, and hydrogen-fuelled vehicles, introduced around 2000. These latter two types of vehicles are examples of ultra low emission vehicles (ULEVs). They emit extremely low levels of harmful gases compared to existing diesel and petrol vehicles. California's Air Resources Board defines a ULEV as *a vehicle that emits 50 per cent less pollution emissions than the average for new cars released in that model year.* We follow that definition here, and include CO_2 as a harmful gas in the examined emissions.[1]

In this chapter we analyse car firm perspectives on hydrogen. We examine the framing of firms by studying their belief systems and actual engagement. The central question is: Which underlying beliefs significantly drive the engagement in the development of hydrogen technology?[2] A second question concerns whether we can map out diversity between firms. We thus 'step into the shoes' of the car firms and analyse their social, business, technological and environmental considerations. Various scholars and industry strategists as well as policy-makers acknowledge hydrogen as an important future energy carrier for transportation.

In the section below entitled 'Firm perspectives on car engines' we introduce how carmakers view car engines in general. After explaining our methodology, we then present the results of our survey to identify the present-day beliefs of global carmakers towards the development of hydrogen propulsion technology (see the section entitled 'Results: engagement and beliefs towards hydrogen technology'). We conclude by discussing the outcomes, also in the light of additional semi-structured interviews.

The outcomes suggest that different firms deal differently with hydrogen vehicle development. Broadly, three groups of firms can be distinguished concerning their engagement in hydrogen: uncertain firms, unwilling firms and optimistic firms. Further, we find that the level of technological and organizational capabilities seems most strongly correlated with the actual engagement of firms in hydrogen technology development. Finally, we address the relationship of actor beliefs and actor framing, offering a final conclusion.

Firm perspectives on car engines

Carmakers are confronted with a range of uncertainties around car engines. What are the future expectations of the various alternative technologies? How strict will future emissions legislation be? How should one enhance technological capabilities around alternative technologies: collaborate with suppliers, build strategic alliances with competitors or perform corporate R&D? How much money should one invest in alternative technologies? When will the market be ready for product launches? Is there a 'business case'? When are prototypes good enough to be launched? These are all questions with no straightforward answers for companies.

How, then, do carmakers deal with alternative propulsion technologies? Inter-firm competition in R&D on new engine technologies is intense (Molot, 2008). There is major competition over petrol and diesel technology, as well as hybrid-electric and hydrogen-powered vehicles. The high cost of this type of research, and the search for competitive advantage, sometimes push car firms into strategic alliances with each other, as well as with suppliers. For cleaner technologies, several partnerships have emerged, often partly government-sponsored (such as California's Fuel Cell Partnership). Though research consortia are important in the initial stages of research on new and costly technologies, inter-firm competition for first-mover advantages imposes limits on the extent of corporate co-operation around R&D.

In this chapter we focus on firm framing towards hydrogen engine technologies as an example of an ULEV. Most of the major (top five) car assemblers are working to produce hydrogen-powered vehicles and doing so in their home countries, generally at sites close to their head offices. Daimler's fuel cell (FC) R&D is based in Germany; Ford's and GM's research is in the USA, and Toyota's base is in Japan. GM, which began co-operation with the US Department of Energy on alternative fuels in 1989–90, was the first assembler to begin research on hydrogen vehicles. GM has invested considerable resources over a long period of time in the search for a viable fuel cell vehicle (FCV) and has alliances with a number of companies, such as Dow Chemical. In recent years, some demonstration projects with fleets of consumer vehicles have been launched in Washington, DC and California in the USA, and Berlin in Germany, among others, where GM and Daimler are participating. Toyota and Honda are also investing quite substantially in R&D around FCVs[3] (Van den Hoed, 2005; Molot, 2008).

Some companies regard hybrid technology as an important transitional step on an indefinite time path to FCVs. Toyota expects car assemblers that master hybrids will have significant cost and learning advantages over competitors when commercializing FCs (Molot, 2008).

Again, because of the significant first-mover advantage (which will accrue to the assembler that first produces a reliable and reasonably priced FCV), it is not surprising that carmakers undertake this kind of leading-edge research in-house. In describing the 'strict partitioning' between research undertaken in-house and that conducted in research partnerships, one expert used the term 'firewall' (Avadikyan and Larrue, 2003). Disruptive technologies are, by definition, uncertain. Despite the huge amount invested in R&D to develop FCVs, a number of uncertainties around cost, infrastructure, hydrogen production and storage remain.

How strong is the dedication of firms to hydrogen technology development? Are firms investing to prepare for market introduction as soon as possible? Or are firms engaging only slightly as a minimal insurance to avoid missing the boat when it sails? These latter matters have stayed unresolved. In the next section we describe hydrogen technology through the eyes of carmakers, and analyse their technological frames.

Methodology

We analyse carmakers' perspectives on hydrogen by assessing the belief systems of the firms' decision-makers. We employ an empirical approach based on a behavioural model, a theory of reasoned action, designed to understand human social behaviour on the basis of the underlying intentions, attitudes, subjective norms and behavioural control (Ajzen, 1988; Ajzen, 1991). The three domains are derived in a range of empirical studies initiated by Ajzen, concerned with the understanding of human social behaviour, finding an explanatory reliability of up to 91 per cent of the variance of behaviour (Ajzen, 1991; Ajzen and Krebs, 1994; Jonas and Doll, 1996). Montalvo (2002, 2006) has shown that this model is also valid for technology development and strategic planning within the firm's context, as strategic planning and technology development are based on goals to be achieved. Montalvo applies the model as a structural model to study the possible determinants of firm innovation behaviour, where he distinguishes nine possible drivers, and concludes that the proposed model is satisfactory.

Although Montalvo's approach was designed to study *behaviour* rather than *frames*, we suppose that both are anticipated by belief systems.[4] Accordingly, we consider the innovation behaviour of car firms as a function of salient *beliefs* that are formed by associating positive or negative connotation with the most relevant aspects of their particular practice of innovation and its implication for stakeholders, activities and so on. With regard to the study of hydrogen, the model suggests that the innovative activities related to hydrogen executed by carmakers are reflected in its decision-makers' willingness to implement hydrogen engine models, which in turn is determined by three domains in their belief-system: (i) their *attitude* towards hydrogen engines; (ii) their *perceived social pressures* to implement hydrogen models; and (iii) their *perceived control* over the implementation process.

The measurement of belief systems is difficult (Malim and Birch, 1998), since they are not directly observable. On the one hand, there is little doubt that holding a certain belief is linked to particular kinds of practices or behaviour. On the other hand, this relationship is far from clear: specific behaviour may contradict a general attitude. To counter this drawback, we should track belief systems for *specific* behaviours (a choice, an action), in which actors are in a 'reasonable' state of mind. For our study, we developed a questionnaire that could identify the various beliefs (and their salience) of car firms with respect to hydrogen propulsion. In this line of enquiry we apply this model, assuming nine possible drivers divided over the three domains: attitude; perceived social pressure; and perceived control.

The *attitude* towards innovation is an indicator of the degree to which relevant decision-makers like or dislike (that is, form positive or negative associations with) the expected direct outcomes of their engagement in innovative activities. Typically, these outcomes refer to expected economic benefits and losses for the innovative entrepreneur and, in the case of an engagement with cleaner technology, also the positive effect of the latter on the environment. Therefore, as the main determinants of the entrepreneurs' perceived attitude towards engaging in the new technology or practice, the model distinguishes between: perceived economic risk/opportunity (ER) and perceived environmental risk (EV).

In contrast to attitude, *perceived social pressure* refers to the positive or negative normative connotations associated with an engagement in the technology. In particular, the model distinguishes:

- the regulatory pressure (RP), referring to the perceived stringency of environmental regulations and standards;

- the market pressure (MP), arising from the perceived attitude of consumers and competitors concerning the technology; and
- the perceived community pressure (CP) from stakeholder groups in the community constituting the social environment of the innovating firm.

The third category of determinants refers to the entrepreneurial decision-maker's *perceived control over the innovation process*. This part of the belief system essentially comprises:

- the perceived technical capabilities (TC) allowing a company to use technological opportunities offered by the market;
- the company's perceived capability to engage in organizational learning (OL);
- perceived capabilities to form strategic alliances (AL) with customers or suppliers; and
- perceived capabilities to use collaboration networks (NW) with research institutions to outsource the acquisition of knowledge needed for the innovation process.

These nine items are potentially salient underlying beliefs related to the willingness to develop the innovation. For our study we designed a survey questionnaire to identity the various beliefs of carmakers with respect to hydrogen engines. By correlating the actual engagement with stated beliefs, we gained an insight regarding the salience of various beliefs.[5]

Results: engagement and beliefs regarding hydrogen technology

This section presents the responses of seven global vehicle manufacturers[6] regarding hydrogen technology. Though the absolute number of this set is not high, the seven firms do represent a significant share of the (around 12) major global car manufacturers. The majority of the respondents were at senior engineering level (see Appendix 1). Therefore, we can be confident that respondents were sufficiently knowledgeable about the firm's strategy on hydrogen use. (The questionnaire is shown in Appendix 2.)

We assume that these people have voiced the vision of their organizations. Regarding the seniority of most respondents and the way we have phrased the questions ('for your organization'), this is plausible. Nevertheless, the response may be biased for two reasons. Apart from personal bias (from the individuals' position or personal opinion), there may be strategic public relations considerations involved that aim to show the firms in a more positive light. At this point it is very difficult to assess to what extent this has played a part with the respondents, but we shall need to take this into account when interpreting the response. We shall therefore discuss the results of the questionnaire survey in the light of two additional interviews (each semi-structured, 90 minutes' duration; see Appendix 1). The combination of questionnaires and interviews has given us sufficient confidence to draw some conclusions from this analysis.

Engagement in developing a hydrogen vehicle

The extent to which a firm is 'engaged in the development of hydrogen propulsion' was a key question asked at both the beginning and end of the questionnaire (a scale of 1–5, ranging through 'not at all, very little, little, moderately, extensively'). From the responses at the beginning of the questionnaire we found that the mean value for the seven respondents was 4.3 for present engagement, and 4.3 for future engagement. In other words, the perceived present engagement is slightly above 'moderate', and, further, we find that firms do not intend to increase their development efforts for hydrogen cars.

To test the stability of the respondents, the same question was posed again at the end of the questionnaire. Here, we found that the engagement in hydrogen vehicle development was 2.5 (present) and 3.2 (future) on average, much lower than the 4.3 that was found for both earlier. It is remarkable that the difference between the scores at the beginning and the end is so great. We suspect that responses at the end of the questionnaire are less biased by possible PR considerations that the respondents might have had in mind at the start of the questionnaire, and because respondents have activated their knowledge regarding their firm's considerations during the course of answering all the questions (completing the questionnaire took about 20 minutes). Finally, with regard to the responses during the interviews at two car manufacturers, we were told that hydrogen technology was only a minor area of development for them and for most of the firms in the sector, and we therefore assume that the mean value of 2.5 is most accurate, suggesting that firms are engaged very little in hydrogen vehicle development.

The distribution of values over the respondents gives an indication of how the mean value was delivered; see Figure 7.1. The distribution of engagement across the car manufacturers shows that the largest share of firms in the dataset (that is, four out of the seven) regard their engagement as 'little'. One firm chose 'moderately' and two chos 'not at all' (one did not respond to this question at the end).

Potential drivers of engagement

Following Montalvo's social-psychological model (Montalvo, 2002, 2006), we consider nine potentially salient beliefs underlying engagement levels, grouped into the three categories detailed in the 'Methodology' section above: attitude towards innovation (EV and ER); perceived social pressures to innovate (CP, MP and RP); and perceived control over the innovation process (TC, OL, AL and NW).

These nine items were scored in the questionnaire by each of the seven respondents (using a scale of 1–7).[7] Table 7.1 shows the mean values.

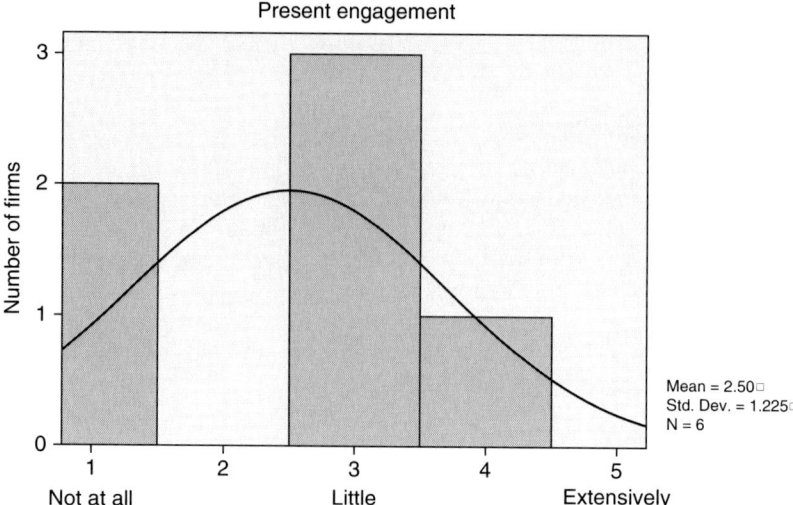

Figure 7.1 Distribution of respondents' present engagement in hydrogen vehicle development

Table 7.1 The mean values of beliefs regarding hydrogen-powered vehicles

Item	Mean
Perceived environmental risk	2.9
Perceived economic risk	2.4
Perceived market pressure	2.4
Perceived community pressure	2.6
Perceived regulatory pressure	2.3
Perceived technological capabilities	3.3
Perceived organizational learning capabilities	3.3
Perceived strategic alliance capabilities	4.4
Perceived networks of collaboration capabilities	4.6

The low values for the salient beliefs observed in Table 7.1 are consistent with the low mean value of the present 'willingness' of the firms (2.5) that we found (see above). Here, we find that the 'attitude' towards the innovation process is slightly negative:

- The economic consequences of developing hydrogen-powered vehicles are perceived to be 'slightly likely' to bring losses (2.4).
- The environmental risks of the current car models (that is, internal combustion engines (ICE)) is perceived to be 'slightly low' (2.9).

The pressure from the social environment to develop hydrogen-powered vehicles is not perceived as high by car firms:

- MP, CP and RP are perceived to be low: between 2.3 and 2.6. These are not likely to encourage the firms towards the development of hydrogen vehicle technology in the short run (one to two years).

The perceived capacity to perform the development of hydrogen-powered vehicles is perceived as weak:

- TC and OL capabilities are 3.3 on average, where 3 is 'slightly low' and 4 is 'uncertain'. AL and NW are about one unit higher (4.4 and 4.6, respectively); that is, between 'uncertain' and 'slightly easy'.

Since the variance in the scores is quite high, a single conclusion regarding the mean values is somewhat risky. Therefore, in addition, we shall examine the answers of the seven companies separately (where actual engagement is scored on a scale of 1–5, whereas beliefs are score on scale of 1–7):

- *Firm 1* is presently engaged very little with hydrogen vehicle development (3) and has few plans to do so within the next 5 to 10 years (3). It is uncertain about the economic consequences of engagement (4). EV is perceived as very low (1). Social pressures are perceived to be very low (1, 2 and 2). Its own TC related to hydrogen is perceived as weak (2), but OL, AL or NW capabilities are uncertain or slightly strong (5). Conclusion: Firm 1 has little ambition regarding hydrogen-powered vehicles and is most optimistic about its NW capabilities (slightly strong).
- *Firm 2* is hardly engaged with hydrogen vehicle development at all (1), and has few plans to do so in the future (3). It would mean big losses for the firm (7). The EV of the current models is uncertain (4). Social pressures are low (2, 1 and 1). Capabilities are very weak (TC: 3; others, 1, 2 and 2). Conclusion: Firm 2 is hardly engaged in hydrogen vehicle development for two possibly related reasons: its current capabilities around hydrogen-powered vehicles are perceived to be weak and engagement would currently mean great losses.
- *Firm 3* is also hardly engaged with hydrogen vehicle development (1), and has no plans to do so in the future (1). It perceives the economic consequences as being very negative (6, moderate losses). The EV of the current models is quite small. Social pressures are small (2, 1, 1). The TC and OL around hydrogen technology are perceived to be very low (both 1). However, the capabilities to form AL and NW are seen as slightly high and fairly high, respectively (6 and 5). Conclusion: similar to Firm 2, Firm 3 is very unwilling to engage in hydrogen vehicle development for two probably related reasons: its current capabilities around hydrogen

technology are very slight, and engagement will currently bring them moderate losses. Firm 3 seems to trust external knowledge sources at the present time.
- *Firm 4* is moderately engaged (4) and has moderate plans (4). However, it still expects moderate losses from its current engagement. The environmental effects are uncertain. Social pressures are uncertain (3, 4, 4). TC and OL are seen as being slightly strong (5), also AL and NW (5 and 6). Conclusion: Firm 4 is reasonable engaged and believes itself to be (slightly) capable of engaging in the development of hydrogen-powered vehicles.
- *Firm 5* is currently engaged very little with hydrogen (3), but has moderate plans for the future (4). Economic opportunities are currently seen as very negative (7). Social pressures are perceived to be very low (1, 1, 1). Technological capabilities are low (2), but the capabilities to build alliances and benefit from networks are slightly and fairly good, respectively (5 and 6). Conclusion: Firm 5 is currently engaged very little with hydrogen technology, but is confident about its network.
- *Firm 6* is at present engaged very little with hydrogen technology (3) but has some plans for the future (4). It is uncertain about all pressures (4), except for environmental risk, which is seen as slightly low (3). Conclusion: Firm 6 is uncertain about hydrogen technology (or at least is not outspoken about it).
- *Firm 7* did not respond to the question on engagement at the end of the questionnaire, but the response to that question at the beginning suggests they are moderately engaged (4). It sees minimal economic losses (5). The environmental risk of the current models is perceived to be quite low (2). Social pressures are uncertain (all 4). TC and OL are perceived to be uncertain; only forming alliances are perceived to be slightly easy. Conclusion: Firm 7 is optimistic, most probably because of its confidence in alliances, and its expectation of some economic returns.

Three conclusions can be drawn from these discussions:

- Conclusion 1: we find three subgroups in the population of firms:
  o Subgroup 1: these firms are somewhat engaged with hydrogen technology (Firms 1, 5 and 6).
  o Subgroup 2: these firms are hardly engaged (firms 2 and 3), probably for two related reasons: their current capabilities around hydrogen-powered vehicles are perceived to be weak, and engagement will currently mean great or moderate losses.
  o Subgroup 3: these firms are moderately engaged (firms 4 and 7). In one case, this is because of its slightly strong capabilities; while the other seems to have great confidence in alliances. Both acknowledge current losses, but these are apparently seen as necessary investments.

- Conclusion 2: the (perceived) level of technological capabilities TC and organizational learning capabilities (OL) seems to have a high correlation with the engagement level. The two non-engaging companies have weak capabilities, and one of the two optimistic companies has high capabilities (for the other firm, the response value is missing). From the three companies with little engagement, two are uncertain about their capabilities.
- Conclusion 3: five out of the seven firms see their capabilities for AL and NW (that is, attaining the required knowledge from outside) as being higher than their own current TC and OL. It seems as though a majority of the firms choose to rely on external technological knowledge sources for their potential hydrogen vehicle development.

Discussion and conclusion

What does this analysis of belief systems clarify about technological framing by car firms? As described above in the section entitled 'Methodology', we supposed that knowledge and beliefs *become manifest* in the frame: the specific hydrogen-related belief-structure of firms shapes their framing of hydrogen technology. This delivers a specific appraisal of the technology, which can be summarized as willingness (that is, as behavioural intention). Subsequently, willingness will shape (though not determine) the firm's behaviour with respect to the technology (see Figure 7.2).

In this chapter we have correlated *specific* beliefs with *specific* engagement, namely regarding the development of hydrogen technology by carmakers. We have found differences in engagement levels, and identified three subgroups. We found that these differences correlate most strongly to different levels of (perceived) technological and organizational capabilities. However, though firms diverge in the appraisal of hydrogen technology, neither from

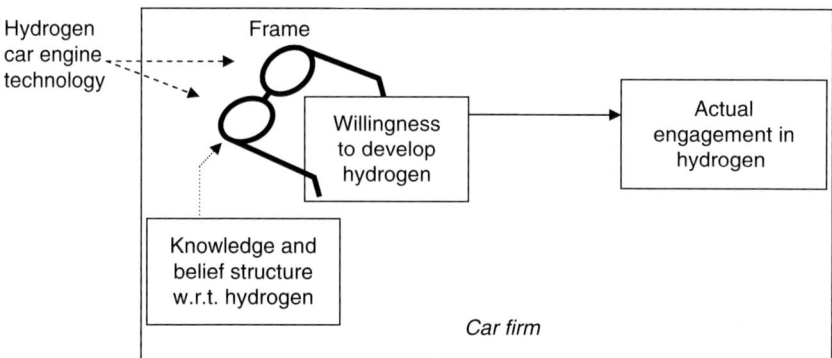

Figure 7.2 Relationship between frame, willingness and belief-structure regarding a specific technology

the questionnaires, nor from additional interviews (see Appendix 1), did we find evidence that frame-*structures* of car firms towards hydrogen technology differ significantly; that is, regarding the weights they attach to economic opportunity, environmental risk and so on. All firms are part of the same global market, with fierce competition. All firms emphasize *business opportunities*, both immediate and in the future, when considering the development of hydrogen technology. A major question for business is: do consumers expect and value cleaner engines (personal communication at Audi and Peugeot)? A second point is that their relative positions regarding competitors is important. Further, social expectations and pressures to engage with novel technologies, such as hydrogen, affect the R&D expenditures of firms to some extent, whereas emissions regulation is a boundary condition that needs to be met.

The more detailed level of communication in the interviews suggested that there are two technological frames related to a technology, or two sub-frames: an *R&D technology frame* and a *product launch technology frame*. The R&D frame evaluates the development of technology for future product launches. The execution of R&D delivers engine prototypes, most importantly as concept cars, which may be shown at international motor shows without any direct intention to put them in production. This frame appraises the extent to which enhancing technological competences is important to the firm.

Second, the product launch technology frame relates to the technology on near-future products. This frame appraises the extent to which innovative technology will be incorporated in new products. Each new car or engine is seen as a project with revenues and costs, the revenues are from expected sales, and costs from product engineering and manufacture. Each project (product) needs to achieve a certain *return on investment* in order to be approved. As long as production investments in a new product technology have no profitable outlook, a firm will usually wait until launching the product.

None of the car firms surveyed considered launching hydrogen-powered vehicles before 2010. All firms perform R&D on hydrogen technology at least to some extent, though the amount varies significantly. Firms that consider themselves to be more strongly engaged in hydrogen vehicle development are those that also perceive their technological and organizational competences to be higher.

Appendix 1: Stakeholder consultation

This appendix provides the details on the consultation of the stakeholders. It describes which stakeholders were involved in the interviews and questionnaires, what their roles in the organization were, and how they were contacted.

Questionnaire survey respondents

We contacted respondents at the Electric and Hydrogen Vehicle Symposium (EVS 21) in Monaco in the spring of 2005.[8] At this type of symposium, company engineers present

Table A7.1 Respondents to the questionnaire surveys

Organization	Respondent (anonymized)	Position
Audi	R. v. D.	Chief engineer
Daimler	Anonymous	Unknown
Honda	Thomas B.	Unknown
Nissan	Yukimasa B.	Engineer
Peugeot	Mr B.	Engineer (Alternative energies)
Renault	Fabien B.	Team Leader
Saab	Tommy L.	Project leader (Engineering)

the latest developments of their firm with regard to electric and hydrogen technology. Some people we made contact with completed the questionnaires on the spot, and some asked for it to be completed by a colleague and returned later. A number of requests were later rejected by email, referring to the confidential nature of the information. Seven questionnaires were returned completed (see Table A7.1). Not all of the respondents have indicated their positions in their organization, but the majority were, as observed in Table A7.1, at senior engineering level. Therefore, we can be confident that respondents were sufficiently knowledgeable about the firm's strategies regarding hydrogen technology. (The questionnaire is printed below in Appendix 2.)

We assume that the people surveyed to have voiced the vision of their organization (see related comments in the 'Results' sub-section above). The way we validated the questionnaire survey was by interpreting the data in the light of (a few) stakeholder interviews that had been performed earlier (see next section). Therefore we can be fairly confident about the responses to the survey questions.

Stakeholder interviews

We contacted a small number of carmakers in Germany and France, requesting an interview regarding hybrid and hydrogen car technology. Typically, we were put through to senior engineers, and it was often hard to get a reply from a specific person and, when someone was prepared to offer an interview, to find time in their busy diaries. Some people who were contacted did not want to participate in an interview (for example, Volkswagen), generally because of the strategic sensitivity of the information. Nevertheless, the three interviewees we did meet (see Table A7.2) were fairly open about the conflict of interest between their business and environmental concerns. Also, their level of seniority suggests that we can be confident that they were sufficiently knowledgeable about the firm's strategy on hydrogen. We applied a semi-structured interview format; see the next sub-section. This interview format gave us the opportunity to ask for elaboration and examples, and to give the interviewees other information (for example, data on market trends or patent information). Therefore, we found this interview method to be slightly less vulnerable to the personal bias of the interviewee and the strategic or public relations considerations of the organization in comparison to the questionnaire survey, and the responses were very useful for interpreting the survey outcomes.

Format for interview with car (or truck) manufacturer
Attitude
- What do you see as the advantages/gains/benefits from your (company's) engagement in the adoption (or development) of hydrogen-powered vehicles?

Table A7.2 Respondents to our interview surveys

Organization	Interviewee	Position	Date
Peugeot–Citroën	Dénis Depuis	Senior research engineer	Summer 2004
Audi	Ingrid Paulus	Head of transport and environment	Summer 2004
DAF Trucks	Peter van der Heijden	Manager performance	Summer 2004

- What do you see as the disadvantages/drawbacks of your (company's) engagement in the adoption (or development) of hydrogen-powered vehicles?
  - If necessary: Is there anything else, either positive or negative, that you associate with the adoption (or development) of hydrogen-powered vehicles for your company?
- What might be the effect on the environment related to the adoption?

Social pressures

- Are there any people, organizations or institutions whom you think want/are pushing you/your company to engage in the adoption (or development) of hydrogen-powered vehicles?
- Are there any people or institutions that you think oppose your company adopting (or developing) hydrogen-powered vehicles?
- Does anybody else come to mind when you think about your company adopting (or developing) hydrogen-powered vehicles?
- Do you have any examples of a policy instrument or measures put in place to encourage the adoption or development of hydrogen-powered vehicles?
  - If necessary: Do you have any examples of barriers/difficulties presented by people/institutions to the development/uptake of hydrogen-powered vehicles?

Control of innovation

- What kinds of skills or abilities do you think you/your company need in order to adopt (or develop) hydrogen-powered vehicles?
  - If necessary ask further about: What experience, information, additional resources in terms of time/money, do you think your company needs to engage hydrogen-powered vehicles?
- Are there any particular circumstances/opportunities you think your company relies on to develop hydrogen-powered vehicles?
- Are there any constraints you think are preventing your company from developing hydrogen-powered vehicles?

Appendix 2: Questionnaire[9]

General information requested

Organization name
European member state within which the organization is based
Respondent's name
Position in organization
Date of interview/survey

Specific information on hydrogen requested
See Table A7.3.

Table A7.3 Hydrogen-powered car questions

1. To what extent is your firm presently engaged in developing a fuel cell car?	Scale 1–5: Not at all – Extensively
2. To what extent does your firm have existing plans to develop a fuel cell car?	Scale 1–5: Not at all – Extensively
3. The environmental effects generated by the usage of our current car models are likely to be:	Scale 1–7: Great losses – Great benefits
4. In general it can be said that the signals (demand) that we perceive from the marketplace (for example, customers, suppliers and competitors) tell us that we should develop and sell a fuel cell car are:	Scale 1–7*
5. In general, the pressure from the community (local and global NGOs, mass media, unions, etc.) that this firm faces to develop and sell a fuel cell car is:	Scale 1–7*
6. There are several regulatory institutions (e.g. the EU and national authorities) pushing us to develop and sell a fuel cell car:	Scale 1–7*
7. We find that the level of state of the art knowledge available in the marketplace for our firm to engage in the development of a fuel cell car is:	Scale 1–7*
8. Our firm has the necessary organizational capabilities to reshape our organizational structures to develop and implement a fuel cell car:	Scale 1–7*
9. Our firm finds the performance of strategic alliances (with suppliers, customers or competitors) to develop a fuel cell car:	Scale 1–7*
10. Establishing networks of collaboration to acquire know-how to develop a fuel cell car for our firm is:	Scale 1–7*
11. For our firm, the venture of developing a fuel cell car seems to imply economically:	Scale 1–7: Great losses – Great benefits
12. After evaluating the outcomes, pressures and capabilities to develop a fuel cell car, what is the likelihood that your firm will engage in the development of this new product:	Scale 1–7*

Note: *Scale 1–7 is as follows: 1 = very unlikely/not at all; 2 = fairly unlikely; 3 = slightly unlikely; 4 = uncertain; 5 = slightly likely; 6 = fairly likely; 7 = very/extensively likely.

Notes

1. Obviously, emissions during the production of the fuel should be incorporated in the calculation.
2. When the term 'hydrogen engines' is used in this chapter, two types of engines are meant: fuel cell vehicles (FCVs), and IC engines with hydrogen as fuel. For the

survey in the section 'Results: engagement and beliefs about hydrogen technology', however, we have only included hydrogen fuel cells.
3. Up to the year 2000, Daimler invested approximately US$1 billion in FCVs. By 2004, an estimated US$6–10 billions had been spent by the automotive industry alone to research and develop FCVs.
4. We regard knowledge and belief systems as the underlying basis of the actor frame, or, in other words, knowledge and beliefs *become manifest* in the frame.
5. For practical reasons it was not possible to include the various alternative innovation options (such as, for car engines, petrol and diesel technology, hybrid-electric and natural gas). Therefore we are working under the (rather strong) assumption that beliefs and opportunities related to the innovation we investigated can be studied independently of beliefs and opportunities related to alternatives.
6. These are: Daimler, Honda, Nissan, Renault, Peugeot, Saab and Audi.
7. Or, more precisely: 1 = very unlikely/not at all; 2 = fairly unlikely; 3 = slightly unlikely; 4 = uncertain; 5 = slightly likely; 6 = fairly likely; 7 = very/extensively likely.
8. Though the survey was conducted in 2005, the focus in this chapter is on the correlation of engagement and motivations in a disruptive technology (and less on engagement in absolute terms), something that is not likely to be very time-dependent, and we therefore argue that the results are still relevant and valid.
9. The original questionnaire was much longer and was part of a European FP6 project. We have left out many detailed questions that were not included in our analysis.

References

Ajzen, I. (1988) *Attitudes, Personality, and Behavior.* Chicago: Dorsey Press.
Ajzen, I. (1991) 'The Theory of Planned Behavior', *Organizational Behavior and Human Decision Process*, 50(2):179–211.
Ajzen, I. and Krebs, D. (1994) 'Attitude and Measurement: Implications for Survey Research', in I. Borg and P. Mohler (eds), *Trends and Perspectives in Empirical Social Research.* New York: Walter de Gruyter, pp. 250–62.
Avadikyan, A. and Larrue, P. (2003) 'The Partnership for a New Generation of Vehicles and the US DoE Transportation Fuel Cells Programme', in A. Avadikyan (ed.) *The Economic Dynamics of Fuel Cell Technologies.* Heidelberg: Springer-Verlag, pp. 133–58.
Chanaron, J.-J. and Teske, J. (2007) 'Hybrid Vehicles: A Temporary Step', *International Journal of Automotive Technology and Management*, 7(4): 268–88.
Cowan, R. and Hulten, S. (1996) 'Escaping Lock-in: The Case of the Electric Vehicle', *Technological Forecasting and Social Change*, 53(1): 61–80.
Jonas, K. and Doll, J. (1996) 'A Critical Evaluation of the Theory of Reasoned Action and the Theory of Planned Behaviour', *Zeitschrift für Sozialpsychologie*, 27(1): 18–31.
Lave, L. and MacLean, H. (2002) 'An Environmental–Economic Evaluation of Hybrid Electric Vehicles: Toyota's Prius vs. Its Conventional Internal Combustion Engine Corolla', *Transportation Research Part D: Transport and Environment*, 7(2): 155–62.
Malim, T. and Birch, A. (1998) *Introductory Psychology.* London: Macmillan.
Molot, M. (2008) 'The Race to Develop Fuel Cells: Possible Lessons of the Canadian Experience for Developing Countries', in L. K. Mytelka and G. Boyle (eds), *Making*

Choices About Hydrogen: Transport Issues for Developing Countries. Tokyo: UNU Press/ Ottawa: IDRC Press.

Montalvo, C. (2002) *Environmental Policy and Technological Innovation*. Cheltenham, UK: Edward Elgar.

Montalvo, C. (2006) 'What Triggers Change and Innovation?' *Technovation*, 26(3): 312–23.

Van den Hoed, R. (2005) 'Commitment to Fuel Cell Technology? How to Interpret Carmakers' Efforts in this Radical Technology', *Journal of Power Sources*, 141(2): 265–71.

8
CNG Cars: Sustainable Mobility Is within Reach

Andrea Stocchetti and Giuseppe Volpato

This chapter describes the opportunities related to the diffusion of compressed natural gas (CNG) as an alternative fuel for vehicles. Indeed, the economic and environmental benefits of CNG are likely to make this fuel the most effective option for the rapid achievement of sustainability targets in road transport.

Despite carmakers' efforts towards the improvement of the efficiency and cleanliness of their products, cars still remain among the main contributors to air pollution in Europe. Despite the average CO_2 emissions from new cars having dropped by 18.5 per cent since the start of the 2000s (from 172.2 gCO_2/km in 2000 to 140.3 in 2010), in the EU about 20 per cent of total CO_2 emissions comes from road transport (EEA, 2011). The overall pollution generated by cars is increasing because there are greater numbers of cars on the road and many of them are getting old. According to ANFAC (2010), the average age of European cars on the road is just over 8 years, and 34 per cent of cars (that is, about 70.4 million cars) are more than 10 years old. In the USA, the average age of cars at the time of writing is over 10 years (NADA, 2011), and this also applies to the rest of the world, with very few exceptions. The greater proportion of the cars in use in Europe and across the world were built before the introduction of strict anti-pollution standards, and it is likely that half of the cars on the road across the world do not even comply with out-of-date standards such as Euro 3. This means that the actual amount of greenhouse gases (GHG) and other harmful pollutants is affected much more by market dynamics and consumer behaviour than by the technological improvement of traditional petrol and diesel engines.

Therefore, the urgent need to attain a rapid and significant reduction in pollution (and specifically in GHG) is beyond question. The ultimate solution to overcome the problem of pollution will be provided by zero emissions vehicles (ZEVs); however, at the time of writing, such vehicles are facing economic and technical barriers that will delay their diffusion significantly. We claim that CNG technology presents, together with lower emissions, a major benefit in that it can be applied to existing cars with a relatively small and rapidly recoverable investment. In our view, such features make CNG a possible intermediate stage on the road towards sustainable mobility.

A crucial stage in the long race towards zero emissions vehicles

Carmakers' efforts regarding innovations leading to the energy efficiency of vehicles are intense; nevertheless, at the present rate of improvement, the general reduction of pollutants is far too low. EU policies have been decisive in pushing manufacturers to take drastic action against noxious emissions; increasingly stringent standards and regulations have been introduced during the 2000s. In December 2008, the European Parliament launched a measure which mandates carmakers to cut the CO_2 emissions of new models, and imposes monetary penalties on those exceeding allowable limits (T&E, 2008). The improvements derived from the gradual introduction of the Euro emissions standards, together with the definition of a target for the average CO_2 for the vehicles produced by European carmakers, has undoubtedly brought significant technological improvement. Between 1997[1] and 2010, the average emissions of new cars sold in Europe by the ten major producers decreased on average by 24 per cent (see Table 8.1).

But together with regulations it is also important to address vehicle demand towards those solutions that might drastically cut the level of pollutants in the short term. This is of crucial importance because of the high percentage of older vehicles on the road.

On the other hand, the persistence of obsolete vehicles frustrates the technological efforts of carmakers towards the improvement of traditional engines, especially regarding the most harmful emissions (carbon monoxide (CO), volatile hydrocarbons (HC), oxides of nitrogen (NOx) and particulate matter (PM)). Of course, when a new car is sold it is not always the case

Table 8.1 Average CO_2 emissions of cars sold in Europe, by group, 1997–2010 (g/km)

	2010	1997	1997/2010 (%)
BMW	148	216	−31.5
Daimler	161	223	−27.8
Fiat	126	169	−25.4
PSA	131	175	−25.1
Ford	137	180	−23.9
Opel/Vauxhall	139	180	−22.8
Toyota	130	163	−20.2
Renault	139	173	−19.7
Volkswagen	143	170	−15.9
Average	139	183	−24.0

Source: T&E (2011).

that an old (and more polluting) car is scrapped. In those markets where the mobility rate is increasing, because of demographic growth and/or to an increase in disposable income, old cars often remain on the road as a second or third car, or are sold on the used car market. As a consequence, there is no correspondence between the number of new cars sold and the number scrapped. Thus the number of cars on the road tends to grow constantly, and to get older, unless specific incentives for scrapping are introduced.

In this regard, even when fully-functional ZEVs are available at a price similar to existing cars and capable of equal performance, it will take a long time to eliminate the problem of pollution. Indeed, assuming that:

- at that time, traditional cars will no longer be for sale;
- both the world supply of cars and the global demand for new cars will be at the present level (in fact the number of cars is increasing year on year, while demand for new cars has decreased in recent years); and
- for each ZEV purchased an old car is scrapped, even within the framework of these very optimistic and unrealistic hypotheses, it would take no fewer than 16 years to replace the present world stock of vehicles.

Therefore, given the perspective of ZEV development, it is reasonable to assume that the world stock of cars will continue to be a significant source of pollution at least until 2050, unless an intermediate stage on the long road towards ZEVs is found.

The solution lies in relatively inexpensive, rapidly deployable technologies, compatible with existing vehicles (even the oldest ones), which do not require major infrastructures and do not lead to significant disadvantages or to dramatic changes in people's lifestyles.

We think that a possible intermediate stage between present technologies and ZEV could be represented by the diffusion of compressed natural gas (CNG) powered engines. At present, among alternative fuels only CNG seems to have the required features, with the further benefit of also improving the economic sustainability of the industry. Indeed, according to Enerdata (2009), in Europe about 50 per cent of oil use is devoted to road transport; since oil imports to Europe amount to around €318.8 billion a year (European Commission – D.G. Energy), the oil bill generated by road transport is more than the value added of the whole European automotive industry, which is estimated in €140 billion (EU, 2009).

The peculiar features of CNG as a fuel make it one of the most suitable solutions for a rapid improvement in the sustainability of individual mobility, waiting for zero-emissions technologies to become reliable, easy-to-use and affordable enough to constitute a competitive solution from the point of view of customer expectations as well as in the eyes of carmakers.

Natural gas (NG) is in fact a mixture of gases extracted from natural fossil deposits containing a variable percentage of methane (between 80 per cent and 98 per cent) plus other elements. 'Natural gas' and 'methane' are commonly used as synonyms, though they are different gases. Methane is also present as the main component in the so-called 'biogas', a family of gases containing between 50 per cent and 80 per cent of methane, together with water, CO_2, and hydrogen sulphide.

In our view, a massive introduction of this fuel is possible within a short time and will provide immediate advantages, for the following reasons:

- CNG is less costly and less polluting than conventional fuels;
- CNG can be used with common petrol engines with a relatively inexpensive adaptation;
- many OEM volume producers are already offering one or more CNG dual-fuel-powered products; that is, cars with a dual-fuelling system that can switch from petrol to CNG at any moment when being driven;
- in countries that already have a network of methane pipelines, the required infrastructure can be developed rapidly and with minor investment;
- proven reserves of NG in the world can meet the need for fuel for at least the next 50 years, thus for a much longer time than is expected for conventional fuels; and
- in the long-term, CNG could be included among renewable resources, in that it will be obtained from biogas.

Given these features, and the importance of setting up initiatives that are able immediately to produce concrete results, we claim that the diffusion of CNG as a vehicle fuel is probably the best among possible short-term solutions.

Reducing emissions through conversion to CNG: an example from Italy

The ecological benefits of CNG have long been known (Gas Research Institute, 1987; Liew and Liew, 1995; Di Pascoli *et al.*, 2001). Yedla and Shrestha (2003) found that CNG cars are the most effective solutions for reducing pollution in congested cities with reference to various parameters, including emission reduction potential, operating costs, availability of the technology, the adaptability of the option, and barriers to implementation. In such an environment, Goyal (2003) demonstrated that CNG contributes significantly to the reduction of all main pollutants. It also offers comparative benefits compared with traditional technologies when life-cycle costs and external costs are taken into consideration (Lave *et al.*, 2000; Roder *et al.*, 2003; Ogden *et al.*, 2004), though Ogden *et al.* (2004) noticed that the advantages of such a specific aspect might disappear in those countries where a major shift to CNG cars would have an impact on the infrastructure and the import of NG.

An important aspect to take into consideration is the greenhouse effect of methane, which is about 20–23 times greater than CO_2. The management of the whole cycle of NG in relation to the losses occurring during extraction and transport is therefore a serious issue. Such losses are estimated at between 0.2 per cent and 0.7 per cent. In this regard, NG total CO_2 emissions (cycle and fuel) are likely to be less than those of petrol and diesel fuels (Onufrio, 2005; European Commission, 2007: Appendix 1 'Description of individual processes and detailed input data'), but the actual ratio depends on the source of the NG and on the efficiency of the well-to-wheel (WTW) supply path. Gifford and Brown (2011) claim that, in the WTW analysis, CNG performs better than both of the traditional fuels in hybrid engines. Moreover, in the specific case of methane originating from biogas produced from waste and manure, the reduction of GHG would be even greater, since burning methane produces less GHG than leaving the biomass to decompose naturally.

The ecological benefits of CNG derive, of course, from the nature of methane itself. The methane molecule consists of a tetrahedral structure in which a carbon atom binds to four hydrogen atoms; this shows the excellent characteristics of methane, which in the presence of oxygen produces the highest amount of heat per unit mass.[2]

Methane also has a high octane index, which allows for a higher knock resistance and therefore greater efficiency of the engine.[3] These technical features mean that CNG has a greater combustion heat (see Table 8.2) and considerably lower emissions of all the main pollutants, even compared to the most recent cars (see Table 8.3).

The environmental performance of CNG engines are particularly noticeable in comparison with older cars (Euro 0–3), but in comparison with more recent cars the results of various studies do not agree.[4] Specifically, while in the NEDC[5] tests CNG cars perform better than either petrol or diesel (see Table 8.3), some tests show that in a slow-speed urban cycle, Euro 4 petrol engines perform better regarding main pollutants apart from CO_2 (see Table 8.4). In any case, the majority of research agrees that, in common conditions, CNG presents lower emissions in comparison with conventional fuels, even when compared with more recent cars (see, for example, Zhang et al., 2010). On average, the CO_2 reduction is estimated to be –23 per cent in comparison with gasoline, –9.4 per cent compared to diesel, and –12.5 per cent compared to LPG.

Table 8.2 Octane index and heat of combustion comparison of fuels

	Octane (RON)	Heat of combustion (MJ/kg)
Petrol	95–100	44.0
Diesel	–	43.3
CNG	130	47.7

Notes: RON = research octane number; CNG = compressed natural gas.
Source: Mahla et al. (2010); Bakar et al. (2002).

Table 8.3 Average pollutant emissions by fuels, in relation to Euro 5 standards (NEDC cycle)

	NMHC		NOx		PM	
	g/km	Index	g/km	Index	g/km	Index
Petrol	0.068	100.0	0.06	100.0	0.003	100.0
Diesel	0.005	73.5	0.18	300.0	0.005	100.0
LPG	0.055	80.9	0.04	66.7	0.002	66.7
CNG	0.015	22.1	0.03	50.0	0.001	33.3

Notes: NEDC = New European Driving Cycle; NMHC = non-methane hydrocarbons; NOx = oxides of nitrogen; PM = particulate matter; LPG = liquified petroleum gas; CNG = compressed natural gas.
Source: EU (2007); Allegrini et al. (2007).

Table 8.4 Average emissions of different fuels in an urban situation, measured by Euro standards

		NOx (g/km)	VOC (g/km)	PM10 (g/km)	CO_2 (g/kWh)
Euro standard 0	Petrol	1.79	2.04	0.04	243
	Diesel	0.63	0.23	0.26	235
	CNG	0.18	0.01	0.009	170
Euro standard 1	Petrol	0.26	0.26	0.04	220
	Diesel	0.55	0.08	0.073	235
	CNG	0.18	0.01	0.009	170
Euro standard 2	Petrol	0.09	0.05	0.011	220
	Diesel	0.55	0.08	0.073	235
	CNG	0.18	0.01	0.009	170
Euro standard 3	Petrol	0.06	0.036	0.008	220
	Diesel	0.37	0.023	0.014	235
	CNG	0.15	0.009	0.009	170
Euro standard 4	Petrol	0.03	0.007	0.004	220
	Diesel	0.12	0.019	0.008	235
	CNG	0.14	0.008	0.009	170

Notes: NOx = oxides of nitrogen; VOC = volatile organic compounds; PM10 = particulate matter; CO_2 = carbon dioxide.
Source: Allegrini et al. (2007).

To better exemplify the potential CNG advantages we calculated the CO_2 reduction that would take place in a situation where a certain number of older cars were equipped with a CNG propulsion device. According to our estimates, incentives along these lines would be more effective than supporting the purchase of new cars against the scrapping of older ones. Apart from the evident difference in costs between the two options (purchase of a new car versus installing a CNG system), the data we collected show that installing a CNG system in a Euro 0 car produces a reduction in CO_2 that is greater than replacing the same car with a Euro 5 car. Specifically, the substitution of older cars by new CNG-powered cars is on average almost twice as effective in CO_2 reduction as the substitution of Euro 0 with Euro 5 vehicles (see Table 8.5).

Our aim is therefore to estimate the reduction of CO_2 that would occur among cars in Italy if a proportion of the oldest cars were converted to CNG power. Since emissions data from various sources are not identical, we have chosen to use those related to use at low speeds in an urban situation. These values are conservative compared to the values derived from the New European Driving Cycle (NEDC) and similar cycles. According to Allegrini *et al.* (2007), who compared CNG with Euro 0–4 in an urban situation at an average speed of 25 km/h, CNG outdoes both petrol and diesel (that is, has better performance on all main pollution parameters) up to Euro 1 cars. CNG still performs better than diesel up to Euro 3 and above with regard to CO_2. The CO_2 reduction in CNG engines compared to the Euro 4 standard is around 22.7 per cent for petrol and about 27.6 per cent for diesel (see Table 8.4). Most recent petrol vehicles at a low speed, however, present much lower emissions of NOx and PM10 (particulate matter smaller than 10 micrometres in diameter – which are pollutants responsible for serious health hazards because of their ability to reach the lower respiratory tract).

We have chosen to use these conservative data to obtain an unambiguous indication in relation to the short-term effectiveness of a policy of conversion to CNG in an attempt to reduce pollutants and greenhouse gases. In our simulation we take into consideration a possible policy to encourage the conversion of older petrol-driven cars (Euro 0–4) into dual-fuel CNG cars. Starting from these data, we weighted the contribution of each class of vehicle on the basis of an average distribution between classes. Specifically, we focus here on CO_2.

Table 8.6 shows the composition of cars in Italy by class and fuel; and Table 8.7 shows the average CO_2 emissions of Euro 0–4 cars on the road in Italy.

Table 8.5 CO_2 reduction by substitution and by 'methanization' (g/km)

Vehicle class	Euro 0–Euro 5 substitution		Euro 0 'methanization'	
	Petrol	Diesel	Petrol	Diesel
<1400 cc	30	20	61	56
1400–2000 cc	42	29	84	77
>2000 cc	60	39	116	106

Source: Data from Allegrini *et al.* (2007); ACI (2011b); EU (2007).

Table 8.6 Numbers of Euro 0–4 cars on the road in Italy by class, fuel type and Euro standard, 2009

		Class			Total by fuel	
		<1400 cc	1400–2000 cc	>2000 cc		
Euro 0	Petrol	3,087,970	676,635	94,930	3,859,535	4,561,370
	Diesel	96,102	322,788	225,499	644,389	
	CNG	33,369	23,080	997	57,446	
Euro 1	Petrol	1,421,250	516,073	31,557	1,968,880	2,374,573
	Diesel	10,291	253,002	106,820	370,113	
	CNG	15,779	19,390	411	35,580	
Euro 2	Petrol	4,584,016	1,242,956	70,233	5,897,205	7,962,379
	Diesel	9,195	1,529,351	412,397	1,950,943	
	CNG	58,297	55,255	1,039	114,591	
Euro 3	Petrol	2,963,351	558,112	70,659	3,592,122	8,134,181
	Diesel	560,552	3,169,465	739,570	4,469,587	
	CNG	28,349	43,414	709	72,472	
Euro 4	Petrol	4,423,677	853,837	155,916	5,433,430	11,446,192
	Diesel	1,846,534	3,182,220	673,550	5,702,304	
	CNG	222,288	86,088	2,082	310,458	
Total vehicles, Italy						34,479,055

Source: ACI (2011a).

At the end of 2009, more than 4.5 million 'pre-Euro' cars (12.4 per cent of the whole number of registered cars) were in use in Italy. The weighted average (w.a.) of CO_2 emissions from this portion of the cars on the road in Italy is 173 g/km (178 g/km for Euro 1), decreasing to 147 g/km for Euro 4 (down15 per cent), while the CNG w.a. emissions for the Euro 0 cars is 137 g/km (even lower than the overall Euro 4 average emissions) and decreasing to 108 g/km for Euro 4 methane cars on the road (down 21.2 per cent).

We have estimated that the conversion of 1 per cent of the Euro 0–3 cars in Italy could decrease CO_2 emissions by 8.96 tonnes per km on average. If all the oldest diesel and petrol cars (Euro 0) in Italy could be converted to CNG the reduction would be 149.8 tonnes per km. The relevance of such a reduction is huge; assuming that each car travels on an average 12,250 km per year, as reported by ACI (2011a), the overall expected CO_2 reduction would be 1,835,050 tonnes, 14.2 per cent of the total CO_2 reduction (13,200,000 tonnes) that the EU Commission expects Italy to meet in order to bring Europe in line with the Kyoto standards by 2012.

The advantages could be even greater if policies were fine-tuned to intervene specifically regarding cars of a higher class (>2000 cc), for which the environmental benefits of CNG are more evident.

Table 8.7 Average CO_2 emissions of Italian Euro 0–4 cars, by standard, fuel type and class

		Class		
		<1400 cc	1400–2000 cc	>2000 cc
Euro 0	Petrol	129.6	178.3	245.1
	Diesel	125.4	172.4	237.0
	Dual fuel	99.8	137.3	188.7
Euro 1	Petrol	149.8	205.9	283.1
	Diesel	160	220.0	302.5
	Dual fuel	115.7	159.1	218.7
Euro 2	Petrol	139.9	192.3	264.5
	Diesel	149.4	205.5	282.5
	Dual fuel	108.0	148.6	204.3
Euro 3	Petrol	117.3	161.2	221.7
	Diesel	125.3	172.3	236.9
	Dual fuel	90.6	124.5	171.3
Euro 4	Petrol	118.8	163.3	224.6
	Diesel	126.9	174.5	239.9
	Dual fuel	91.7	126.1	173.5

Source: Data from Allegrini et al. (2007); ACI (2011b); EU (2007).

In this context, it is noteworthy that in Italy there are already about 761,000 dual-fuel (gasoline + CNG) cars in use. In our simulation we assumed a 6 times increase in the existing number of CNG cars; for the feasibility of such a policy it is crucial to evaluate constraints and opportunities related to availability, costs, transport and distribution of this fuel on a large scale.

The economic perspective: CNG availability and cost

Apart from the environmental benefit, CNG would also secure a significant economic benefit, since this resource is: (i) widely available; (ii) replaceable by biogas, which is a renewable source of energy; (iii) less costly; and (iv) easier to transport and distribute than conventional fuels.

Natural gas reserves are much more spatially extended than oil reserves. In 2009, known reserves of NG were estimated at 178 trillion cubic metres. Worldwide consumption at that time was 3.02 trillion cubic metres, which is expected to rise to 4.32 trillion cubic metres by 2030 (EIA, 2011).[6] Known global reserves have been increasing since the 1980s and at the present rate of consumption would last for over 40 years.

The geographical distribution of NG is less concentrated than that of oil (see Tables 8.8 and 8.9), and the refining process of gas is less complex, less costly and more energy efficient.

Table 8.8 Geographical distribution worldwide, known reserves of NG, 2008

Area	natural gas reserves (%)
North America	4.94
Asia and Oceania	6.88
Central and South America	4.26
Europe	2.70
Eurasia	31.88
Middle East	41.44
Africa	7.90

Source: Data from EIA (2011).

Table 8.9 Geographical distribution worldwide, known reserves of NG: top 20 countries, 2009

Country	Share (%)
Russia	26.7
Iran	15.8
Qatar	14.2
USA	4.2
Saudi Arabia	4.1
UAE	3.4
Nigeria	2.9
Venezuela	2.7
Algeria	2.5
Iraq	1.7
Indonesia	1.7
Turkmenistan	1.5
Kazakhstan	1.4
Malaysia	1.3
Norway	1.3
China	1.3
Uzbekistan	1.0
Kuwait	1.0
Egypt	0.9
Canada	0.9
Total top 20 countries	**88.0**

Source: EIA (2011).

Recent technological advances in this area allow NG to be transported safely and efficiently worldwide. CNG can then be distributed by pipelines or by gas carrier ships, which are already widely used in Europe and the USA and are being developed in other countries. Italy, though already crossed by pipelines aims in this way to diversify supplies. The supply through gas ships needs a transformer station to turn the CNG into LNG (liquefied

Table 8.10 Natural gas production, by geographical area, 2009

Area	Share (%)
North America	26.6
Eurasia	24.2
United States	19.7
Russia	19.4
Asia and Oceania	14.4
Middle East	13.5
Europe	9.9
Africa	6.7
Canada	5.3
Central and South America	4.6

Source: EIA (2011).

natural gas) at the port in the producing country. The turning of the gas into liquid form by cooling to–minus 162°C reduces the volume by 600 times and makes it feasible to transport NG by ship. At the destination harbour the LNG is heated, which returns it to its gaseous state, using a procedure called 'regasification'. This gives an opportunity for additional energy recovery during the gas heating and expansion, thus improving the energy balance of the overall process.

It is noteworthy that new technologies have a liquefaction capacity significantly greater than that of the previous generation of systems; moreover, new LNG tankers hold almost twice the tonnage with respect to carriers available only few years ago. This allows a reduction in transportation costs to such an extent that gas tankers have now become competitive with the construction of new pipelines, particularly for distances greater than 4,000 km.

According to various sources (Exxon, 2009; BP global website), in 2006 the global trade flows of LNG was around 150 billion cubic metres per annum, equivalent to 5 per cent of the total NG trade. The share of this transport is expected to rise to 15 per cent of global NG trade, equivalent to 720 billion cubic metres (BP, 2011), while cost per unit of the overall supply chain will continue to decrease. Supplies will come mainly from the Middle East, Africa and Australia (see Table 8.10), and will be consumed primarily in North America, Europe and the Asia Pacific region. These structural advantages are expected to have an effect on the pump price. In fact, even at the time of writing there is a significant price disparity between methane and both petrol and diesel. Table 8.11 compares fuel prices as at February 2011 in Europe and some major countries.

The data show that the price of CNG is significantly lower than petrol and diesel. The average petrol price in 2010 was €1.54 per litre, while diesel price was €1.42 per lt. The average price of CNG per kg was €1.07, but again the price of CNG at an equivalent energy level is definitely lower: €0.68 against

Table 8.11 CNG pump prices compared to petrol and diesel with equal energy content, various countries, 2011

Country	CNG price equivalent per litre gasoline	Petrol (€/litre)	CNG price equivalent as % of petrol price	CNG price equivalent per litre diesel	Diesel (€/litre)	CNG price equivalent as % of diesel price
Austria	0.55	1.37	39.8	0.63	1.33	47.0
Belgium	0.68	1.64	41.2	0.78	1.47	52.7
Finland	0.84	1.59	52.5	0.96	1.39	69.0
France	0.57	1.52	37.4	0.65	1.34	48.7
Germany	0.64	1.55	41.1	0.73	1.4	52.2
Greece	0.6	1.71	35.0	0.69	1.47	46.9
Iceland*	0.69	1.46	47.2	0.79	1.49	53.0
Italy	0.6	1.6	37.5	0.66	1.47	44.6
Luxembourg	0.39	1.33	29.6	0.45	1.19	37.9
Netherlands	0.47	1.67	28.4	0.54	1.36	40.0
Norway	0.85	1.83	46.5	0.97	1.77	55.1
Portugal	0.63	1.58	39.7	0.72	1.38	52.1
Spain	0.72	1.35	53.3	0.94	1.28	73.7
Sweden	1.15	1.57	73.3	1.49	1.5	99.0
Switzerland*	0.8	1.26	63.6	0.94	1.33	71.0
United Kingdom	0.73	1.55	46.9	0.83	1.6	52.0
Avg. Western Europe	**0.68**	**1.54**	**44.6**	**0.8**	**1.42**	**55.9**
Bosnia and Herzegovina*	0.31	1.13	27.7	0.36	1.12	32.1
Croatia*	0.53	1.22	43.3	0.61	1.13	53.6
Czech Republic	0.66	1.46	45.6	0.76	1.44	52.9
Estonia*	0.45	1.26	35.7	0.51	1.25	40.9
Hungary	0.56	1.45	38.3	0.64	1.41	45.2
Latvia	0.45	1.32	34.0	0.51	1.3	39.6
Lithuania	0.82	1.36	60.2	0.93	1.26	74.4
Moldova*	0.22	0.93	24.1	0.26	0.84	30.5

(continued)

Table 8.11 Continued

Country	CNG price equivalent per litre gasoline	Petrol (€/litre)	CNG price equivalent as % of petrol price	CNG price equivalent per litre diesel	Diesel (€/litre)	CNG price equivalent as % of diesel price
Poland	0.55	1.32	41.3	0.63	1.27	49.4
Russia*	0.22	0.72	31.1	0.26	0.56	45.8
Serbia *	0.52	1.14	45.6	0.6	1.18	50.4
Slovakia	0.7	1.46	47.7	0.8	1.34	59.4
Slovenia	0.73	1.29	56.4	0.83	1.24	67.1
Ukraine*	0.38	0.81	46.5	0.43	0.76	56.7
Avg. Eastern Europe	**0.51**	**1.2**	**41.2**	**0.58**	**1.15**	**49.9**
Argentina**	0.19	0.46	40.9	0.22	0.45	47.9
Bangladesh**	0.16	0.49	32.9	0.18	0.34	54.3
Belarus	0.24	0.55	44.0	0.28	0.55	50.4
Chile	0.2	0.56	35.2	0.23	0.34	66.4
China**	0.3	0.56	54.4	0.35	0.5	69.8
Egypt	0.05	0.12	44.8	0.06	0.15	41.0
India**	0.24	0.76	31.8	0.28	0.53	52.3
Indonesia**	0.21	0.44	46.8	0.24	0.61	38.7
Iran	0.19	0.29	67.2	0.22	0.11	205.2
Japan**	0.52	1.31	39.9	0.7	1.19	58.6
Kyrgyzstan**	0.23	0.6	38.8	0.27	0.7	38.1
Liechtenstein**	0.77	0.92	83.8	0.88	1.09	81.1
Malaysia**	0.16	0.46	35.2	0.18	0.46	40.3
Mexico**	0.18	0.43	41.7	0.21	0.34	60.4
Singapore **	0.46	0.92	50.5	0.53	0.86	61.8
Tajikistan**	0.19	0.55	34.2	0.22	0.45	47.9
Thailand	0.19	0.87	21.6	0.22	0.74	29.1
Uzbekistan**	0.1	0.34	29.0	0.11	0.34	33.2

Notes: *data as at 2010; **data as at 2008.
Source: NGVA Europe (2011).

petrol and €0.80 against diesel. Thus the average cost saving per km of CNG is minus 53 per cent compared to petrol and minus 44 per cent compared to diesel (see Table 8.12).

For example, in the case of an annual mileage of 20,000 km with a 'D' segment car consuming 7 litres per 100 km, the average yearly saving in Western Europe is about €1140 compared to a petrol engine and about €875 compared to diesel. In Europe, the average payback period of installing the CNG system in the petrol engine (estimated at between €2,500 to €3,000) is less than three years (between 2.2 and 2.6 years), while in some countries this drops to just over two years. Considering smaller mileages, the average payback period of a CNG system still remains below the average age of European cars on the road (8.2 years) (see Figure 8.1).

At present prices, from the car owner's point of view, the purchase of or adaptation to a CNG-powered car is definitely convenient in both Western and Eastern Europe, though both excise duties and the oligopolistic market structure of gas distribution could have a significant impact on final prices in the future. In this regard the study 'Well-To-Wheels Report' developed on behalf of the European Commission claims that: 'historically the price

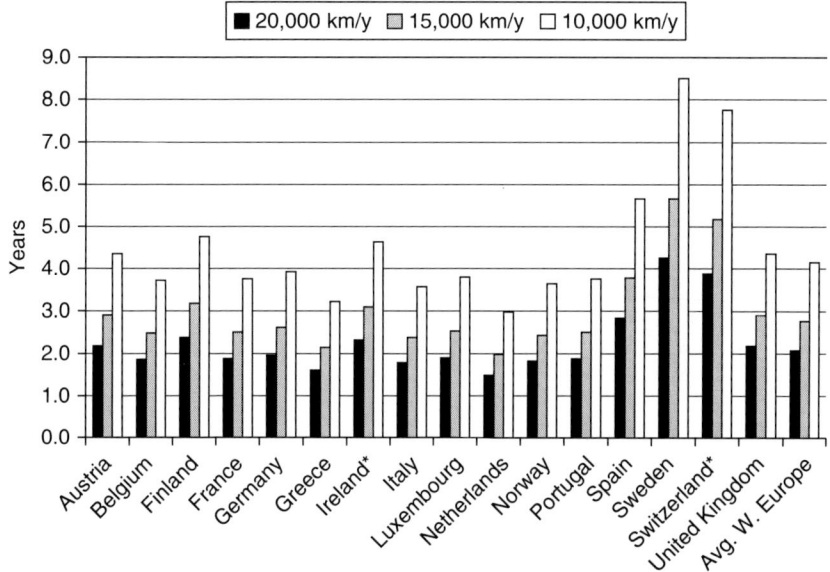

Figure 8.1 Average payback period[1] of a CNG system in a petrol-driven car at various mileages per year

Note: 1. Estimated cost of CNG system €2,500; average consumption 7 lt/100 km. *Data as at 2010.

Source: Data from NGVA (2011).

of natural gas has been loosely linked to that of crude oil, trading in Europe at around 60 to 80% of North Sea crude oil on an energy content basis' (European Commission, 2007).

At the time of writing there is no NG market, in a technical sense – that is, a (virtual) place where offer and demand meet and establish the price for all the market operators. Contracts for NG supply derive from bilateral agreements where the supplier can fix the price according to the functional value of NG (for example, calorific value, energy content and so on) within an oligopolistic framework. Such oligopoly is reinforced by the presence of company-owned pipeline networks. In the future, the development of biogas and the diffusion of regasifiers would break the oligopolistic structure of the production market, thus creating perhaps the setting for the creation of an actual NG market and for a generalized decrease in NG prices.

The diffusion of CNG vehicles

The proportion of CNG vehicles (CNGVs) in different countries is very varied. There are countries where CNGVs are in the majority and others in which this fuel system is virtually absent. According to statistics reported by NGVA Europe (2011), at the end of 2010 the circulating number of CNGVs comprised almost 13.5 million units, about 1.4 per cent of the total worldwide number of cars on the roads (see Table 8.12). The majority of these vehicles were cars or light commercial vehicles, but there are also a considerable number of medium-sized (medium duty), and heavy (high duty) commercial vehicles, including buses.

Of course, a crucial role in the development of CNG-powered vehicles is the availability of an adequate network of CNG filling stations. It is no coincidence that the top four countries across the world are also equipped with an extensive network of distribution points.

In the EU-27 the diffusion of CNG passenger cars is generally low, but for heavy duty vehicles (HDV) and buses it is higher than in other areas. Table 8.13 shows that Italy, with 761,340 CNG-powered vehicles on the road, was the top European adopter in 2011.

Among the countries with high vehicle mobility, only Germany is about to exceed 100,000 CNG-powered units. However, a growing interest by consumers and public administrations in several European countries is emerging regarding CNG vehicles. This trend has been further enhanced by the international economic crisis that in recent years has pushed policymakers from almost all the European countries to promote policies to support demand through incentives for scrapping cars. Such incentives have promoted not only the sale of smaller cars but also cars powered by LPG and CNG. This happened mainly in Germany, France and Italy, in favour of dual fuel cars (petrol and LPG, or petrol and CNG).

Table 8.12 Natural gas vehicles and number of CNG refuelling stations, selected countries

Country	Total no. vehicles on road	Total NGVs	Percentage	CNG refuelling stations	Reference year
Pakistan	3,276,000	2,850,500	87.01	3,300	2010
Bangladesh	293,472	200,000	68.15	600	2010
Armenia	329,596	101,352	30.75	303	2008
Iran	12,182,219	2,605,364	21.39	1,690	2011
Bolivia	821,400	140,400	17.09	156	2010
Argentina	12,568,528	1,460,769	11.62	1,894	2011
Colombia	2,917,530	334,876	11.48	651	2011
Peru	1,505,296	114,186	7.59	158	2011
India	14,554,000	1,100,000	7.56	683	2010
Uzbekistan	1,700,000	120,000	7.06	71	2010
Egypt	2,373,723	139,804	5.89	129	2010
Brazil	35,120,203	1,684,474	4.80	1,719	2011
Ukraine	7,558,100	200,019	2.65	294	2009
Thailand	11,827,911	267,735	2.26	444	2011
Bulgaria	2,733,000	61,623	2.25	95	2010
Italy	40,894,491	761,340	1.86	858	2011
Venezuela	5,000,000	90,000	1.80	220	2011
China	42,564,512	600,000	1.41	2,100	2011
Moldova	528,240	5,000	0.95	14	2007
Sweden	4,802,668	36,380	0.76	166	2011
Singapore	782,359	5,428	0.69	4	2011
Malaysia	7,419,643	48,992	0.66	167	2011
Georgia	533,716	3,000	0.56	42	2008
Chile	2,339,983	8,173	0.35	15	2010
Russia	35,455,227	86,000	0.24	244	2011
Germany	44,240,519	94,890	0.21	900	2011
Switzerland	4,444,529	9,494	0.21	129	2010
Belarus	2,721,018	5,500	0.20	23	2008
Korea	16,438,968	30,443	0.19	178	2011

(continued)

Table 8.12 Continued

Country	Total no. vehicles on road	Total NGVs	Percentage	CNG refuelling stations	Reference year
Dominican Republic	1,208,352	1,614	0.13	3	2011
Austria	4,673,347	5,910	0.13	210	2011
UAE	1,557,488	1,296	0.08	18	2011
Canada	20,491,639	14,205	0.07	81	2010
Luxembourg	356,982	234	0.07	8	2011
Czech Republic	5,134,682	3,075	0.06	49	2011
Japan	75,448,380	40,429	0.05	333	2011
Netherlands	8,882,000	4,300	0.05	150	2011
Latvia	1,061,540	500	0.05	1	2008
Slovakia	1,798,098	823	0.05	11	2010
USA	248,164,738	112,000	0.05	1,100	2011
France	37,212,000	13,500	0.04	300	2011
Turkey	10,190,335	3,339	0.03	14	2010
Finland	3,331,670	970	0.03	18	2011
Serbia	1,791,253	519	0.03	8	2010
Australia	14,725,803	2,825	0.02	47	2009
Norway	2,855,258	545	0.02	10	2011
Mexico	27,442,436	4,830	0.02	3	2011
Indonesia	18,465,568	2,550	0.01	6	2008
Spain	27,613,145	3,051	0.01	48	2011
Poland	19,093,899	2,082	0.01	47	2011
Portugal	5,757,400	504	0.01	5	2009
Greece	6,152,872	520	0.01	3	2010
New Zealand	3,099,708	201	0.01	14	2010
Algeria	3,175,445	125	0.00	3	2005
Belgium	6,270,150	241	0.00	10	2011
United Kingdom	34,457,011	220	0.00	5	2010
World total	**949,475,440**	**13,449,411**	–	**19,679**	

Source: NGVA Europe (2011).

Table 8.13 Natural gas vehicles in the European Union-27

Country	All road vehicles	Total NGVs	NG buses	NG HDV	NG buses and HDV share of NG vehicles (%)	Reference year
Austria	4,673,347	5,910	133	6	2.4	2011
Belgium	6,270,150	241	0	6	2.5	2011
Bulgaria	2,733,000	61,623	105	11	0.2	2010
Czech Republic	5,134,682	3,075	326	41	11.9	2011
Denmark	2,646,306	0	0	0	0.0	2010
Estonia	639,472	69	6	3	13.0	2011
Finland	3,331,670	970	85	15	10.3	2011
France	37,212,000	13,500	2,200	1,100	24.4	2011
Germany	44,240,519	94,890	1,590	1,200	2.9	2011
Greece	6,152,872	520	412	108	100.0	2010
Hungary	3,441,323	87	83	0	95.4	2010
Ireland	2,278,199	3	0	0	0.0	2010
Italy	40,894,491	761,340	2,300	1,200	0.5	2011
Latvia	1,061,540	500	10	187	39.4	2008
Lithuania	1,889,209	185	110	0	59.5	2011
Luxembourg	356,982	234	35	0	15.0	2011
Netherlands	8,882,000	4,300	590	180	17.9	2011
Poland	19,093,899	2,082	276	4	13.4	2011
Portugal	5,757,400	504	354	54	81.0	2009
Slovakia	1,798,098	823	334	60	47.9	2010
Slovenia	1,167,223	8	0	0	0.00	2011
Spain	27,613,145	3,051	1,405	1,028	79.7	2011
Sweden	4,802,668	36,380	1,725	1,080	7.7	2011
United Kingdom	34,457,011	220	0	150	68.2	2010
Total Europe 27	266,527,206	990,515	12,079	6,433	1.9	

Note: HDV = heavy duty vehicle.
Source: NGVA Europe (2011).

Future technological opportunities for the development of CNG vehicles

The use of CNG systems in vehicles should not be considered the ultimate solution to the problem of pollution generated by road transport; nevertheless, it will for many years be the most sustainable solution. Moreover, there is an interest in developing the demand for CNG-powered vehicles for the corresponding industrial sector. The technology associated with this kind of engine is able to develop further, producing interesting innovations from economic, productive and ecological points of view.

The first aspect concerns the spread of CNG over the next generation of petrol engines, and to diesel engines. The search for more economical and less polluting engines is prompting the automotive industry towards turbocharged engines of small displacement that do not sacrifice performances and are 'fun to drive'. Thanks to its high knocking resistance, natural gas is the ideal fuel to use in the new, downsized and turbocharged engine platforms, where boosting is mandatory to realize low-end torque associated with better fuel economy. CNG kits compatible with new-generation engines are already available and, if the distribution of natural gas grows enough to make feasible cars powered exclusively by CNG, it will be possible to exploit the high octane of natural gas to increase the compression of engines and achieve higher specific power.

Another promising technological trajectory is that of the CNG REEV (Range Extended Electric Vehicle); that is, cars with electric engines in which a natural gas engine is coupled to a generator that recharges the batteries. Such a vehicle would be a purely electric vehicle from the point of view of driving performance, but with a longer range thanks to the recharge system. Moreover, the internal CNG engine can also be used for heating (or cooling) of the passenger compartment, a comfort option that would rapidly drain the batteries of any present-day electric car. Of course, emissions would not be zero but they could be significantly reduced with respect to a hybrid car, in that the combustion engine would not be used for traction and could therefore be set up to work constantly at an optimal regime of rotations per minute (rpm).

In connection with this advance, there is also the possibility of mixing hydrogen with CNG. If hydrogen becomes available at competitive prices it would immediately be possible to use a mix of CNG (70 per cent) and hydrogen (30 per cent) using current engine technology. The advantage would be a further reduction of CO_2 emissions, but probably the most interesting aspect will come from the fact that this type of change would trigger the conditions for a broader use of hydrogen. The diffusion of hydrogen for transport suffers the same problem as CNG: the inadequacy of the distribution network discourages the use of hydrogen-powered vehicles, and this in turn prevents the expansion of the network. The use of a CNG–hydrogen mixture would be the first step in creating the conditions for a wider application of hydrogen and would facilitate its diffusion.

As for the future of CNG as a fuel, another important aspect is related to the possibility of producing CNG from biogas. Biogas contains almost pure methane (biomethane) and is produced from biomass comprising lignocellulosic (straw and wood) matter, other crops or organic waste, using a variety of technologies. Biomethane is chemically almost identical to natural gas and is fully interchangeable with it. Biomethane produced from waste offers a more favourable GHG balance than any other fuel (including hydrogen produced by renewable power). If the biomethane is produced from manure and organic waste it is CO_2 neutral and, moreover, reduces the overall GHG impact, since it avoids the leakages of methane, ammonia and nitrous oxide (N_2O) deriving from the natural decomposition of these biomasses (Ahlvik and Brandberg, 2011). Furthermore, the biogas production from crops and biomass is the most land-efficient process compared to other biofuels; specifically, 3.21 times more efficient than the production of biodiesel from rapeseed, 3 times more efficient than the production of ethanol from cereals, and 1.36 more efficient than ethanol from sugar cane. The biogas 'oil-equivalent' production potentiality of 1 hectare of land is 4,500 litres, corresponding to about 3,537 kg of methane and to a mileage of between 63,000 and 70,700 km (NGVA Europe, 2010).

The use of biogas is increasing in Europe. In Sweden, about 25 plants for biogas production are in operation at the time of writing, and many more are planned. Biomethane now accounts for more than 50 per cent of all methane used in vehicles in Sweden. Switzerland followed the Swedish initiative in 1998 and biomethane now accounts for about 30–40 per cent of all methane used in vehicles. In France, the city of Lille operates a fleet of more than 300 NG buses where biomethane makes up 50 per cent of the fuel used. Both Germany and Austria are developing programmes ensuring that by 2020 biomethane will account for 20 per cent of all methane used in vehicles.[7] In the 2030 scenario, biomass could provide a contribution of approximately 15–16 per cent of the energy base in the European Union (Gerini et al., 2008).

Finally, among other possible developments worth mentioning there is the use of methane in its liquid state. Liquefied natural gas (LNG) has the considerable advantage of allowing a much higher range with the same size of cylinder, while maintaining the economic and ecological benefits of CNG. LNG would thus be a particularly interesting solution for heavy duty vehicles.

In 2003, an Economic Commission for Europe/United Nations report emphasized the strategic importance of 'blue corridors' (intended as highways or inter-state roads equipped with CNG and LNG refuelling stations) as a crucial step to compensate for the constant increase in road freight transport in Europe (Economic Commission for Europe/United Nations, 2003). The technologies necessary to implement LNG vehicles have already been acquired, but once again, problems stem from the proximity of the points of supply, which in the case of LNG is proximity to a regasification plant that receives the liquid methane from gas carriers ships. This could supply the vehicles directly, thus avoiding the double process of converting the liquid to gas and back again.

Policy implications

The research carried out shows that there are many economic and ecological benefits arising from an expansion of the number of vehicles fuelled with CNG – benefits that should further increase over time, as a result of technological improvements associated with investments in technology and the diffusion of this type of vehicle. The next step in technological development could be the development of electric vehicles with range extenders working with a mixture of compressed biogas and hydrogen.

Probably the most promising field of application for CNG is in urban environments, especially in big cities where the levels of noise and harmful emissions, particularly PM, pose a serious threat to the health of the population. In such a context the positive environmental effects would be immediately evident and the presence of a large number of CNG cars in a specific area would shorten the payback period of investments in the CNG distribution network.

The use of CNG is also highly recommended for HDVs and buses operating in urban areas. Since the technical solutions are already available on the market, a policy of incentives in this field should be addressed mainly towards the final user and towards investment in CNG refuelling stations. On the one hand, a policy of incentives would hardly be successful in the absence of an adequate distribution network, and on the other, spontaneous investment in CNG refuelling stations depends on the number of CNG cars on the road. In such context, the infrastructural investments must be made before or at the same time as policies in support of CNG cars. The required investment is relatively small in those countries where NG pipelines are already in place. In countries that do not have such an infrastructure it is possible to stimulate the creation of plants producing biomethane from biomass, using technologies and experience that are already working and reliable, such as in the case of the Spanish company FCC, and long-established installations in Sweden and Switzerland. Coastal areas could be served through gas carrier ships.

In any case, we believe that a policy of incentives addressed towards the conversion of older cars (that is, those compliant with Euro 0–3 standards) to CNG is far more sustainable than a policy of incentives encouraging the purchase of new conventional cars, both from economic and an environmental points of view. Unless it is possible to reduce the total number of cars on the road and, specifically, to increase the numbers scrapped with respect to the number of new car registrations, the persistence of older cars on the road will cancel out technological improvements for many years to come.

At the same time, the development of an adequate CNG retail network would probably stimulate the R&D of specific CNG-powered vehicles, including hybrid CNG–electric vehicles that would probably be less polluting after hydrogen and purely electric vehicles. At the time of writing, technological research in this field has not been as focused as, for example, in the field of electric cars and batteries.

Notes

1. This was the first year in which carmakers were required to provide CO_2 emissions data on the basis of an official test cycle.
2. The calorific value (or heat of combustion) measures the energy that becomes available when a fuel is burned; it provides the basis for calculating the thermal efficiency of an engine using that fuel. Energy content can be expressed in megajoules per kilogram (MJ/kg) or per litre (MJ/l). One cubic metre of methane is equivalent to about 1.1 litres of petrol, and one kilogram of methane is equivalent to about 1.5 litres of petrol. See Ahlvik and Brandberg (2011).
3. In spark-ignition internal combustion engines, the fuel's knock resistance is expressed by the research octane number (RON). The maximum allowable compression ratio of an engine, and hence its efficiency, depends on the knock resistance of the fuel, since a fuel with too low a RON will knock at high loads. Thus the higher the octane number, the more knock-resistance and the greater the efficiency of the engine.
4. In this chapter we refer to conventional cars to which are applied a subsidiary system consisting of a CNG cylinder and fuel injection system. These cars can use both petrol and CNG, and change from one to the other during normal driving. At the time of writing, such systems are applicable only to a wide range of petrol engines; the adaptation of a diesel engine is still too costly in respect to the advantages. However, there are recent reports that a new system for diesel engines has already been tested and will be commercialized soon by the Landi Renzo company, the world leader in the production of such systems, as the company has announced the launch of a CNG injection system for diesel engines. Of course, the engines specifically designed to run only on CNG are even more efficient.
5. The New European Driving Cycle (NEDC) is the homologation test used to evaluate car emissions; it is a 20-minute sequence of accelerations, decelerations and constant speed paths that is intended to simulate normal driving conditions.
6. EIA statistics are provided in cubic feet (1 cubic metre = 35,314667 cubic feet; 1 metric tonne = 48.700 cubic feet).
7. Persson *et al.* (2006) present a number of cases of the production of biogas for motor fuel, and a survey of the most significant of biogas plants in the world.

References

ACI (2011a) *Annuario Statistico*. Rome: Automobile Club d'Italia.
ACI (2011b) 'Autoritratto 2010', Report on Italian vehicle statistics. Available at: http://www.aci.it/sezione-istituzioale/studi-e-ricertche/dati-e-statistiche/autoritratto-2010.html.
Ahlvik, P. and Brandberg, A. (2011) 'Well-to-Wheel Efficiency for Alternative Fuels from Natural Gas Biomass'. Borlänge, Sweden: Swedish National Road Administration.

Allegrini, I., Bertuccio, L., Parenti, A. and Pascalizi, F. (2007) 'Benefici ambientali del metano per autotrazione', Research Report. Euromobility/CNR Institute of Atmospheric Pollution Research.

ANFAC (Asociación española de fabricantes de automóviles y camiones) (2010) *European Motor Vehicle Park 2008*, March.

Bakar, R.A., Sera, M. A. and Mun, W. H. (2002) 'Towards the Implementation of CNG Engines: A Literature Review Approach to Problems and Solutions', BSME-ASME International Conference on Thermal Engineering, 31 December 2001–2 January 2002, Dhaka, Bangladesh.

BP (2011) 'BP Statistical Review of World Energy', June. Available at: http://www.bp.com/statisticalreview.

Di Pascoli, S., Femia, A. and Luzzati, T. (2001) 'Natural Gas, Cars and the Environment. A (Relatively) 'Clean' and Cheap Fuel Looking for Users', *Ecological Economics*, 38(2): 179–189.

Economic Commission for Europe/United Nations (2003) *Blue Corridor Project on the Use of Natural Gas as a Motor Fuel in International Freight and Passenger Traffic*. New York/Geneva: United Nations.

EEA (European Environmental Agency) (2008) *Greenhouse Gas Emission Trends and Projections in Europe*. Copenhagen: European Environmental Agency.

EEA (European Environmental Agency) (2011) *Monitoring CO_2 Emissions from New Passenger Cars in EU27*. Copenhagen: European Environmental Agency, June.

EIA (US Energy Information Administration) (2011) "Natural Gas Annual 2010. Available at: http://205.254.135.7/naturalgas/annual/pdf/nga10.pdf.

Enerdata (2009) *The Impact of Lower Oil Consumption in Europe on World Oil Prices*. Brussels: European Federation for Transport and Environment.

European Commission (2007) *Well-to-Wheels Report*, Version 2c, Joint Research Centre, Commission of the European Communities, Brussels.

European Commission (2009) *Responding to the Crisis in the European Automotive Industry*, COM(2009)104 final. Brussels: Commission of the European Communities.

European Commission - Energy (2010) 'Registration of Crude Oil Imports and Deliveries in the European Union' (EU27), Market Observatory: EU Crude Oil Imports. Available at: http://ec.europa.eu/energy/observatory/oil/import_export_en.htm.

EU (European Union) (2007) 'Regulation (EC) No. 715/2007 of the European Parliament and of the Council, 20 June 2007', *Official Journal of the European Union*, L171/1, 29 June.

Exxon (2009) *È arrivato in Italia il terminale di rigassificazione Adriatic LNG* Rome: Argomenti Esso.

Gas Research Institute (1987) *Environmental Benefits of CNG-fueled Vehicles – Topical Report*. Chicago, IL: Gas Research Institute.

Gerini, A., Novella, P., Ziosi, M. and Cornetti, G. (2008) 'Potential and Perspectives on Natural Gas and Biomethane Propulsion', in C. Stan and G. Cipolla (eds), *Alternative Propulsion Systems for Automobiles II*. Renningen, Germany: Expert Verlag.

Gifford, J. D. and Brown, R. C. (2011) 'Four Economies of Sustainable Automotive Transportation', *Biofuels, Bioproducts and Biorefining*, 5(3): 293–304.

Goyal, S. P. (2003) 'Present Scenario of Air Quality in Delhi: A Case Study of CNG Implementation', *Atmospheric Environment*, 37(38): 5423–31.

Lave, L., MacLean, H., Hendrickson, C. and Lankey, R. (2000) 'Life-cycle Analysis of Alternative Automobile Fuel/Propulsion Technologies', *Environmental Science and Technology*, 34(17): 3598–605.

Liew, C. J. and Liew, C. K. (1995) 'The Use of Compressed Natural Gas (CNG) in Motor Vehicles and Its Effect on Employment and Air Quality', *The Annals of Regional Science*, 29(3): 315–34.

Mahla, S. K., Das, L. M. and Babu, M. K. G. (2010) 'Effects of EGR on Performance and Emission Characteristics of Natural Gas Fueled Diesel Engines', *Jordan Journal of Mechanical and Industrial Engineering*, 4(4): 523–8.

NADA (2011) 'State of the Industry Report', *NADA DATA*. Available at: www.nada.org.

NGVA Europe (2009) *CNG Trucks in Urban Garbage Collection. The Successful Case of the FCC Fleet in Madrid*. Madrid/Brussels: NGVA.

NGVA Europe (2010) *Fact Sheet: Biomethane Production Potential in the EU-27 + EFTA Countries, Compared with Other Biofuels*. Madrid/Brussels: NGVA.

NGVA Europe (2011) 'Statistics 2011'. Available at: http://www.ngvaeurope.eu.

Ogden, J. M., Williams, R. E. and Larson, E. D. (2004) 'Societal Lifecycle Costs of Cars with Alternative Fuels/Engines', *Energy Policy*, 32(1): 7–27.

Onufrio, G. (2005). 'Aspetti ambientali del ciclo del gas naturale' (Environmental aspects of the natural gas cycle), Paper presented at 'Riprendere la strada per Kyoto: il ruolo del gas naturale', Rome: ISSI – Istituto Sviluppo Sostenibile Italia.

Persson, M., Jönson, O. and Wellinger, A. (2006) *Biogas Upgrading to Vehicle Fuel Standards and Grid Injection*. IEA Bioenergy.

Roder, A., Kypreos, S. and Wokaun, A. (2003) 'Evaluating Technologies in Automobile Markets', *International Journal of Sustainable Development*, 6(4): 417–35.

T&E (European Federation for Transport and Environment) (2011) *How Clean Are Europe's Cars? An Analysis of Carmaker Progress towards EU CO_2 targets in 2010*. Brussels: European Federation for Transport and Environment.

T&E (European Federation for Transport and Environment) (2008) *Reducing CO_2 Emissions from New Cars: A Study of Major Car Manufacturers' Progress*. Brussels: European Federation for Transport and Environment.

Yedla, S. and Shrestha, R. M. (2003) 'Multi-criteria Approach for the Selection of Alternative Options for Environmentally Sustainable Transport System in Delhi', *Transportation Research Part A: Policy and Practice*, 37(8): 717–29.

Zhang, C.-H., Xie, Y.-L., Wang, F.-S., Ma, Z.-Y., Qi, D.-H., Qiu, Z.-W. (2010) 'Emission Comparison of Light-Duty In-Use Flexible-Fuel Vehicles Fuelled with Gasoline and Compressed Natural Gas Based on the ECE 15 Driving Cycle', *Proceedings of the Institution of Mechanical Engineers, Part D: Journal of Automobile Engineering*, 225(1): 90–8.

9
Institutional, Technological and Commercial Innovations in the Brazilian Ethanol and Automotive Industries[1]

Marcos Amatucci and Eduardo Eugênio Spers

> Cicero demands of historians, first, that we tell true stories. I intend fully to perform my duty in this occasion, by giving you a homely piece of narrative economic history in which 'one damn thing follows another'.
> David (1985)

In Brazil, the most promising sources of renewable energy include ethanol and, more recently, biodiesel. Brazil has been producing ethanol for fuel since 1980, and the production of biodiesel began in 2003. Biofuels are expected to bring significant environmental, social and economic advantages as well as innovations such as flex-fuel engines for cars and motorcycles. This chapter describes the institutional, technological and commercial aspects of Brazilian ethanol and automotive system innovations.

Demand for sustainable fuels has caused the rise of innovative organizations, the development of more efficient systems/technologies and commercialization, resulting in the diversification of supply sources, particularly with regard to environmentally friendly, clean and renewable energies. Several countries are seeking to reduce their dependence on and use of petrol oil fuels, either by replacing them with another source or another product, or by adding other fuels that decrease pollution. We focus here on the Brazilian approach to this sustainable strategy, describing how the associated innovation has spread throughout the automotive industry and the ethanol value chain, and the roles of the governance structure and the national innovation system in this transition.

This chapter is divided into four parts. The next section defines the concepts and describes the models that were used to analyse the Brazilian experience: the system of innovation, institutional and path dependence analysis. The second section addresses innovation with regard to the ethanol chain, and the third section describes innovations in the Brazilian automotive industry.

Finally, the fourth section contains remarks regarding the lessons that can be drawn from the Brazilian experience.

System innovation and the institutional theoretical approaches

The fully matured economics of ethanol fuel in Brazil is a system innovation that has been conceptualized by Geels *et al.* (2004) as the transition of one socio-technical system to another on the level of societal functions (for example, communications, or energy and transport in our case). The system innovation approach accounts for technological innovation, institutional change and the path-dependent sequence of events and is therefore a suitable subject for analysis. According to these authors, the system innovation can be characterized by four main features: (i) supply-side and demand-side co-evolutionary development of technology, knowledge and industry structure; (ii) large-scale changes involving the structural elements of a societal socio-technical system; (iii) multi-actor processes involving a wide range of social groups; and (iv) a long time scale in the order of several decades. We may connect the Brazilian situation to these elements as follows:

- ethanol leads to innovations in the agricultural, processing and automotive industries by providing consumers with a less expensive fuel;
- from the distribution system of the fuel, via petrol stations all over the country, to engine manufacturing, which involves a whole supply chain (even Mercosur-imported cars projected to run in Brazil have flex-fuel engines), the entire transportation socio-technical system has been affected;
- on the agribusiness side, sugar-cane mills, which concentrate their production and research in south-east and north-east Brazil, have reached an economy of scale and spread the fuel technology to highly productive farms and ethanol production plants;
- the entire ethanol value chain is integrated and influenced by several social actors involving research support and sustainability programmes; and
- ethanol production can be traced back to the 1930s, when it was influenced by the progressive involvement of social actors and social change.

Using the system innovation as a framework, this report highlights some of the foundations of our analysis, specifically the institutional and path-dependence approaches.

An institutional, technological and commercial innovation in a production value chain concerns three main economic and political forces at different levels: the institutional environment and the organizations (macro level); markets (meso level); and the companies (micro level).

At the macro level, North (1990) defined institutions as 'the rules of the game' and organizations as 'the players'. The oil crises in the 1970s provided the institutional incentive for the Brazilian military government to institute a programme called PROALCOOL ('*Pro-alcohol*', see the sub-section below entitled 'Innovation in Brazilian ethanol commercialization'). The main goal of this programme, which was supported by the law and public investment, was to find alternatives to fossil fuels. Similarly, at the meso (market) and micro (company) levels, Williamson (1993) argued that the institutional environment provides fundamental rules (property rights and the law) that induce the formation of different types of organizations that, in turn, structure the institutional arrangement. With the incentives of PROALCOOL, several agricultural industries began to produce ethanol, and the automotive industry began to develop engines that could work exclusively with this new fuel.

Olson (1965) defined an organization as a group of individuals with common interests. Organizations develop collective actions that maximize the value of the founders and try to increase competitiveness. Brazilian examples of organizations in the ethanol value chain are UNICA[2] (for ethanol production) and ANFAVEA[3] (for carmakers).

Public and private organizations also contribute to the competitiveness of the environment. With regard to innovation in ethanol production, there are both private research organizations such as CTC[4] and public organizations such as EMBRAPA.[5] Together, they could change significantly the standards that dictate competitive advantage in the market. In other words, changes in individual strategies can alter the competitive environment of a market and, consequently, the institutional environment. How the organizations interact is particularly important to the competitive dynamics (Olson, 1965), as further discussed in the sub section below entitled 'Innovation in Brazilian ethanol commercialization'.

The system innovation that resulted in a mature market economy for ethanol fuel involved several institutional changes. In this regard, Alston *et al.* (1996) discussed two main forces that drive institutional change: (i) the symbiotic relationship between institutions and organizations that defines the structure of incentives (for example, in sugar cane production, the strong co-ordination led by the automotive industry and the vertical integration of the ethanol industry made possible the intense symbiotic relationship in those chains); and (ii) the feedback that drives the perception and reactions of organizations towards new opportunities (for example, in Brazil, government incentives, the right signals to the market and the reputation of PROALCOOL paved the way towards long-term investments in innovations).

This chapter focuses on three main institutional changes in the ethanol value chain: the rapid growth of the ethanol sugar cane industry; the emergence of the ethanol engine and its infrastructure; and flex-power technology.

Given that the sugar cane economy represents the first established non-extractive production mode in the country since the Portuguese arrived in the New World, it is clear that the influence of a path-dependent sequence of events determinates the adoption of this particular fuel solution. As David (1985) defined it, 'a path-dependent sequence of economic changes is one of which important influences upon the eventual outcome can be exerted by remote events' (p. 332). These remote events include chance elements. David further emphasized that a path-dependent history is sometimes the only way we can understand the logic of an existing situation. Indeed, David's three-determinant elements for path dependence formation: *technical interrelatedness, economies of scale* and *quasi-irreversibility of investment,* seem to be the very pillars upon which system innovation takes place.

We have developed a model, based on similarities in the institutions and technological and commercial innovations in the Brazilian ethanol industry, which considers the evolution from technological innovation to market adoption, with institutional changes governing the process. The model is summarized in Figure 9.1.

To describe the technological, commercial and institutional innovations for the ethanol and automotive industries we propose the following complementary model. The institutional and organizational environments show how the changes provide a sustainable structure of governance at the macro level (institutional innovations). Also at the macro level, we incorporate the concept of path dependence as a longitudinal approach. In the meso and micro levels, we analyse which innovations (technological innovations) occur in each industry in the automotive (suppliers and makers) and ethanol (sugar cane and ethanol production) chains. We choose the main transaction in each chain (TA1, TA2, TE1 and TE2) to describe the commercial innovations (see sub-sections 'Innovation in Brazilian ethanol commercialization' and 'Commercial innovations in the Brazilian automotive sector' below). Figure 9.2 further illustrates this model.

The longitudinal analysis of the Brazilian system of innovation is discussed in the next section.

Innovation within the ethanol chain

The system innovation assumes a pre-existing governance structure involving several different agents (see Figure 9.2). The sugar cane and ethanol

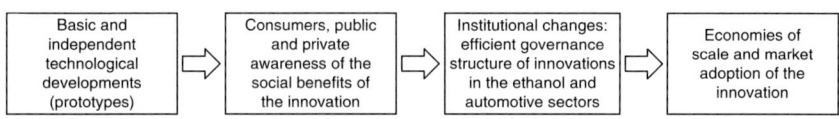

Figure 9.1 Institutional changes and governance structure mediating the market adoption of a technological innovation

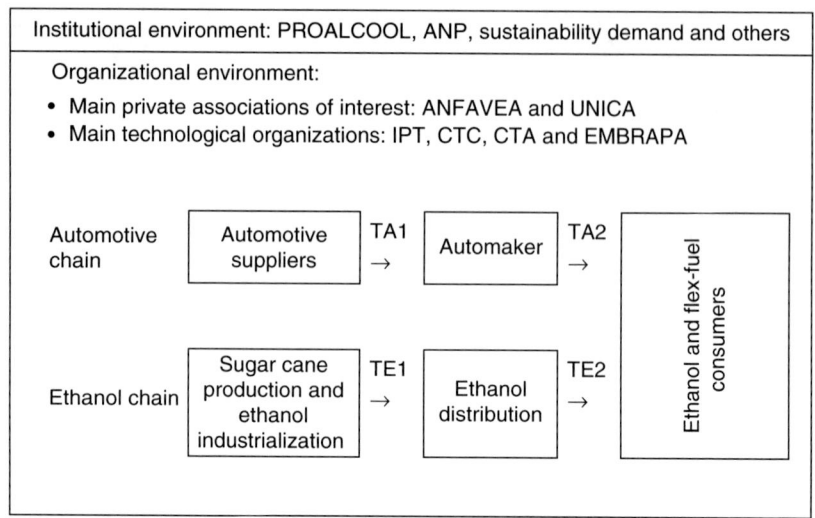

Figure 9.2 The Brazilian ethanol and automotive innovation system
Notes: ANP = National Agency for Petroleum, Natural Gas and Biofuels; ANFAVEA = National Association of Motor Vehicle Manufacturers; UNICA = Sugar Cane Industry Association; IPT = Technological Research Institute of the University of São Paulo; CTC = Sugarcane Technology Center; CTA = General Command for Aerospace Technology; EMBRAPA = Brazilian Enterprise for Agricultural Research.
TA1: Transaction between automotive suppliers and carmakers.
TA2: Transaction between carmaker and the consumer.
TE1: Transaction between ethanol industrialization and distribution.
TE2: Transaction between ethanol distribution and the consumer.

industry agents co-ordinate the knowledge, development and the flow from the agricultural sector, and the information and skills spread throughout the automotive sector, to support the flex-fuel technology. Both innovations are described below.

Path dependence and innovation in the Brazilian sugar cane sector

The sugar cane plant (*Saccharum officinarum*) was first found in India, from where it went to China and Persia (in the time of Alexander the Great), and then to Syria, Egypt and Sicily. The Portuguese *infante*,[6] D. Henrique, introduced it to the Madeira Islands during the fourteenth century, giving Portugal an important source of revenue during subsequent centuries (Souto Maior, 1967).

Its discovery by the Portuguese in 1500 immersed Brazil in the mercantilist economy, using Portuguese navigational expertise to promote intense commerce with India (mainly for specialties such as sugar that sold for a good price in Europe). The first economic activity in Brazil involved the extraction of an ashy (*verzino* in Italian, *brésil* in French, *brasil* in Portuguese) wood used as the source of a red dye for colouring fabrics; it was called '*pau-brasil*' (ashy wood) and gave the country its name. Sugar cane plantations were introduced between 1516 and 1525 to Brazil, constituting the first colonial non-extractive productive activity. The production of sugar cane employed a technology already dominated by the Portuguese, namely the slave-operated central *engenho* (machinery that processed the sugar cane to produce sugar), located in the middle of the plantation. Sugar production provided Portugal with more wealth than all its other colonial activities, including mining. By 1570, Brazil had approximately 60 *engenhos*, most of them in the north-east of the country. Between 1570 and 1710, Brazil exported an average of 28,000 tons a year, reaching the peak of 58,800 tons in 1600 (Vicentino and Dorigo, 1998).

In addition to the main *engenho* used for sugar production, a smaller *engenho* (or *engenhoca*) was used for the production of sugar cane liquor (*cachaça*), the first alcoholic by-product of sugar cane production and an industry that was particularly important to the slave trade in Africa.

By the end of the seventeenth century, competition from Central America and the West Indies caused sugar prices to drop to commodity levels, and the importance of Brazilian sugar cane production to the Portuguese economy decreased accordingly. But despite the competition and two major setbacks – the war against the Dutch settlements that destroyed many assets in the north-east and the massive workforce transfer to gold mining – sugar production was never interrupted.

Institutional and technological innovation in Brazilian ethanol production

Some important organizations with different origins have been important to the governance structure of the Brazilian system of ethanol innovation. EMBRAPA and CTC are good examples. EMBRAPA is the federal and public agency of innovation for the agricultural sector. Individual specialized centres exist in almost every region of Brazil. The EMBRAPA-CENARGEN[7] has conducted research into the adaptation of agricultural species to different regions of the country, and houses a genetic collection for use in developing genetically modified crops, including sugar cane. CTC is an example of private investment in agricultural innovation. Sustained by private companies, it supports research into cellulose, ethanol, biotechnology, agronomy, benchmarking, rural and industrial mechanization, sugar cane and ethanol production, and bioenergy.

The 1975 law (No. 75,593) after which the PROALCOOL programme was initiated, established price guarantees to ethanol producers based on corresponding sugar prices. Several subsequent laws renewed the incentive until the ethanol market was established and deregulated in 1999. From the consumer point of view, prices were guaranteed to be lower than those of petrol, a situation established by governmental subsidies when production costs exceeded petrol production prices. Consumer prices were deregulated in 1996, putting an end to this incentive (Puerto Rico, 2007). At the time of writing, ethanol production prices vary significantly depending on the producing region. Furthermore, they float according to international sugar prices, which compete for ethanol raw materials in a free market. Ethanol production has generally become less expensive than petrol in the south/south-east of the country (mainly São Paulo, which is the main consumption centre).

Currently in Brazil, 48 per cent of the total energy generated comes from renewable sources, and sugar cane supplies 15 per cent of the total country's domestic energy, mostly for automobile flex-fuel engines, (OECD, 2011). The country is not only the world's largest producer of sugar cane ethanol but has also become a leader in innovation technologies and processes for ethanol production. The state of São Paulo has the largest concentration of sugar cane ethanol production sites in the country.

In ethanol production, the governance structure has been based on a strong relationship between agriculture and industry, which has solved the problem of co-ordination and provided support for rapid growth in the scale of production. The sugar cane industry has used this economy of scale to invest in agricultural and industry innovations. Most of this innovation has been driven not only by efficiency and/or productivity goals but also by concerns about sustainability.

The agricultural or agronomic innovations that were conducted by the sugar cane industry and several research institutes and universities include direct seeding and rotations (with other cultures) every four years. This sustainable innovation helps to maintain/conserve both soil and water quality. In addition, thanks to the introduction of mechanical engines, the crop-burning process used in human harvesting is no longer necessary.

Some research associations are owned by ethanol organizations, but the government itself still invests in promoting innovation, especially in the agricultural sector. UNICA, the largest private Brazilian association of ethanol producers, contributes significantly to the governance structure of the industry's innovation, and supports communication and the promotion of the product in other countries (UNICA, 2011).

The production of ethanol has also led to promising technological innovations. One of the newest developments in ethanol production is called 'second-generation' ethanol. Because plant cells are surrounded by cellulose, a polymer that is very hard to penetrate, developing an economically viable means of breaking down the cellulose is essential to the production of second-generation ethanol, which can be extracted from any biomass (not only from sugar cane). Once implemented, this technology will increase biofuel production without requiring an increase in the size of the sugar cane plantations themselves. Another example of technology that is being developed employs enzymes found in termites and ruminant digestive systems to decompose the cellulose. An alternative innovation involves the use of specific acids to break down the fibres.

The commercial side of the ethanol chain involves innovation challenges regarding price oscillations. The prices of petrol and diesel are controlled by the government, which therefore also controls the percentage of ethanol mixed with the petrol. Because the price of ethanol is less controlled, depending more on the laws of supply and demand. For example, since early 2011, the state of São Paulo has witnessed a 19 per cent increase in the price of ethanol. According to the National Agency for Petroleum, Natural Gas and Biofuels (ANP) (ANP, 2011), the average price of a litre of ethanol fuel in this state has increased to R$1.99 (compared with R$1.67 in December 2010). Consequently, at the time of writing, petrol is priced more competitively than ethanol in 25 states and in the Federal District. While showing an exceptional situation that tends to return to balance, this example reveals the nature of the price oscillations problem that needs to be addressed.

To reduce this problem, 30 sugar and ethanol plants, responsible for processing 65 million tons of cane (12 per cent of the annual volume processed at the centre/south region of Brazil), attempted to anticipate the 2011/12 crop of sugar cane (UNICA, 2011). Given the high price of sugar on the international market, the industry has decided to produce sugar instead of ethanol. Another potential commercial innovation involves the future stock-exchange prices for ethanol. Between 7.1 million and 8.9 million tons of sugar produced in April 2011 already have a fixed price on the New York Stock Exchange.

The ethanol market is also dependent on other international institutional environments. For example, the USA is attempting to counterbalance the growing economic power of China by making room for biofuel development and production. Another example is the current US tariff on ethanol imports, which could be reduced. In December 2010, though, the US Congress passed the renewal tariffs on ethanol import (AEB, 2011).

Innovation in Brazilian ethanol commercialization

Brazil has mixed ethanol and petrol for fuel use since the 1930s. Law 17,917 (of 20 February 1931) established the use of 5 per cent of ethanol in all imported petrol. In 1948, Law 25,174 extended the 5 per cent mix to Brazilian-produced petrol. During the Second World War, the percentage of ethanol reached higher levels (Flores, 2010). Therefore, mixing alcohol with petrol has been a historical national solution in Brazil to international fuel shortages.

The 1973 oil crisis hit the Brazilian economy, an economy that was already indebted and suffering from high inflation and a weak currency.

At that time, Brazilian oil production was far below demand, forcing the country to rely on oil imports for consumption and growth. Table 9.1 shows the main data pertaining to the Brazilian economy from 1972 to 1979, highlighting the country's great economic dependence on oil imports.

The Brazilian growth strategy was based on accruing external debt as a way of capitalizing the economy, increasing international reserves and financing external trade imbalances. However, the relative weight of oil to (total) imports rose with the devaluation of the Brazilian currency, the rise of international oil prices and the growth of the Brazilian economy itself, which demanded increased oil consumption. Furthermore, because the 1973 oil crisis impacted worldwide, external markets were not suitable for Brazilian export growth. Though the devaluation of the currency fostered exports, it did so at the expense of imports. The bottom line was that oil represented more than one-third of total imports by 1979.

The military government put forward the traditional idea of substituting petrol by ethanol as a fuel, thus instigating the PROALCOOL programme. The programme was officially announced by President Ernesto Geisel in his televised speech on 9 October 1975 (Geisel, 1976) and sanctioned by Law 76,593 on 14 November 1975. Between 1975 and 1979, the programme consisted of fostering ethanol production for fuel mixtures – a traditional Brazilian response to petrol supply crises. The PROALCOOL programme aimed first to drive the production of dehydrated ethanol (99.5 per cent pure ethanol), adding it to petrol at a rate that would not require engine modifications (a proportion between 10 per cent and 15 per cent, or E10 and E15, respectively). Later research showed that the ethanol content could be as much as 25 per cent and still require only minor engine adaptations. The resulting fuel is called *gasohol*. This approach remains in effect at the time of writing; Brazilian petrol has been 25 per cent ethanol (E25) since 2007; the exact proportion may vary slightly according to ethanol supply levels.

In 1978, with the country on the verge of petrol rationing that would decelerate its expensive economic growth, the automotive industry representatives headed a national agreement and produced 250,000 vehicles in 1980 powered exclusively by ethanol, 300,000 in 1981 and 350,000 in 1982. Agribusinesses produced enough ethanol to power these vehicles, in exchange for a governmental promise not to put rationing into effect. The protocol was signed on 19 September 1979, successfully halting the movement to petrol rationing (Garnero, 1980). Part of the deal involved a government tax incentive for cars powered by the new engine, resulting in 5 per cent less in IPI, the tax on industrial products (Puerto Rico, 2007). Product distribution across Brazil's 8,500,000 sq. km was delegated to Petrobras, the Brazilian state petrol company that enjoyed a petrol *distribution* monopoly at that time (though retail was free marketed). That solved the distribution problem, but at the retail level there was another one: how to store and sell ethanol immediately in every petrol station in

Table 9.1 Brazilian economic growth, evolution of external debt and the importance of the oil account, 1972–9, US$ millions

Year	GDP[3]	Annual growth[3] (%)	Gross external debt[1]	Liquid external debt[1]	Exports[2]	Imports[2]	Oil imports[2]	Oil/imports[4] (%)
1972	58,539	12	9,500	5,300	3,991	4,232	469	11.1
1973	79,279	14	12,600	6,200	6,199	6,192	769	12.4
1974	105,136	9	17,200	11,900	7,591	12,641	2,961	23.4
1975	123,709	5	21,200	17,200	8,669	12,210	3,100	25.4
1976	152,678	10	26,000	19,400	10,128	12,346	3,354	27.2
1977	176,171	5	32,000	24,700	12,139	11,999	3,660	30.5
1978	200,801	3	43,500	31,600	12,658	13,683	4,093	29.9
1979	224,969	7	49,900	40,200	15,244	17,961	6,188	34.5

Sources: [1]Adapted from Bank of Brazil and Banco Central in Lacerda et al., 2000, p. 114.
[2]Adapted from Garnero, 1980, p. 28.
[3]Authors' compilation with data from the World Bank.
[4]Oil imports over exterior trade balance.

the country. The creative idea was to abandon the commercialization of premium ('blue') petrol and to keep only the regular ('yellow') petrol at the pumps, using the premium petrol tanks to store the ethanol. By this means, every Brazilian petrol station could, almost instantaneously, commercialize the new fuel because every filling station would have at least one petrol pump and an ethanol pump.

The consumer price of ethanol was determined by the producers' value plus the distribution cost. With this price strategy, the final cost of ethanol was higher than that of petrol. To compensate, the government fixed the final ethanol price at 65 per cent of the price of petrol (the ethanol engine had a 20 per cent higher consumption than a regular petrol engine). During the Iraq–Iran War in 1982, the Brazilian government fixed the ethanol price at 59 per cent of that of petrol to provide an incentive for ethanol consumption (Puerto Rico, 2007, p. 75).

These developments led to the interaction between the two organizations/areas, namely automobile manufacturing and the sugar cane agribusiness.

Innovation in the automotive chain

This section explores the same innovations and path dependence (as above) in the automotive chain, in which the two main innovations were the creation of the ethanol car and the flex-fuel engine.

Path dependence in the automotive chain

The automotive industry was one of the first industries to internationalize and arrived in Brazil early in this process. Ford entered the market in 1919, followed by GM (General Motors) in 1925. It was difficult for Brazil, a late bloomer as far as industrialization was concerned, to afford to support the automotive industry, evidenced by the fact that gasohol (a mixture of gasoline and alcohol) had been a means of saving foreign currency since the 1930s. After the Second World War, the automotive industry was chosen as the leading industrialization sector by President Juscelino Kubitschek, who instituted the GEIA (Automobile Industry Executive Group), a government committee designed to recommend policies to attract foreign carmakers to Brazil, while committing them contractually to gradually enhance local input. The first automobile multinational to accept the Brazilian government's invitation was the German Volkswagen (VW), who began local production in 1953; Fiat followed in 1974.

A growing automobile industrialization process requires that the main transportation system is the road. Consequently, to transport people and goods throughout Brazilian's 8.5 million sq. km, Brazilians have become fuel-dependent. Therefore, when the country was hit by the first international oil crisis, it was the industrialization of the country and its required mode of transport that pushed Brazil towards an alternative to imported oil.

Institutional and technological innovation in the automotive industry

Ethanol-engine technology has been made possible through long-term research projects at the General Command for Aerospace Technology (CTA), in São José dos Campos, São Paulo. Established in 1947, CTA has had a crucial role in the development of Brazilian gasohol and aerospace programmes, producing not only the ethanol engine but also founding a world-class aircraft manufacturer, Embraer (which was later privatized). According to Dahlman and Frischtak (1993), of the dozens of Brazilian military technical institutes that were built in this period, CTA was the most important. Urbano E. Stumpf, professor and engineer at CTA, furthered the work of Eduardo Sabino de Oliveira and Lauro de Barros Siciliano to develop the first Brazilian modern prototypes of ethanol-powered engines in the automotive industry (Teixeira, 2005).

Private companies also played an important part in the innovation process. The first exclusively ethanol-powered passenger car produced in Brazil was a modified 147 model from Fiat in 1979. As discussed in the sub-section 'Innovation in Brazilian ethanol commercialization', a protocol signed in the same year by President Figueiredo and automotive industry representatives – mainly ANFAVEA: Associação Nacional dos Fabricantes de Veículos Automotores (National Association of Motor Vehicle Manufacturers) – invited all the other automakers to join the programme.

Also, in conjunction with governmental and independent research institutes and universities, multinational corporations established engineering centres in their Brazilian subsidiaries, encouraged by the availability of well-trained engineers with comparatively less expensive labour costs, and the local market characteristics (Dias and Salerno, 2004; Queiroz and Carvalho, 2005; Consoni and Quadros, 2006; Amatucci and Bernardes, 2007, 2009; Balcet and Consoni, 2007; Quadros and Consoni, 2009). This is particularly the case in the automobile business. According to Amatucci and Bernardes (2009), product development in the Brazilian subsidiaries of international carmakers resulted from a local need to adapt the product, the presence of a Porter's diamond (Porter, 1990) in the local industry, the availability of a specialized workforce, and the investment of headquarters in training, expatriation and the use of standardized product development software. Not only did the main car such as GM, Fiat, Volkswagen, Ford and Renault support local product development capabilities, but Brazilian subsidiaries of vehicle parts suppliers also did so (take, for example, the cases of Robert Bosch, Magneti Marelli, Delphi and many others). Indeed, Magneti Marelli's product development capabilities in Brazil were responsible for the important development of second-generation flex-fuel technology (based on software).

In addition to technological innovations in both public and private organizations, institutional changes in the economic and political environments pushed the automobile market, as a whole, to adopt the flex-fuel technology. By adopting this technology, Brazilian consumers could choose their fuel combinations (using petrol and ethanol) depending on their own value perceptions (Amatucci and Spers, 2010). Analysis of the flex-fuel engine technology adopted by Brazilian carmakers reveals parallels between this process and the US experience.

In the USA, carmakers have been developing electronic fuel injection for use with biologically-sourced fuels since the middle 1980s, a persistent hangover from the petrol crisis of the 1970s (Teixeira, 2005; Yu et al., 2009). Through the Alternative Motor Fuels Act of 1988, the California Air Resources Board set out the technical parameters for a dual-fuel engine and offered tax incentives to support its adoption; and the Clean Air Act Amendments from 1990 gave the US Environment Protection Agency (EPA) the power to control toxic emissions. Under this governance structure, the automotive industry offered the ethanol engine alternative to consumers in California. Between 1992 and 2005, approximately 2.5 million such cars were sold in the USA, despite the fact that there was not enough ethanol to fuel them all (Teixeira, 2005).

Early dual-fuel technology used in the USA was developed by Robert Bosch, and relied on a dielectric sensor in the fuel feed line. This took advantage of the dielectrical differences between the two fuels to provide the necessary adjustments to the burning parameters. This 'first generation' dual-fuel technology (called 'dual-fuel' rather than 'flex-fuel' because it does not allow any mixing of the two fuels but works with one at a time) added costs to the regular electronic injection command device, but it was well adapted to the specific US anhydrite ethanol.

In Brazil, early ethanol research also began as a response to the petrol crisis of the 1970s. However, before electronic injection became available, the automotive industry worked with two different engines – a regular petrol engine and an ethanol-only engine. This double standard shortened scale efficiency, as the physical characteristics of ethanol demanded several engine modifications, such as corrosion-resistant materials, as well as different regulations and compression rates. Thus the differences between the two engines were considerable, and involved separate production lines.

According to Yu et al. (2009), the flex-fuel innovation in Brazil was born out of previous technological development, mainly electronic injection and its application to ethanol engines. Despite the efforts of the technology supplier to persuade Brazilian carmakers to adopt the Robert Bosch technology used in USA, this did not happen in Brazil. For the Brazilian market, the additional costs associated with the dielectric sensor prevented the local carmakers from acquiring the innovation.

It was the Brazilian subsidiary of Magneti Marelli that introduced the 'second generation' of flex-fuel technology, the SFS (software fuel sensor).

The new technology works on a feedback basis, processing information about burnt gases (from the escape system) that is being collected via the regular electronic injection system. The SFS system is completely software based, and for all practical purposes, apart from the fact that the electronic module processor had to be upgraded to receive some 100,000 new code lines, it has no marginal costs when compared to the traditional injection controller. Therefore, once the R&D is amortized, each additional device adds no extra costs to the engine.

That was not enough for the industry to adopt the new technology, however. At first, the carmakers were not very enthusiastic about this new technology either. The innovations were presented to the major Brazilian sectors of the global carmakers, but it took some time before the market evolved to accept the new technology. Apparently, the carmakers and their consumers were still resentful about their past experiences with pure alcohol engines (sales of these relied on government tax incentives, but they suffered from ethanol supply shortages) caused by the technical and economic problems of the incipient development of the ethanol industry.

Adoption of the new technology was delayed until an event at IPT (Technological Research Institute, São Paulo state), where Bosch and Magneti Marelli allowed them the opportunity to present their technology to representatives of the industry and Brazilian government representatives, to show that the situation had changed. With IPT entering the game, the new governance structure of innovation received government attention, and the incentives provided to pure-ethanol engines were extended to (future) flex-fuel engines. As this technology allows customers to use either petrol or ethanol, the environmental gains from the subsidies were not achieved, but the Brazilian government was investing in the idea that consumers would behave in a pro-environment fashion. It was only then that the carmakers adopted the innovation fully.

Thus, evidence suggests that the early process in California and the more recent developments in Brazil had followed a similar pattern. The Robert Bosch dual-fuel technology of the 1980s did not take off until the State of California and US federal incentives were put in effect. Only then did the American automotive industry adopt the innovation, leading to the sale of an impressive number of dual-fuel engines in the American market even when there was not enough ethanol available for all of these cars.

Similarly, it took the Brazilian government's extension of tax incentives for pure alcohol engines to the new flex-fuel technology to kick-start the industry. The first to adopt the Magneti Marelli innovation in Brazil was Brazilian VW with its Gol Power 1.6 Total Flex, in 2003.

Another parallel can be traced regarding technological innovation development and its availability before its adoption. During the technological development phase, suppliers in both cases had worked without support from either industry or the government. In Brazil, the Brazilian branches of Bosch (from 1988) and Magneti Marelli (from the mid-1990s) developed prototypes of the flex-fuel, attempting to lower costs related to the expensive sensor that had been necessary to balance the fuel mixture. With this kind of governance structure, the lack of support was so dramatic that the suppliers had to buy their cars on the open market to develop and test the prototypes. Some technicians tested the new devices in their own cars.

In Brazil, the new technology attracted new players, both in the supply industry and among the carmakers themselves (for example, GM preferred an in-house solution to provide their own flex-fuel devices). Therefore, the flex-fuel technology market is almost the textbook pure competitive market. With competition and scale, the new technology gained a place in the market, and in 2010 the Brazilian government dropped the incentives on flex-fuel vehicles. The Brazilian flex-fuel fleet grew exponentially, as shown in Figure 9.3.

Commercial innovations in the Brazilian automotive sector

The production of ethanol-only powered vehicles reached its peak in the middle of the 1980s, when more than 90 per cent of passenger-car production was focused on ethanol-powered engines. In 1989, however, ethanol demand surpassed its supply, causing shortages. Long queues appeared at petrol stations, causing mass consumer abandonment of the technology. This history continues to influence consumer behaviour even today.

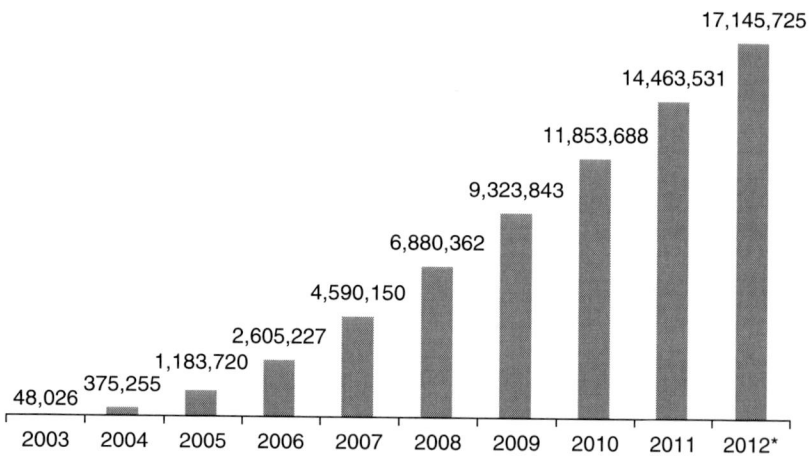

Figure 9.3 Brazilian flex-fuel fleet evolution: numbers of cars and light trucks using the technology, 2003–12
Note: *Data for 2012 projected.
Source: MMA (2011).

Eventually, international oil prices dropped to 'normal' levels, which eliminated the economic advantages of ethanol.

As a result, sales of new ethanol cars dropped to 4 per cent in 1994 and 0.56 per cent in 1996. Some small workshops offered 'reconverting' services to turn ethanol engines 'back' to petrol engines during this period. These macroeconomic factors, allied to technical issues characteristic of innovative processes (weaker performance compared to traditional engines, lower efficiency, difficulties with cold starts, corrosion problems and so on), caused consumers to consider ethanol engines as inferior (Kremer and Fachetti, 2000).

Even years later, with a reliable ethanol supply and fully developed technology (enhanced kilometres per litre, although still 30 per cent less efficient than today's petrol engines, but delivering more horsepower than an equivalent petrol engine), the market has remained hesitant about adopting exclusively ethanol-powered engines. Ethanol-only cars gradually lost value in the secondary market (used cars), costing less than their petrol peers, and practically disappeared from assembly lines.

The flex-fuel technology eliminated the market's fears about fuel shortages and the inconveniences of international oil price fluctuations. The VW Gol 1.6 litre, powered by injection technology from Magneti Marelli, the first hybrid produced on a large scale in Brazil, was rapidly imitated by other players (Mello *et al.*, 2005). The latest figures, from March 2009, indicate that 87 per cent of all new vehicle number plates are registered to cars with flex-fuel engines (Automotive Business, 2009).

This new technology changed the process of choosing between ethanol and petrol. Before the flex-fuel engine, deciding which type of car to buy was a one-off, long-term decision; now, the decision is which fuel to use – a daily and non-permanent decision that carries no consequence in terms of vehicle depreciation or resale pricing. It is purely a matter of immediate price and sustainability.

As flex-fuel technology has provided scale gains both to the industry and to suppliers, the incentives have become unnecessary, and, in consequence, were eliminated in 2010. However, the Brazilian government continues to use tax incentives to intervene in the market, and the ethanol market is subject to strong economic oscillations. Therefore, it would not be surprising if the incentives return.

Final remarks

This brief historical report of the Brazilian ethanol experience, from its production as a fuel to mix with petrol to the flex-fuel technology, brings to light a system innovation or social function transition that can be broken into three parts: the ethanol production increase to meet the PROALCOOL goals of gasohol availability; the adoption of pure ethanol and the pure ethanol engine by both the industry and the market (included in this phase is distribution problem-solving); and, after the end of Phase II, a third phase consisting of the adoption of flex-fuel technology. These three phases are summarized below.

Phase I: Gasohol Ex[8]

Each phase influenced the following phase, and the first phase was strongly path dependent on the Brazilian history of sugar cane plantations and previous ethanol/petrol mixtures as a way of saving foreign currency. The use of ethanol as a fuel, well known in the Brazilian economy, was 'recycled' to respond to the oil crises of the 1970s. So, while being an important part of the systemic transformation, the first phase was not really an innovation (unless one considers the perfecting of the mixture and ethanol quality) in the technological or the social sense.

Phase II: Pure ethanol engines

The second phase (and beyond), even with the inheritance from the first phase and previous Brazilian history, had both technological and social innovations that involved regulations, private companies and other social entities, such as class associations, being widely adopted by Brazilian consumers. This solution dwindled because of fluctuations in availability that severely penalized owners of the locked-in, pure ethanol-driven vehicles.

Phase III and beyond: flex-fuel

The third phase also involved technological innovation and social mobilization in order to bring commercial viability. With the flex-fuel car, the consumer could choose the best fuel with regard to price and sustainability. This flexibility has 'unlocked' consumers to deal with supply oscillations more smoothly. With the Brazilian economic growth of the 2000s, which brought a large proportion of the formerly excluded population into the market, a record number of cars were sold. In addition to this economic growth, the price of sugar was also attractive. Investments were, and continue to be, necessary to keep up with this internal market-consumption increase and the external demand for sustainable fuels. The Brazilian system of innovation still provides incentives for innovation. Some future examples are: cellulose ethanol, ethanol for aircraft, diesel from ethanol, ethanol lubricants, genetically modified sugar cane plants, and new sources of ethanol, among others. The automotive industry is also investing in improving the efficiency of flex-fuel engines.

Four common elements of the three phases

All three phases have some common elements, without which each phase would not exist. First is the strong perception of a social problem that must be solved urgently; that is, a belief that things cannot continue as they are. Second is the availability of a technology which, when combined socially with logistics solutions (in other words, when put in motion), can make the social solution viable. It seems that the technology often already exists; the difficult part is to put it in motion. The third common element is that the social arrangement (to put the technology in motion to address the perceived problem) needs to be catalysed by the action of one or more entrepreneurs. By the same token, governmental incentives are necessary in the early stages of the transition; however, they can be withdrawn when scale, productivity and consumer confidence achieve normal market levels.

An essential part of the social transition is market adoption of the new technology. A fourth common element to all the above phases is the initial fragility of the situation, which requires the customer to walk an extra mile, paying a higher price while tolerating inferior performance, until the technology has been perfected. Table 9.2. illustrates all the phases that were supported by entrepreneurship pushing institutional change. Future commercial innovations demand supranational organizations and an institutional environment that supports global ethanol consumption: for example, an initiative to foster Africa's sugar cane economy to provide an additional source of ethanol.

In conclusion, the institutional changes regarding ethanol as a fuel in Brazil provide a stable governance structure to mediate the development of all phases of the production chain from sugar cane to the automobile; and economic agents such as UNICA and the Brazilian government are striving for international institutional changes to extend the local production chain into a global one.

Table 9.2 Phases of the use of ethanol as a fuel

Innovation	Phase I: Gasohol (Ex)	Phase II: pure ethanol engines	Phase III: flex-fuel	Future
Commercial	Creativity to solve logistic problems	Economy of scale provided by sugar cane plantations	Competition with sugar and energy	Access to global market
Technological	Ethanol engine prototypes (FIAT/CTA)	The 'alcohol car' and agribusiness technology	The 'flex-fuel' car	Ethanol from cellulose
Institutional	Oil crisis and import/export imbalance; PROALCOOL programme	Tax incentives	Sustainable demands	Supranational institutions

Notes

1. We would like to express our gratitude for the information provided by product development executives from Bosch and from Magneti Marelli, respectively the first generation (hardware-based) and the second generation (software-based) flex-fuel technology pioneer providers; for the information provided by two executives from FTP (Fiat Powertrain), flex-fuel technology adopters and ethanol engine pioneers; and for the assistance provided by a former ANFAVEA and Brazilian VW president, the social entrepreneur who strongly influenced the adoption of ethanol as fuel in the Brazilian automobile market and who served as the ethanol retail distribution idealizer.
2. União da Indústria de Cana-de-Açúcar (Sugarcane and Ethanol Industry Association) For more information about UNICA see: http://english.unica.com.br.
3. Associação Nacional dos Fabricantes de Veículos Automotores (National Association of Motor Vehicle Manufacturers). For more information about ANFAVEA, see: http://www.anfavea.com.br/index.html.
4. Centro de Tecnologia Canavieira (Sugarcane Technology Center). For more information about CTC, see: http://www.ctcanavieira.com.br/site/index.php.
5. Empresa Brasileira de Pesquisa Agropecuária (Brazilian Agriculture Research Organisation). Further described in the section 'Institutional and technological innovation in Brazilian ethanol production'. For more information about EMBRAPA, see: http://www.embrapa.br/english.
6. An *Infante* was the Portuguese or Spanish legitimate son of a king who was not the heir to the throne.
7. For more information about EMBRAPA–CENARGEN (Genetic Resources and Biotechnology Centre for Plants), see www.cenargen.embrapa.br (in Portuguese).
8. Ex = different proportions of ethanol in gasohol: E5, E10, E25 and so on.

References

AEB (Associação Brasileira de Comércio Exterior – Brazilian Foreign Trade Association) (2011) 'Making Algae Biodiesel at Home'. Available at: http://www.making-biodiesel-books.com/algae-biodiesel-SPform.html; accessed 15 August 2011.

Alston, L. J., Eggertsson, T. and North, D. C. (1996) *Empirical Studies in Institutional Change*. Cambridge, UK: Cambridge University Press.

Amatucci, M. and Bernardes, R. C. (2007) 'O Novo Papel das Subsidiárias de Países Emergentes na Inovação em Empresas Multinacionais: O caso da General Motors do Brasil. RAI', *Revista de Administração e Inovação*, São Paulo, 4(3): 5–16.

Amatucci, M. and Bernardes, R. C. (2009) 'Formação de competências para o desenvolvimento de produtos em subsidiárias brasileiras de montadoras de veículos' *Produção*, ,19(2): 359–75.

Amatucci, M. and Spers, E. E. (2010) 'The Brazilian Biofuel Alternative', *International Journal of Automotive Technology and Management*, 10(1): 37–55.

ANP (Agência Nacional do Petróleo, Gás Natural e Biocombustíveis – National Agency for Oil, Natural Gas and Biofuels) (2011) 'Biofuels in Brazil'. Available at: http://www.anp.gov.br/?id=470; accessed 10 February 2011).

Automotive Business (2009) 'Veículos flex representam 86.7% dos emplacamentos', Internet document; available at www.automotivebusiness.com.br/noticia_det.asp?id_noticia=3026 (accessed 20 March 2009).

Balcet, G. and Consoni, F. L. (2007) 'Global Technology and Knowledge Management: Product Development in the Brazilian Car Industry', *International Journal of Automotive Technology and Management*, 7(2/3): 135–52.

Consoni, F. L. and Quadros, R. (2006) 'From Adaptation to Complete Vehicle Design: A Case Study of Product Development Capabilities on a Carmaker in Brazil', *International Journal of Technology Management*, 36(1/2/3): 91–107.

Dahlman, C. J. and Frischtak, C. R. (1993) 'National Systems Supporting Technical Advance in Industry: The Brazilian Experience', in R. R. Nelson (ed.), *National Innovation Systems – A Comparative Analysis*. Oxford, UK: Oxford University Press.

David, P. A. (1985) 'Clio and the Economics of QWERTY', *The American Economic Review*, 75(2): 332–7.

Dias, A. V. C. and Salerno, M. S. (2004) 'International Division of Labour in Product Development Activities: Towards A Selective Decentralisation?', *International Journal of Automotive Technology and Management*, 4(2/3): 223–39.

Flores, N. (2010) 'Política governamental de energia substitutiva – cana de açúcar. Do pró-álcool ao etanol' MSc. dissertation (Economic history), Faculdade de Filosofia, Letras e Ciências Humanas, Universidade de São Paulo, Brazil.

Garnero, M. (1980) *Energia: o futuro é hoje*. São Paulo, Brazil: ANFAVEA.

Geels, F. W., Elzen, B. and Green, K. (2004) 'General Introduction: System Innovation and Transitions to Sustainability', in B. Elzen, F. W. Geels and K. Green (eds), *System Innovation and the Transition to Sustainability*. Northampton, MA: Edward Elgar.

Geisel, E. (1976) 'Discursos', *Assessoria de Imprensa da Presidência da República*, 2: 191–219.

Kremer, F. G. and Fachetti, A. (2000) *Alcohol as Automotive Fuel – Brazilian Experience*. SAE International (Society of Automotive Engineers) Technical paper No. 2000-01-1965.

Lacerda, A. C., Bocchi, J. I., Rego, J. M., Borges, M. A. and Marques, R. M. (2000) *Economia brasileira*. São Paulo: Saraiva.

Mello, A. M., Vasconcellos, L. H. R. and Marx, R. (2005) 'Estariam as montadoras abrindo mão de suas competências essenciais no desenvolvimento de motores? Um estudo de caso do primeiro veículo nacional bicombustível', Proceedings of the XXIX ENANPAD (on CD-ROM). Brasília: ANPAD.

MMA (2011) *1º Inventário nacional de emissões atmosféricas por veículos automotores rodoviários*. Ministério do Meio Ambiente (Ministry of the Environment), Brazil.

North, D. C. (1990) *Institutions, Institutional Change and Economic Performance*. Cambridge, UK: Cambridge University Press.

OECD (Organisation for Economic Co-operation and Development) (2011) 'Biofuels', OECD–FAO Agricultural Outlook 2011–2020. Available at: http://www.agri-outlook.org/dataoecd/23/56/48178823.pdf; accessed 12 August 2011).

Olson, M. (1965) *Logic of Collective Action: Public Goods and the Theory of Groups*. Cambridge, MA: Harvard University Press, p. 201.

Porter, M. E. (1990) *The Competitive Advantage of Nations*. New York: The Free Press.

Puerto Rico, J. A. (2007) 'Programa de biocombustíveis no Brasil e na Colômbia: uma análise da implantação, resultados e perspectivas'. MSc. dissertation. São Paulo, Brazil: Universidade de São Paulo, Programa Interunidades de Pós-Graduação em Energia.

Quadros, R. and Consoni, F. L. (2009) 'Innovation Capabilities in the Brazilian Automobile Industry: A Study of Vehicles Assemblers' Technological Strategies and Policy Recommendations', *International Journal of Learning, Innovation and Development*, 2(1/2): 53–75.

Queiroz, S. and Carvalho, R. Q. (2005) 'Empresas multinacionais e inovação tecnológica no Brasil', *São Paulo em Perspectiva*, 19(2): 51–9.

Souto Maior, A. (1967) *História do Brasil*. São Paulo, Brazil: Cia. Editora Nacional.
Teixeira, E. C. (2005) *O desenvolvimento da tecnologia Flex-fuel no Brasil*. São Paulo, Brazil: Instituto DNA. Available at http://146.164.33.61/silviocarlos/PF%2008/Rodrigo%20Faria%20PF%2007/Cap2/FLEXFUEL.pdf; accessed 28 January 2011.
UNICA (União da Indústria da Cana-de-Açúcar – Sugar Cane and Ethanol Industry Association) (2011). 'Etanol e Bioeletricidade: A cana-de-açúcar no futuro da matriz energética'. Available at: http://www.unica.com.br/downloads/estudosmatrizenergetica/pdf/livro-etanol-bioeletricidade.pdf; accessed 18 February 2011.
Vicentino, C. and Dorigo, G. (1998) *História do Brasil*. São Paulo: Editora Scipione.
Williamson, O. E. (1993) 'Transaction Cost Economics and Organisation Theory', *Industrial Corporate Change*, 2(1): 107–56.
Yu, A. S. O., Nascimento, P. T. S., Nigro, F. E. B., Frederick, B. W. B., Varandas Junior, A., Vieira, S. F. A. and Rocha, R. L. (2009) 'The Evolution of Flex-Fuel Technology in Brazil: The Bosch Case', in Proceedings of PICMET '10, Portland International Center for Management of Engineering and Technology (PICMET), Portland State University, Portland, Oregon, USA, pp. 1–11.

10
New Forms of Vehicle Maker–Supplier Interdependence? The Case of Electric Motor Development for Heavy Hybrid Vehicles

Dedy Sushandoyo, Thomas Magnusson and Christian Berggren

Hybrid powertrains have become a key technology for environmentally friendly passenger cars. Almost all of the major carmakers have incorporated this technology in their products, and some of them have enjoyed significant sales in recent years. Several heavy vehicle makers (such as Scania, Volvo Trucks, Daimler) and MAN, have also developed hybrid electric buses and trucks; however, at the time of writing, the costs of these vehicles remain very high, therefore launching commercially viable heavy hybrid electric vehicles is still a very challenging task.

The difficulties of developing cost-effective energy storage present a well-known hurdle; but there are several other engineering obstacles – for example, those related to the development of electric motors. Electric motors are already in use in hybrid passenger vehicles; however, the demands for performance, reliability and robustness in heavy hybrid vehicles are an order of magnitude higher. In addition to these stringent demands, heavy hybrid vehicles require much more powerful electric motors, and such motors currently available are typically designed for stationary power generation, not for vehicle applications.

Electrical motors are vital components in hybrid powertrains, which strongly influence their performance and therefore need to be integrated within the whole vehicle development (Eriksson, 2007). Such a demand could be an argument for producing electric motors in-house, as Toyota is currently doing. This strategy requires huge investments and very substantial internal economies of scale, far from the volumes produced by heavy vehicle manufacturers. Relative to the passenger car market, the market for heavy hybrid electric vehicles is expected to be quite small, therefore in-house production will suffer from diseconomies of scale. For this reason, it seems that developing appropriate partnerships with electric motor suppliers is a positive factor for manufacturers in the development of heavy hybrid vehicles.

An important challenge here is that firms in the electro-technical industry have never engaged in the commercial production of electric motors for heavy vehicle applications. There are significant differences between firms in the traditional electro-technical industry and those in the vehicle industry; for example, established firms in the electro-technical industry are used to specify the products themselves and then to sell them to their widely diversified customers. In the vehicle industry, however, suppliers typically have to adapt and develop their products to meet the ambitious targets set by the vehicle manufacturers and deliver the products in high volumes, with stringent demands being made regarding delivery, quality and continuous cost reduction. As a result, electro-technical firms perceive collaboration with vehicle manufacturers to be very demanding, which makes them reluctant to involve themselves in this new business (Interview with the manager of parallel hybrid development, Scania, 2008/2011; Interview with Sadarangani, 2010).

The aim of this chapter is to outline the key issues for heavy vehicle manufacturers in their collaboration with suppliers of electric motors. The information was collected via an in-depth study of the hybrid product development conducted by two heavy-vehicle manufacturers with their respective electric motor suppliers: Scania–Voith and Volvo Trucks–Kollmorgen. The two vehicle manufacturers used different types of electric motor for their hybrid powertrains. They also chose different types of architecture for their hybrid vehicles. By analysing the similarities and differences in these two collaborations, the chapter will demonstrate some of the key challenges involved.

The next section presents the hybrid development processes at Volvo and Scania, and the following two sections deal with the technological dynamics of electric drive systems and the development of the electric motors used in Volvo Trucks' and Scania's heavy hybrid electric vehicles. A comparative analysis of hybrid development and supplier collaboration conducted by Volvo Trucks and Scania follows. The final section provides conclusions on how new product development (NPD) collaborations are organized and a discussion on lessons learnt from the study.

The hybrid vehicle development processes at Volvo and Scania

Volvo Trucks' engagement with hybrid vehicle development may be traced back to the mid-1980s, when the company began a theoretical study of hybrids, followed by the development of demonstrators and tests of various technologies, including mechanical hybrids, hydraulic hybrids, and series and parallel electric hybrids. In 2002, the company began an advanced engineering project to investigate which type of hybrid solution would offer the best combination of fuel efficiency, robustness and cost-effectiveness. 'We ruled out quite a few configurations of hybrids because they were too expensive, and ended up with what is known as a parallel hybrid system, in which you use one electric motor and an automatically-geared mechanical

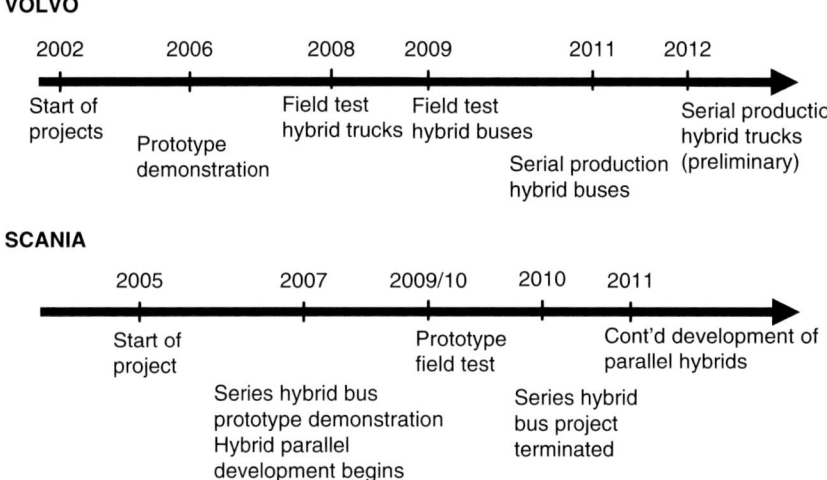

Figure 10.1 Timelines of Volvo's and Scania's hybrid vehicle developments

gearbox,' said Anders Kroon, Director of Hybrid Technology at Volvo Powertrain, the Volvo Group's in-house producer of powertrain systems (Volvo, 2009). In 2006, Volvo demonstrated parallel hybrid prototypes, and in the same year, the CEO took the decision to develop these further. In 2008, Volvo began field tests of their hybrid vehicles, followed by the serial production of hybrid buses in 2011, and the company is planning to produce hybrid trucks in 2012 (see Figure 10.1).

Since the 1980s, Scania has been engaged in engineering studies related to two types of hybrid technology: series hybrid and parallel hybrid (Scania, 2008). In 2005, Scania's top management took the decision to develop a series hybrid bus concept, and in 2007, the prototype was presented to the public. In late 2007, a new unit dealing with the development of parallel hybrid trucks was created within the company. In 2009–10, Scania carried out field tests on six series hybrid buses in Stockholm, but eventually decided to terminate the project. In its place, Scania decided to focus on the development of parallel hybrid configurations for both buses and trucks. This development has been given no public launch date, hence the project is beyond the scope of this chapter. An overview of the developments at both companies is shown in Figure 10.1.

Technological dynamics of electrical drive systems

An electrical drive system consists mainly of electric motors, power electronics and a control system. An electric motor is a complex product, the development of which requires both scientific and practical know-how. The former includes areas such as electromagnetic theory, electrical engineering, materials technology, mechanical engineering, solid mechanics, fluid mechanics, thermodynamics, control theory and power electronics. The latter consists of areas such as computer-aided engineering, manufacturing technology and test methods (Eriksson, 2007).

The development activities related to electric motors and power electronics for hybrid and electric vehicle applications have grown significantly since the mid-1990s. Figure 10.2 shows the growth in inventive activities related to power electronics, electric motors/generators and control systems for electric vehicle and hybrid electric vehicle applications, drawn from granted patents as listed in the US Patent and Trademark Office (USPTO) database. Inventions in power electronics include converters, inverters and rectifiers. The patents on electric motors/generators in the figure are mainly related to the structures of the electric motors (for example, winding, structure, permanent magnets, rotors and stators). It seems that the increasing number of patents around 1995/6 are related to preparations to introduce hybrid passenger cars to the market. Numbers of inventions related to control systems seem rather low. This may be explained partially by legal problems related to the patenting of software (EPO, 2005). However, there is a growing interest among vehicle manufacturers in patenting control systems (for example, algorithms) related to hybrid vehicle applications (Control engineer, Scania, 2010).

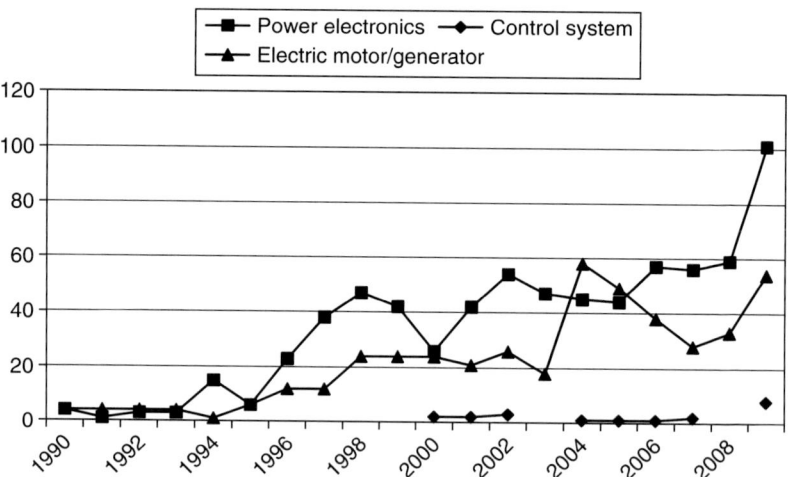

Figure 10.2 Patents granted related to power electronics, electric motors/generators and control systems for hybrid and electric vehicle applications
Source: USPTO, 1990–2009.

The development of the electric motors for hybrid vehicles at Volvo and Scania

Hybrid technology and electric motor technology

Based on their component configuration, hybrid powertrains can be classified into *series* and *parallel hybrid* systems. In a series hybrid system, an engine is coupled to an electric generator to produce the power, which drives an electric propulsion motor. Hence the power capacity has to be sufficient to provide the maximum energy required by the vehicle, which means that powerful electrical equipment is required. Such equipment may also provide optimal capacity for recuperating braking energy (Guzzella and Sciarretta, 2007). This type of hybrid system is also used in extended-range electric cars, such as the GM Volt.

A parallel hybrid vehicle can be propelled either by an engine and an electric motor simultaneously, or by only one of them. This means that the maximum power required may be shared between the engine and the electric motor. As a result, a combination of a downsized engine and a relatively small electric motor is possible. In a hybrid parallel system, the power produced by the engine and the motor is transmitted via mechanical couplings (including clutch and gears) to propel the vehicle (Guzzella and Sciarretta, 2007). The VW Touareg Hybrid is an example of a vehicle with a parallel hybrid system.

While a parallel hybrid vehicle typically consists of only one electric motor, a series hybrid vehicle requires a generator and an electric motor (Guzzella and Sciarretta, 2007). Moreover, a series hybrid powertrain typically requires a more powerful electric motor than does a parallel hybrid. Hence the electrical components are more dominant in a series hybrid powertrain.

Electric motors can be categorized as direct current (DC) or alternating current (AC) machines. As a result of their higher power density and efficiency, the majority of hybrid vehicles use AC rather than DC motors (Guzzella and Sciarretta, 2007). There are typically two types of AC electric motors: asynchronous and synchronous. The asynchronous motor used to be preferred in demanding applications because of its robustness, simple controls and low cost. However, the trend is moving away from asynchronous to synchronous machines equipped with permanent magnets (Neudorfer and Franz, 2010). These permanent-magnet motors provide high efficiency, superior power density and torque-to-inertia ratio, making them a frequent choice for hybrid vehicles (Wallmark, 2006). The Toyota Prius, for example, uses such an electric motor (Eriksson, 2007). Since these motors also offer high efficiency at lower speed ranges, this type is also suitable for hybrid city bus applications (Neudorfer and Franz, 2010).

Electric motors may also be classified by their operating principle as *radial*, *axial* or *transverse* flux machines. Of these, radial is the most commonly used type (Eriksson, 2007). Both Volvo and Scania equip their hybrid vehicles with permanent-magnet synchronous motors, but with different technologies: radial and transverse flux motors, respectively.

Volvo–Kollmorgen electric motor development

Sourcing and knowledge development

Volvo's hybrid powertrain is a parallel hybrid configuration (see Figure 10.3), with a radial flux 3-phase permanent magnet synchronous electric motor. Volvo developed and produces the engine, the clutch, the transmission and the hybrid power control unit (HPCU) in-house. The HPCU is the brain of the system and 'regulates engagement and disengagement of electric and diesel power as needed, as well as gear changing modes and battery recharging' (Volvo website, 2010). The electric motor, energy storage system and power electronics are obtained from suppliers. The electric motor was developed in conjunction with, and manufactured by, the supplier. The power electronics and software for the electric motor drive system have been developed by the supplier alone.

In 2005, Volvo presented its hybrid vehicle prototypes, indicating that the company had developed a significant architectural knowledge base. In addition, Volvo had developed new component knowledge by encouraging selected employees to pursue doctoral studies related to key components (for example, energy storage). To speed up the development, Volvo recruited a university professor with a special competence in hybrid systems and electric motors. Furthermore, since the standards/specifications related to the high voltage system had not yet been established in the vehicle industry,

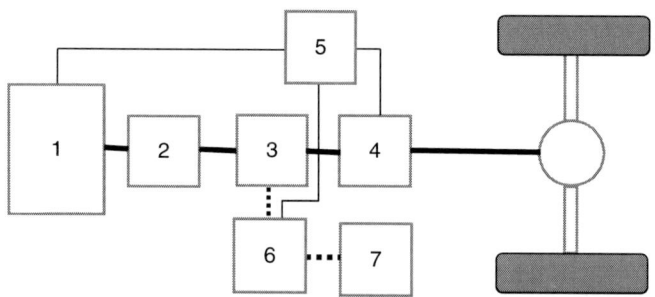

(1) Engine; (2) Clutch; (3) Electric motor/generator; (4) Transmission;
(5) Electronic control unit; (6) Energy converter AC/DC; (7) Energy storage

Key note: Bold lines: mechanical link; dash lines: electrical link; solid lines: control link

Figure 10.3 Volvo's parallel hybrid configuration
Source: Adapted from Volvo (2007).

Volvo also developed competences in defining these technical specifications. The company gained patents for a hybrid drive system in 1997 and for an electric motor in 2005. In 2008 and 2009, Volvo submitted patent applications to the USPTO related to an energy converter and a hybrid drive system, respectively.

Danaher Motion, through its subsidiary Kollmorgen, was chosen to supply the motors, including their control units (that is, the electric motor's control system) for Volvo's hybrid vehicles. The Group's business is organized under five brands: Kollmorgen, Thomson, MEI, Dover and Portescap. Kollmorgen's expertise is in magnetic, mechanical and control technologies (Kollmorgen, 2010). For hybrid heavy vehicle applications, Kollmorgen offers electric motors, inverters and software. Danaher Motion has six relevant patents, mainly in power electronics and control systems, obtained between 2006 and 2009.

Volvo–Kollmorgen collaboration

Volvo's business package team (BPT) is the interface between Volvo and its potential suppliers. The team is also responsible for co-ordinating purchasing, R&D and manufacturing within Volvo from the early phase of the NPD process. Related to Volvo's hybrid vehicle development, the main tasks of the team are to choose technologies, select suppliers and evaluate components with regard to quality, delivery and cost. Some members of the team are assigned responsibility for key components. For example, six team members are responsible for the energy storage system and four are responsible for the motor drive system (Interview with the BPT manager, 2009). After the general technical specifications are issued, Volvo begins the search for an electric motor supplier. The selected supplier then works closely with the team to define detailed technical specifications, which are then communicated to the line organization for approval (that is, the sub-system manager for transmission and hybrids). The team members responsible for purchasing then evaluate the approved specifications.

In Volvo's hybrid powertrain, the diesel engine is mounted behind the rear wheels. Since the engine is coupled directly to the electric motor, the packaging of the drive train in such a limited space is technically demanding. During the development of the electric motors, one challenge was to keep the driveline compact and sufficiently light. The verification and validation of procured components are typically the suppliers' responsibility. In this collaboration, Volvo took on some of the tasks, such as the component vibration test, which is normally carried out by the supplier. In Volvo's hybrid powertrain, the electric motor is placed between the engine and the transmission, and the shaft of the electric motor is connected tightly to the shafts of Volvo's components. To carry out a vibration test, information on the characteristics of the engine and transmission are needed. Since the diesel engine and transmission system are produced in-house by Volvo (2007), and the supplier had no access to these, the supplier is unable to perform this vibration test. Further, this electric motor supplier was not familiar with components made from aluminium. Hence Volvo, with its experience in aluminium housing for the transmission and clutch took on the responsibility of also handling the load for the electric motor casing.

During the collaboration, Volvo assigned some of its employees to be responsible for the collaboration, and set clear delivery dates and time schedules. Face-to-face meetings were arranged when necessary. For example, in order to make sure that the components fitted Volvo's system, the company sent an engineer to work with the supplier's engineers for a couple of weeks. On this occasion, Volvo's engineer was accompanied by staff from the company's production and aftermarket departments.

Scania–Voith electric motor development

Sourcing and knowledge development

The configuration of Scania's series hybrid powertrain is shown in Figure 10.4. Scania obtained the electric motor and generator from Voith, whose transverse flux motor was chosen for its technical performance: delivering high power and torque densities as well as high efficiency (Overgaard and Folkesson, 2007). Voith also delivered a drive system that controlled the generator, electric motor and energy storage. The supercapacitor cells of the energy storage were obtained from Maxwell, while the engine and the overall hybrid management system were developed and produced in-house by Scania.

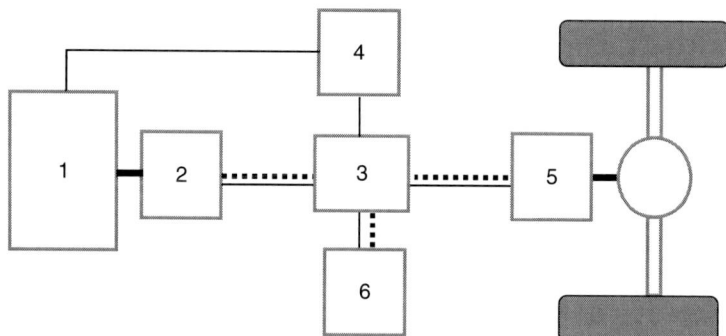

(1) Engine; (2) Generator; (3) Power electronics and 650 V grid;
(4) Hybrid management system; (5) Motor; (6) Energy storage

Key note: Bold lines: mechanical link; dashed lines: electrical link; solid lines: control link

Figure 10.4 Scania's series hybrid configuration
Source: Adapted from Scania (2009) and information from interviews.

To develop the necessary architectural and component knowledge bases, Scania encouraged selected employees to pursue doctoral studies. During the development of the prototype, one employee was doing doctoral research on hybrid energy systems. Two other employees were pursuing doctoral studies on the impact of introducing high-voltage systems into Scania's products, and on diesel control optimization. The company is very strong in diesel engine technology; however, it has no significant knowledge base in electric motors. In 2001, the USPTO granted a patent to Scania for a (parallel) hybrid electric drive. Scania also submitted a patent application to the European Patent Office (EPO) in 2010 for a heating system specifically designed for a hybrid vehicle application. Scania has no granted patents or patent applications in electric motors or generators.

The Voith Group comprises four divisions: Voith Paper, Voith Turbo, Voith Hydro and Voith Industrial Services. The Voith Turbo division traditionally produces automatic transmissions, retarders, turbochargers and vibration dampers for road vehicle applications. Voith Turbo has recently developed two hybrid solutions for bus applications: DIWAhybrid and ElvoDrive. The DIWAhybrid is based on an asynchronous motor using relatively mature technology (that is, radial flux technology) and is intended for a parallel hybrid bus. The ElvoDrive is based on Voith's transverse flux machine (TFM) technology and proposed for a series hybrid bus. The TFM technology is relatively new (Lange, 2010). In terms of electric motors and electric drive systems for hybrid vehicle applications, Voith has been granted 29 patents from the USPTO and the EPO. These patents can be classified in the areas of electric motors/generators (for example, cooling, rotor, stator, armature and electric motor drive system), hybrid vehicle systems and control methods.

Scania–Voith collaboration

The first informal contact between Scania and Voith occurred in 2004. Both shared similar visions for a hybrid bus, such as: (i) a distrust of the suggested electric drives for hybrid buses at the time, such as wheel-hub motors; (ii) both of them thought that a hybrid powertrain should be simple and offer optimum fuel economy; and (iii) because of their robustness and high power density, they preferred supercapacitors over batteries for energy storage.

Both firms also found that the competencies of the firms complemented each other. Voith had very strong technical competencies in making electric motors but no experience in implementing them in hybrid applications. Further, Voith's TFM electric motor was suitable for Scania's series hybrid concept for city bus application. To begin the collaboration, they decided to develop a full-scale prototype. Here, the collaboration was not a conventional supplier–vehicle manufacturer relationship, but rather each party contributed its own components and expertise.

When the project was initiated, both Scania and Voith created small, dedicated teams. Both teams worked beyond their firms' core business units, making the members of these teams feel, as described by a Voith engineer, like a small family. As a result, according to this Voith engineer, when technical problems arose, they had a willingness to help, rather than to blame each other. There were no specific task divisions within each team, so the engineers and the project leader often had to do jobs beyond the remit of their conventional tasks. Furthermore, there was no detailed time schedule. Sometime Scania's and Voith's engineers met frequently over a period of several weeks, particularly when a problem arose, but then perhaps they would not meet again for a couple of months. They also communicated by telephone, sending emails and exchanging data.

Faced with great technological challenges, as described by the Voith engineer, the engineers from both sides worked closely and promoted open communication. Such communication led to a high level of trust among them. As a result, Voith was willing to report in detail about a problem and let Scania know whether or not their company had sufficient know-how to overcome the difficulty; Scania was similarly open with Voith.

A comparative analysis of the development of electric motor applications

Knowledge bases

Both of the vehicle manufacturers studied were active in developing their architectural knowledge bases related to hybrid powertrains. However, they differed in the development of component knowledge. Volvo developed significant knowledge about electric motors, whereas Scania had more limited activities related to knowledge development in this regard. These differences may be explained by the different degrees of technology novelty among the components involved. The TFM technology was invented a few years before the Scania–Voith collaboration began, and the supplier protected it through patents, whereas Volvo selected a more mature technology.

As can be seen from Table 10.1, the suppliers have adopted different technological strategies. Kollmorgen has no patents for electric motors or hybrid drive systems, perhaps because the technology and manufacturing related to these types of machines are relatively mature. In contrast, Voith is very active in developing both its architectural and component knowledge bases (see Figure 10.5). This may be because this supplier's components are dominant in a series hybrid system. Further, Voith also delivers both electric motors and hybrid management systems (that is, energy management) for its US customer (Lange, 2010). Developing an architectural knowledge base is thus an advantage.

Technology and application novelty

Synchronous permanent magnet motors produced on a large scale are generally based on radial flux technology (Interview with Sadarangani, 2010; Eriksson, 2007). This means that technology novelty is relatively moderate

Table 10.1 Component and architectural knowledge development

Firms	Component knowledge	Architectural knowledge
Scania	No	Yes, together with the supplier
Voith	Yes, before the collaboration started	Yes, before the collaboration started
Volvo	Yes, before the collaboration started	Yes, before the collaboration started
Kollmorgen	Yes	No

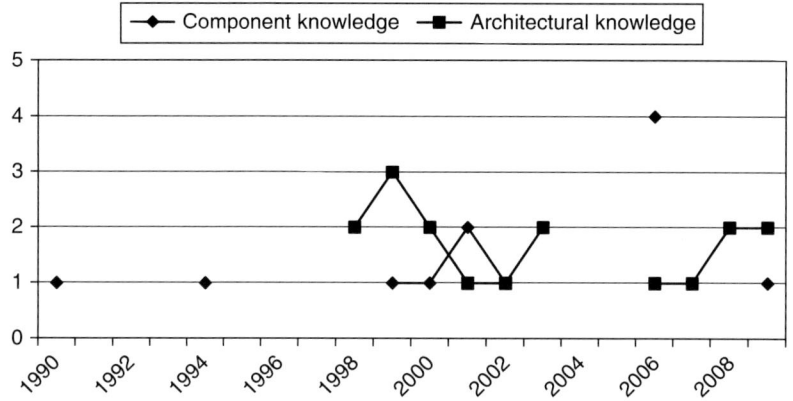

Figure 10.5 Voith's component and architectural knowledge related to electric motors and hybrid drive systems
Sources: Data from USPTO and EPO, 1990–2009.

for Kollmorgen, which had already produced radial flux machines, though not for hybrid vehicle application. At Voith, the technology novelty of its TFM was significantly higher, since it was invented just a few years before the Scania–Voith collaboration began and had never been applied to hybrid vehicles (see Figure 10.6).

When the collaboration with the supplier was initiated, Volvo had already developed vehicle prototypes, implying that Volvo had established an architectural knowledge base. The technology novelty of the radial flux electric motor for Volvo may also be considered as relatively moderate. One of the indications for this is that Volvo had a patent for electric motors before the collaboration was initiated (see Figure 10.7).

In the Scania–Voith collaboration, the prototype was developed by Scania in conjunction with the supplier. This means that the novelty of the series architecture may be categorized as high for Scania. Similarly, the technology novelty of the TFM was high for Scania, as the company had no prior knowledge concerning this type of motor (see Figure 10.7).

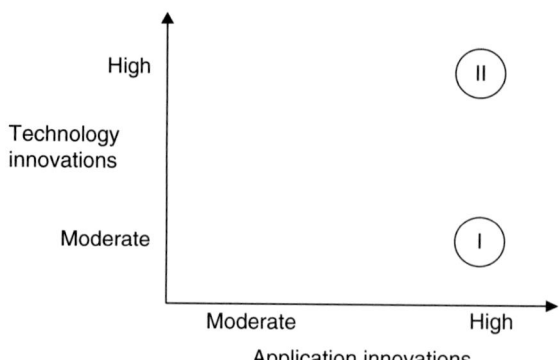

Key: I : The radial flux motor used in Volvo's hybrid powertrain
II : The transverse flux motor used in Scania's hybrid powertrain

Figure 10.6 Technology and application innovations of electric motors relative to suppliers

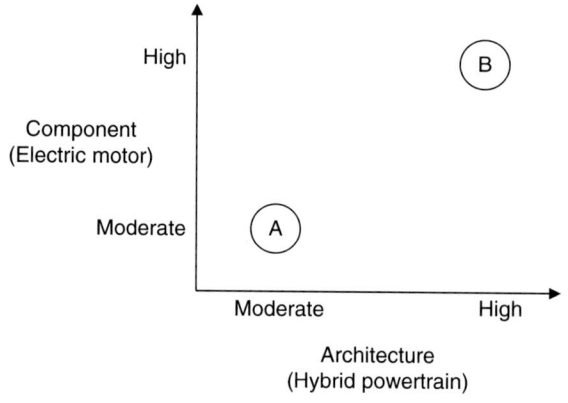

Key: A: The technology and application novelty of the electric motor and hybrid powertrain in the Volvo–Kollmorgen collaboration
B: The technology and application novelty of the electric motor and hybrid powertrain in the Scania–Voith collaboration

Figure 10.7 Technology innovations relative to vehicle manufacturers

Organizing effective collaborations

Volvo began looking for a component supplier after the concept and the general technical specifications had been defined. Thus a strategy of late supplier involvement was adopted, where the supplier had little influence on the overall development effort.

The decision to involve the supplier late in the product's development was justified by the relatively low technology novelty of the component, and the supplier's limited technological knowledge base. In the Scania–Voith case, the collaboration was initiated by developing the prototype jointly at a very early stage. It is obvious from the supplier's patents that Voith has very strong technological capabilities in both TFM technology and hybrid architecture. In contrast, Scania had no component knowledge related to the TFM. This means that component development was the sole responsibility of the supplier. Furthermore, the supplier made a significant contribution in defining the global design of the hybrid drive system.

In the Volvo–Kollmorgen collaboration, the technical specifications were settled early, before the project was initiated. Under these conditions, Volvo could select a supplier under a conventional procurement procedure. After the supplier was selected, both parties defined detailed product specifications and cost-effective solutions. When necessary, ad hoc meetings were arranged. For example, when a problem arose related to the aluminium casing of the electrical motor, Volvo sent its engineer to work with the supplier's engineers. Further, Volvo set a clear delivery date and time schedule for the project. This implies that formal procedural mechanisms were applied.

In the Scania–Voith collaboration, both parties created dedicated teams beyond the firms' core business. Faced with a high level of technology novelty, they decided to build a joint prototype in the early stages of the advanced engineering phase. This is not usual with Scania's standard NPD projects, since developing such a prototype is expensive. The series hybrid also involved a high level of application novelty, so it was decided to conduct field tests with a key Scania customer (SL, Stockholm's public transport provider) under normal traffic conditions. During the critical stages of the development project, the supplier's engineers were also co-located at Scania's site. Similarly, when necessary, Scania's engineers were sent to the supplier's site. Since both parties were focused on solving any technical problems, this collaboration was not organized under a conventional supplier–assembler relationship, but rather was directed towards learning new technologies.

Conclusion: from broad search to focused development

The Volvo Trucks–Kollmorgen and Scania–Voith cases illustrate three essential issues for vehicle manufacturers when collaborating with new suppliers of critical technologies, such as electric motors for the hybrid development process. These three issues are:

- find suppliers with complementary technological knowledge;
- develop an internal technological knowledge base to create sufficient overlap with suppliers; and
- adapt product development collaboration to the knowledge contributions of the organizations involved.

Finding suppliers with complementary knowledge is essential, but not sufficient. Complementary knowledge bases need to be accompanied by a knowledge overlap in critical areas, and collaborative arrangements fitted to the division of knowledge. In the Scania–Voith collaboration, the supplier had strong technological knowledge at both the component and architectural levels. Therefore, the supplier could participate actively and carry a great level of responsibility in the development project. Further, the parties organized the product development in a rather informal way, with open communication and continuous discussion. At times, project integration co-ordination through physical co-location was also necessary. Such co-ordination, combined with the joint development of the hybrid bus prototype, facilitated learning for both parties involved. However, from the perspective of Scania, the supplier retained a major part of the new knowledge, and would thus appropriate a large share of the new value added, if the development project entered full-scale production.

In the Volvo–Kollmorgen collaboration, Volvo had already developed hybrid vehicle prototypes before the collaboration began. Moderate levels of novelty in the component technology implied stable interfaces with other components, and therefore the choice of a supplier with no architectural knowledge was a feasible solution. In addition to architectural knowledge, Volvo had also some component knowledge related to electric motors. This overall position also meant that Volvo would control the greater part of the value added.

In a situation where the technology and application novelty of a new component are moderate, vehicle makers face fewer difficulties in building their own knowledge base. The study illustrates how such component knowledge helped Volvo to organize effective product development collaboration, involving the supplier only at a late stage in the development process. Interactions between the Volvo and Kollmorgen were based on relatively formal procedures, including meetings on an infrequent basis. On certain occasions, however, both parties needed to work face-to-face to make sure that the component could be integrated properly into the system, or to solve occasional problems.

Scania and Volvo exhibit different approaches, both in their development efforts and their supplier relations. The following discussion summarizes the related differences in technological choices and their outcomes.

The case of Scania: the problems of developing two technologies in parallel

Products embodying radically new technologies offer high potential benefits. However, involving such technologies in product development is also associated with a high level of uncertainty, regarding, for example, actual efficiency (fuel saving) and reliability during real driving conditions. In the Scania case, several technical problems related to the electric motor emerged during the development process, which is not unusual during the application of a new technology. However, the field test also demonstrated that the optimal range for a series hybrid application was more limited than previous simulations had indicated. The level of fuel saving of the series hybrid bus was apparently very much a function of the topography of the actual routes taken, and optimal fuel saving could only be obtained in limited topographies (for example, frequent stop-and-go). The bus business at Scania is only a tenth of the truck business, and city buses are only a small part of this bus volume. Further, in cities with deregulated public transportation, where the tender system offers no (or small) incentives for the introduction of new technology, operators typically are reluctant to adopt a novel concept such as a hybrid bus. These conditions implied that the potential market for hybrid city buses was becoming highly segmented, and that a dedicated series hybrid bus would be a narrow niche product. This was particularly serious because of the absence of synergies between different vehicle applications (trucks and buses).

In the later phases of the bus project, Scania also started to develop a parallel hybrid system, and thus was investing in two types of hybrid powertrains simultaneously. The series hybrid configuration would be optimal for city buses in urban stop-and-go traffic, whereas a parallel architecture was intended for later applications, mainly in the truck programme, but it could also be used in urban buses. Moreover, with a parallel hybrid configuration, Scania would maintain a higher share of value added in-house. The dual efforts increased the total development costs and created potential rivalries between the two technologies, such as competition for resources and for gaining support and acceptance from the organization. All these factors contributed to the decision made at Scania during the deep financial crisis in 2008–9 to terminate the series hybrid project and to concentrate only on parallel hybrid development.

The case of Volvo: the advantage of focus

In contrast, Volvo had opted much earlier for a less radical parallel hybrid system for both bus and truck applications. As part of this strategy, Volvo chose electric motors characterized by moderate levels of novelty in the component technology. While this solution offered fewer potential benefits (for example, less capacity in capturing braking energy), the development of such a product typically also entails fewer technical problems. Volvo's parallel hybrid powertrain could also enjoy synergies with advances in the company's conventional components, such as its new computer-controlled gearbox (the I-shift). This made the new product likely to gain support/acceptance from the organization.

Initially Volvo had conducted advanced engineering activities and developed prototypes based on both series and parallel hybrid technologies, in a similar way to Scania. In 2005, however, Volvo decided to focus only on parallel hybrid technology. This focus on one technology helped the company to maintain momentum in the development project. Volvo Trucks was also hit by the global economic crisis in 2008–9, which forced the company to reschedule the launch dates for its hybrid buses. In 2011, however, Volvo was able to start serial production and introduce the parallel hybrid buses to selected markets. Further, Volvo plans to begin serial production of parallel hybrid trucks in 2012.

Technology as a search process with spillover effects

The comparison of hybrid development at Volvo and Scania illustrates how, in a situation of high uncertainty in both technologies and markets, firms typically do not know at the start which technologies will eventually be viable. In such a period, various search processes as well as entrepreneurial activities tend to encourage firms to engage simultaneously in several technological directions. In this case, both Scania and Volvo conducted advanced engineering studies and built hybrid prototypes related to both series and parallel hybrid technology, but the process of technology evaluation led the two vehicle manufacturers to focus their development of components and hybrid architectures on specific configurations (that is, parallel hybrid powertrains and radial flux electric motors). The Scania experience shows how difficult it is to develop different types of complex technological platforms concurrently (that is, both series and parallel hybrid vehicles), to invest in deep collaboration with one technology supplier, to involve customers in field tests, and then to converge eventually on only one technology option, without delaying their planned market launch.

The study also illustrates unintended spillover effects. In the Scania–Voith case, for example, while the hybrid project was terminated, the mutual understanding obtained during the collaboration served as a strong basis on which to continue conducting collaboration in a new business area. The success or failure of a product development project needs to be evaluated, not only in relation to the project itself but also in its wider aspects.

Appendix 1: Data collection

Information presented in this chapter was based on three types of data collection: patents, interviews and documents.

Patents

In this chapter, a patent study provided information on the dynamics of inventive activities, and technological trends related to electric motors, power electronics and control systems for hybrid and electric vehicle applications were gathered from the US Patent and Trademark Office (USPTO). The data were then plotted in respect to the years when the patents were granted (see Figure 10.2). To assess the studied firms' component and architectural knowledge bases, both granted patents and patent applications gathered from the USPTO and the European Patent Office (EPO) databases were used (see Figure 10.5).

Interviews

Twenty semi-structured interviews were conducted to gather information on the products and chosen technologies, the firms' strategies related to knowledge development, and the organization of inter-firm relationships. The interviewees were selected in relation to their involvement in the development projects. From Scania, we interviewed such people as the president of R&D, the head of the bus chassis development division, the head of the vehicle definition division, the manager of parallel hybrid systems, the project leader and a number of key engineers (that is, specialists in hybrid/energy systems, energy storage and control systems). An interview with Voith's project leader was also undertaken (see Appendix 2).

In the case of Volvo, we interviewed the project leader, an energy storage engineer, and the group leader of Volvo's business package team (BPT), as well as a professor of electric motors employed as a key expert on the project (see Appendix 2). The BPT serves as an interface between Volvo and their suppliers, and is responsible for promoting commonality among Volvo's products. The team consists of individuals representing different sections within Volvo, such as engineering, quality control, purchasing, manufacturing and the aftermarket.

To complement and triangulate the data from these interviews, we also conducted five interviews with university professors. These professors were experts in technological fields related to the development of hybrid vehicles, including electric motors, vehicular systems, control systems and energy storage (see Appendix 2).

Documents

Information from the interviews was complemented by information gathered from commercial websites, firms' official websites and other written documents (for example, doctoral theses).

Appendix 2: Interviews

Table A10.1 Interviews at Scania

Date	Position	Interview topics
18 June 2009	Vice President, R&D Department	An overview of the development of the hybrid series bus and the parallel hybrid truck
22 August 2008	Director, Vehicle Definition Division	An overview of the development of the parallel hybrid truck The position of the parallel hybrid truck project within the structure of Scania's line organization
5 September 2008	Head of Division/ Engineering Director, Bus Chassis Development	The strategic importance of the hybrid series bus project for the company Technological and industrial challenges related to the development of the hybrid series

(continued)

Table A10.1 Continued

Date	Position	Interview topics
20 October 2009	Project leader	The history of the development process of the Scania hybrid series concept bus
26 September 2008 23 August 2010	Project leader, Hybrid Series Project	Managing the hybrid series bus project Evaluation of the hybrid series project
26 September 2008	Energy system engineer	The technological challenges and new knowledge bases required to develop the hybrid series bus Managing the project, especially in the early stages
12 November 2008 15 September 2011	Manager, Parallel Hybrid Systems Group	An overview of the technological challenges related to the development of the parallel hybrid truck Organizational adaptations related to the development of the parallel hybrid truck Progress in Scania's parallel hybrid vehicle development
26 November 2008	Manager, Hybrid Components Group	The modularization strategies instituted by Scania related to the development of the parallel hybrid truck
5 May 2009	Energy storage engineer	Technological and industrial challenges related to energy storage
30 April 2010	Control engineer	The development of the hybrid management control system

Table A10.2 Interviews at Volvo

Date	Position	Interview topics
16 April 2007 11 February 2008	Project leader Volvo's main consultant, a professor in electrical engineering	Volvo's hybrid project Technical and industrial challenges related to the development of Volvo's hybrid electric bus
20 April 2009	The team leader for the Business Package Team Hybrid, the engineering leader of the team	Technical and managerial challenges related to the development of Volvo's hybrid parallel bus Sourcing strategies
20 April 2009	Battery engineer	Technical and industrial challenges concerning battery for hybrid heavy vehicle applications

Table A10.3 Interviews at Voith

Date	Position	Interview topics
24 February 2010	Project leader; Technical leader	Voith's electric machines Scania–Voith collaboration

Table A10.4 Interviews with university professors

Date	Position	Interview topics
5 May 2009	Lars Nielsen, Professor of Vehicular Systems, Linköping University	The development of control systems for hybrid electric heavy vehicle applications
25 May 2009	Lennart Josefson, Director of the Swedish Hybrid Vehicle Centre (SHVC)	How collaboration between universities, vehicle makers and suppliers is conducted
25 May 2009	Bo Egardt, Research Co-ordinator, 'System Studies and Tools' at SHVC; Professor of Control Engineering, Chalmers University of Technology, Gothenburg	The development of control systems for hybrid electric heavy vehicle applications Industrial challenges related to the development of the control system
June 2009	Kristina Edström, Professor of Chemistry/Battery Technology, Uppsala University	Technical and industrial challenges in developing batteries for hybrid vehicle applications
30 April 2010	Chandur Sadarangani, Professor of Electric Machines, KTH (Royal Institute of Technology), Stockholm	Technical and industrial challenges in developing electric machines for hybrid vehicle applications

References

Eriksson, S. (2007) 'Electrical Machine Development: A Study of Four Different Machine Types from a Swedish Perspective', Doctoral thesis, KTH (Royal Institute of Technology), Stockholm, Sweden.

Guzella, L. and Sciarretta, A. (2007) *Vehicle Propulsion Systems: Introduction to Modelling and Optimization*. Berlin: Springer Verlag.

Lange, A. (2010) 'Hybrid Drives Development at Voith in Hybrid Vehicles, Electric Vehicles and Energy Management', *Proceedings of the 7th Symposium: Hybrid Vehicles, Electric Vehicles and Energy Management*. Braunschweig, Germany: ITS Niedersachsen.

Neudorfer, H. and Franz, F. (2010) 'Simulation and Economic Evaluation of a Bus with Serial Hybrid Drive in Comparison to a Conventional Diesel Bus', *Proceedings of the 7th Symposium: Hybrid Vehicles, Electric Vehicles and Energy Management*. Braunschweig, Germany: ITS Niedersachsen.

Overgaard, L. and Folkesson, A. (2007) *Scania Hybrid Concept – With Robust Technology into the Future*, 57th UITP World Congress, Helsinki, Finland, 20–24 May.

Wallmark, O. (2006) *Control of Permanent-Magnet Synchronous Machines in Automotive Applications*, Doctoral thesis, Chalmers University of Technology, Gothenburg, Sweden.

References to internet sources

EPO (2005) available at: http://www.epo.org/about-us/press/releases/archive/2005/06072005.html; accessed 29 April 2010.

Kollmorgen (2010) available at: http://www.kollmorgen.com/website/com/eng/industry_solutions/hybrid_solutions.php; accessed 5 April 2010.

Scania (2008) available at: http://www.scania.com/media/feature-stories/technology/hybrid-trucks-on-the-way.aspx; accessed 13 September 2011.

Scania (2009) available at: http://www.scania.com/Images/P09601EN%20Innovative%20hybrid%20bus%20concept_114691.pdf; accessed 30 January 2012.

Volvo (2007) available at: http://www.volvobuses.com/SiteCollectionDocuments/VBC/Global%20site/Buses%20and%20coaches/Components%20fact%20sheets/Volvo_Hybrid_Eng1.pdf; accessed 12 April 2012.

Volvo (2009) available at: http://www.volvotrucks.com/trucks/qatar-market/en-qa/newsmedia/pressreleases/pages/press_article.aspx?pubid=7872; accessed 13 September 2011.

Volvo (2010) available at: http://www.volvogroup.com/group/global/engb/responsibility/envdev/alt_drivelines/hybrid_tech/isam_concept/Pages/hybrid_concept.aspx; accessed on 23 March 2010.

Part III
Surrounding Conditions for the Development of Alternative Vehicles

11
Electric Vehicles and Power Grids: Challenges and Opportunities

Ettore Bompard, Elena Ragazzi and Alberto Tenconi

Fuel price and availability are, of course, relevant to determining the diffusion patterns of vehicles. The example of natural gas (CNG) cars, whose wide distribution was hampered by the poor development of the refuelling station network, is well known. But in the case of electric vehicles it is even more important to consider the question, because of the complex management of the electricity supply.

Taking into account electricity networks and markets is fundamental to evaluating the perspectives and impacts of the development of electric vehicles (EVs). This assessment will examine three different levels. First, the evaluation of *generation capacity*. Will it be enough in case of large and rapid diffusion of EVs? Second, the evaluation of the *distribution network constraints*: will the grid support wide and dispersed connections for battery recharge? Finally, an integrated vision of electricity supply and vehicle energy efficiency is indispensable for a correct evaluation of the *environmental impact*.

As far as the first two questions are concerned, the answer is not simple. It is not only question of producing reliable scenarios regarding EV sales (the yearly flow of new EVs) and numbers of vehicles (the accumulated stock, which must take into account turnover of vehicles), but also predicting daily recharge patterns for batteries (its time distribution as well as the amount of electricity consumed), and finally considering the technological development of the electric energy system. In parallel with the EV paradigm, a new paradigm is appearing within electricity distribution and generation – the smart grid – and many interconnections can be found between the two.

With regard to the environmental aspect, the EV is often considered to be a zero-impact mobility system. This is true at the local level, since the purely electric motor has no CO_2 or polluting emissions, and so EVs represent a solution in situations of acute pollution caused by traffic concentration. But the evaluation it is not so straightforward if we consider the 'well-to-tank' process. Electricity is not a primary source, but rather the output of a complex productive and distribution cycle, based on different primary fuels and on generating plants with different efficiency levels. It is then important to stress that both on the side of energy efficiency and on that of greenhouse gas (GHG) and polluting elements emissions, there can be impacts that tend to be underestimated by the final user.

This is not to deny the environmental interest of EV. The technological progress in electric motors on the one hand, and in generation, transmission and distribution devices on the other, have developed in parallel, and now the energy losses of the internal combustion engine (ICE) – whose energetic efficiency is low, at around 20 per cent – are higher than the aggregate losses of all the phases involved in electricity generation and distribution. But it must be clear that the net environmental advantage depends heavily on generation technologies and on the state of the transmission and distribution network. The environmental impact of EVs, as well as the dependence on energy from abroad, cannot be assessed separately from an evaluation of the electricity supply system and of the primary source mix of countries, matched with the true characteristics of the EV vehicles. The complete scenario concerning the electricity supply side must consider the present situation and the likely evolution of:

- *the generation-plant stock*, with its mix of primary sources and generation technologies, plus the capacity for import;
- *the transmission grid* (some renewables are concentrated at a distance from the consumption location, thus involving a need for investment); and
- *the distribution grid*, which must be enhanced with new devices, guaranteeing increased intelligence and efficiency (smart grid).

In this chapter we shall recall the basic categories of EVs and their refuelling options, and discuss their impact on the electricity system; and then provide a portrait of the likely evolution of electricity grids in the near future, discussing the twofold role that EVs will play in this context – that is, as a load but also as a distributed storage capacity able to interact actively with the network. Finally, we shall outline possible advantages and disadvantages for the electric power systems and distribution grids with a massive deployment of EVs.

Classification of electric vehicles

The transport sector still relies largely on fossil fuels; however, the 'more electric' solutions are being paid more attention in all transport systems:

- *Air transport*: the more electric aircraft (MEA) concept – that is, the wider adoption of electrical systems in preference to established hydraulic, pneumatic and mechanical systems, is now a reality in civil aircraft such as the Airbus A380 or Boeing 787 (Rosero *et al.*, 2007).
- *Sea transport*: the hybrid diesel electric propulsion systems which free naval architects from the 'tyranny of the shaft-line', and offer advantages

in terms of vibration, fuel consumption and maintenance costs, have now been adopted for several cruise ships (King and Ritchey, 2002).
- *Ground transport*: electric and hybrid electric traction systems have traditionally been adopted in railway transportation, but recently have also been gaining popularity in road transport applications, from bicycles to buses, and in a range of vehicles from hybrid electric (HEV) to purely battery electric (BEV) or fuel cell electric (FCEV).

While the 'more electric' air and the sea transport have a negligible or zero impact on the electricity grid, in contrast, ground transport, in many cases, is supplied by the grid. These include the traditional and well-established railway and underground trains, but also the emerging BEVs and the plug-in hybrid electric vehicles (PHEVs), which draw on power from the grid to recharge their battery packs.

The impact on the grid in connection with the railway and underground trains is outside the scope of the chapter, and, in any case, the related power and energy demands can be considered as being settled historically in the grid.

Coming to the focus of the chapter, the 'green' road vehicles panorama includes several items of different sizes, from e-bikes (motor power 300–750 W and battery energy 200–1000 Wh) to e-buses (motor power 100–250 kW and battery energy 50–300 kWh), and with the different architectures of the powertrain (hybrid or purely electric).

From the point of view of the grid, the vehicles with a hybrid powertrain (at least one internal combustion engine and at least one electric motor) can be divided into two main types:

- *charge-sustaining HEVs* (micro-hybrid, mild-hybrid and power-split full-hybrid), where the battery pack essentially acts as power buffer, hence it is not necessary to charge the accumulators connecting the vehicle to the grid. As a consequence, even if it is expected that these kinds of vehicles will be the most diffused 'green vehicles' in the short-to-medium term, their impact on the grid is zero; and
- *charge-depleting HEVs* (blended plug-in, green zone plug-in and extended plug-in), where the battery not only acts as a power buffer, but also contributes to the energy balance and hence it is discharged progressively during the vehicle's use. In consequence, the vehicle must be plugged into the grid to charge the battery (PHEVs).

Vehicles with purely electric powertrains can have different on-board energy sources:

- vehicles supplied by fuel cells obviously do not need to be connected to the grid, even if some studies propose to connect FCEVs to deliver energy to the grid (Lipman *et al.*, 2004); but

- vehicles supplied by batteries (BEVs) obviously *do* need be plugged in to the grid to restore the energy consumed.

In this chapter we shall basically consider the light-duty vehicles that have the most impact regarding the electricity grid, for two reasons: first, these vehicles have high potential for diffusion; and second, the mission profile of these general-purpose vehicles can vary across a wide range of different duties (Boulanger *et al.*, 2011). The energy consumption of this size of vehicle, when running purely in electric mode, ranges from 120–200 Wh/km for a mid-sized saloon car, to 250–300 Wh/km for an SUV or a delivery van. The on-board stored energy ranges from 5–15 kWh in HEVs with 20–60 km pure electric autonomy, to 15–30 kWh in BEVs with 130–170 km autonomy in urban areas.

Refuelling EVs

BEVs and PHEVs need to be 'refuelled' at the end of or during the journey. In the refuelling process the key issue is the energy storage device; that is, the battery. The widely diffused battery technologies are, in order of maturity and inversely for energy density:

- lead-acid battery (25–45 Wh/kg);
- nickel-metal hydrate battery (60–80 Wh/kg); and
- lithium-ion and lithium-polymer batteries (100–150 Wh/kg).

While the different technologies present diverse characteristics and performance, from the point of view of the charging techniques they have similarities:

- when the state of charge is low, the battery can be charged at relative fast rate provided the current does not exceed a specified threshold value; and
- when the battery is nearly full it must be charged more slowly, keeping the voltage no higher than the allowed maximum value.

In all these technologies it is possible to charge the battery at high current (power) rate with the intention of reducing the charging time; the drawback, however, is that puts stress on the battery, and shortens the life of the electrochemical elements.

The large variety of battery charging systems for light-duty vehicles can be divided into two types:

- on-board chargers with a typical power rate of 1–3 kW, usually connected to a single-phase supply; and

- completely independent chargers with a typical power rate of 10–50 kW, usually connected to a three-phase supply.

While the low-power, on-board charger is compatible with a home electricity outlet, the high-power, independent chargers need a dedicated unit or a charging station. The on-board battery-chargers need between 5 and 10 hours, or even more, to totally charge the accumulators (overnight charging); the independent chargers need between 1 and 3 hours to totally charge the accumulators (fast charging), or 20–30 minutes to bring a half-charged battery to full capacity.

A third solution is battery swapping at dedicated switch stations, where the empty battery pack is replaced with a fully charged one; the empty batteries are then recharged for reuse (Lombardi et al., 2010).

Studies are being carried out on prototype electrified lanes (Zell and Bolger, 1982), where the vehicles, with a limited amount of on-board battery, are supplied with power directly via a line; this solution is potentially revolutionary, but its viability is still not proven.

From the grid point of view it is possible to identify two basic options:

- change the battery, which will be charged independently of the vehicle; this option is more consistent with the traditional centralized control of the power grid, since professional providers are more likely to be controlled (with technical constraints and economic incentives) by the distribution system operator (DSO); or
- charge the battery connecting the vehicle to the grid (with different possibilities); this option provides less controllable behaviour, it can have different impacts on the network as a result of different charging features (technical), behaviour of the users (social habits), and tariffs (economic and market issues).

In terms of its potential impact on the grid, a key distinction must be made between conventional ('non-smart') and 'smart' charging:

- In 'non-smart' charging, the power profile is defined only by the technical features of the charging device, and by the individual behaviour of the EV owners. Technical constraints are inherently imposed by the power converter (control strategies, regulation thresholds, protections and so on) while the market information is 'off-line', basically the tariffs plan for peak and off-peak hours.
- 'Smart' charging is based on the availability of two features: the battery charger can communicate with the power system (power grid, DSO, retailer, price signals, remote billing possibilities and so on), and can analyse the information to decide the power profile (level and timing); hence the power profile can be remotely and adaptively controlled in

terms of technical constraints and real-time prices acting as market signals of scarcity.

To sum up, 'non-smart' charging is based on the vehicle's needs, while 'smart' charging may shape the power profile, making the best compromise between the needs of the vehicle and the distribution network.

Smart grids: an emerging paradigm with many interconnections to electric vehicles

Electric power systems are facing a change in the paradigm that will probably be key in their development in the coming years. The traditional paradigm, implemented since the early stages of the electricity industry, is characterized by a small number of large power plants connected to a transmission grid responsible for transferring power to a few nodes of the network whose job is to supply the distribution grids. The task of the distribution system is to supply each individual load at a voltage level (high, medium or low) that decreases with the reduction in size of the parts of the system. This distribution system is mainly passive (not equipped with generation facilities) and the control of the system is mainly centralized and operated by a national/regional utility.

The new paradigm affects mainly the distribution that becomes active with a multitude of small-scale distributed generators, most from renewable sources; and the decision-making context in which, also as a result of the introduction of competition in the electricity industry, various players ('prosumers' – private users both producing and consuming electricity, retailers, distribution system operators and so on) make their own decisions.

All the new issues arising in the distribution systems in terms of distributed generation, markets for electricity and new services, and electricity storage, need to be matched with the reliability and quality requirements of the network while pursuing general efficiency goals (both economic and with regard to energy conservation) and controlling environmental impact. The only way that the central decision-makers (policy-makers and regulators) can optimize overall performance (economic, energy conservation, environmental) of the system is to provide constraints or incentives, mainly economic, on the side of the distributed decision-makers (demand side) and to contrast sources of market power (lack of information, congestion and so on) on the supply side.

To refer to the undefined vista of the future we use the term *smart grids*, in which the technology is embedded into grid assets and involves the decentralized capability of the grid components of sensing, computing and communicating. These ICT applications would provide a stage on which to implement, in both operation and control, new strategies, thus allowing the exploitation of new opportunities.

This dispersed intelligence and real-time communication opportunities will allow an elastic, feasible, diffused and cost-efficient management of battery charging. The EV, while provoking some concerns because refuelling will be supplied by an aged distribution network, provides in a smart grid context brand new opportunities in terms of distributed storage, allowing operators to give up the 'just-in-time' production that has characterized the electricity industry since its early days.

Recent studies (Ragazzi, 2011) on innovation activity have shown that the technology underlying the smart grid scheme is mature, and little further basic research is needed to make smart grids feasible. Nevertheless, a widespread adoption of the smart grid perspective seems unlikely given the present circumstances. Economic operators appear reluctant to invest massively before the institutional and market context becomes clear, and benefits and spillovers are clearly predictable. One of the most relevant question marks concerns the development path of EVs, representing both an incentive and an element of the smart grid scenario.

EV impact on electric energy systems

Scale of the impact

Of course, the impact of EVs on electricity systems depends on the scenarios regarding their diffusion. Forecasts regarding the spread of EVs have come from many sources (International Energy Agency, 2011) but figures are variable, because estimates depend on hypotheses based on different plans and different modes of interaction. Within the automotive industry, forecasts depend on firm strategies and consumer acceptance of different types of EV. But many conditions are dictated by industries close to the car sector, such as ancillary products (batteries) and services (charging systems), or by predictions about prices of alternative transport means and their fuels. Concerning the impact of EV diffusion, but also the EV diffusion pattern related to the greater or lesser economic advantage and rate of return on the investment, it is important to assess the electricity generation mix, the prices for primary inputs for the generation process, and of the electricity supply itself. Finally, the diffusion pattern also depends greatly on the opportunities and ways of use, which in turn are linked to the characteristics of the charging system and hence to the intelligence embedded in the distribution network (smart grid).

To understand how different hypotheses might impact heavily on future scenarios, let us take as an example the University of California, Berkeley's estimates (Becker et al., 2009), which include an assessment of the reduction of GHGs by 2030. Adopting in their model some restrictive hypotheses, such as a charge based on the battery switch, the maintenance of the actual source mix, and the technological level of the electrical system, reductions in GHG range from 8 per cent to 47 per cent. In the case of cars powered with non-polluting sources of electricity, the reduction in GHG would, of course, be much higher (20–69 per cent), but the range is even wider.

One piece of clear evidence coming from scenario analysis is that a massive deployment of electric mobility will come about in the next 10 to 20 years. Deutsche Bank (2008) foresees an increase of the share of EV from 22 per cent in 2015 to 49 per cent in 2020 for the USA, and an even sharper European penetration – from 50 per cent up to 65 per cent in the same time scale – because of stronger fuel economy requirements, with a recorded starting point of 2 per cent in 2007. But only a very small proportion of these vehicles will connect to the electrical network; PHEVs will account for 5 per cent of vehicles in the USA and 2 per cent in Europe, while purely electric vehicles will be 2 per cent and 3 per cent, respectively. The most likely scenario will see hybrid cars (micro, mild and full hybrid) prevailing strongly over other types.

Following the same, rather conservative, forecast, between 2015 and 2020 fewer than 5.5 million plug-in electric cars will be sold on the European market. This indicates that the challenge to the electricity network seems not to be imminent, since the wide diffusion EVs, and above all of purely electric cars, will not happen for some time. But it is also true that the attractiveness of these types of cars is linked firmly to the development of electricity distribution grids and dispersed generation plants, and on the smart management of both. If there were a major swing in the electrical system towards active demand and decentralized management, the attractiveness of purely electric cars would increase and the forecast numbers prove to be too low. This underlines again how important it is to consider the electric car jointly with the system guaranteeing its refuelling.

Technical impacts of EVs on the system

The possible impacts of EVs may be categorized according to different scales.

From a macro perspective, the impacts of EVs are characterized, from an electro-energetic point of view, in terms of the two typical dimensions of power and energy. The EVs, from a load perspective, will shift energy consumption from oil-derivates to electricity, with an impact on the total amount of electricity that has to be supplied and provided for those loads by the power grids at various levels (transmission, distribution). Of course, the power profile according to which that energy is consumed, committing power at different time of the day (see the next sub-section, 'EVs and efficient use of electricity...'), may provide considerably different impacts on the investments and operation of electric energy systems. The source of that energy/power, either from large, centralized thermo-electric power plants or via distributed generators from renewable sources, may make the difference. The needed investments and environmental impacts are based strongly on the chosen option. In the first case, and with no control over charging strategies to the customer (time and power for charging), the power profile might be modified in such a way as to increase the peak-time power, to cause an increase in installed capacity requirements.

From a micro perspective, the impacts of EVs have to be assessed with reference to the low voltage (LV) and medium voltage (MV) distribution grids to which huge numbers of them will be connected. The most likely needs and impacts range from the need for new design approaches to new strategies for the operation of the system, and the assurance of reliability and power quality (voltage profiles, harmonic pollution and so on).

The EVs in electric energy systems might provide both problems and opportunities in achieving the aforementioned performance of the grid, depending on the choices made in terms of distribution grid design and the introduction of 'smart' strategies and technologies for interaction between EVs and the grid. Opportunities are described in the section below dedicated to V2G (vehicle to grid).

From the point of view of the 'problems', if the system goes towards uncontrolled EV charging ('not smart charging'), new loads will need to be supplied, with an increase in power demand during peak hours, which implies the need to dispatch less efficient units with higher electricity costs and to invest in new power plant capacity. If we assume that, in 2020, there will be, at the global level, 8 million BEVs and PHEVs, each with an average battery power of 15 kW, we can estimate the power withdrawn from the network to be around 120 GW, if all are charging simultaneously. Furthermore, assuming around 100 million EVs in 2050, the total capacity needed would be 1,500 GW (International Energy Agency, 2011). These numbers are huge if we consider that, for a country such as Italy, the peak power over the national grid is around 50 GW and the gross capacity of the world's generation plants was about 4,700 GW in 2008 (Terna, 2010a). In addition, the total energy estimated for EVs in 2050 would be around 800 TWh (Roadmap, 2050, 2010), which, again, is a considerable amount of electricity compared to the 20,000 TWh that was the global gross electricity production in 2008. Deciding whether this will become a problem depends both on the charging scenarios and on the charging management described below.

Network problems may also arise, having an impact on the hardware of the power grids, in terms, for example, of an overload of transformers and lines during peak hours, of increased losses and a decrease in reliability with the reduction of power margins, from generation level to distribution. Those concerns would probably drive the system towards new designs for the wiring and network hardware, with a need for expansion that would involve new and higher investments in the network assets. The operation of the distribution grid would become more complicated, especially with the widespread use of V2G that, with distributed generation, will make the power flows bidirectional (from grid to EVs and vice versa). The system needs to be protected against various faults, with proper strategies based on the relay settings. In the new context, batteries will raise the short-circuit currents, and power flows will become bidirectional, so requiring new protective schemes. Present research indicates that, while transformer life can be maintained, service transformers and three-phase primary lines are the most susceptible to issues related to the higher loads caused by PHEVs (Maitra et al., 2009). In particular, investigations at the Pacific Northwest National Laboratory (PNNL) in Washington State, showed an increase in transformer failures as a result of the penetration of PHEVs (Kintner-Meyer et al., 2008). And the integration of PHEVs could cause power losses and voltage deviations in the distribution grid (Clement et al., 2008). In addition, research carried out at Oak Ridge National Laboratory, Tennessee, showed that the reliability in terms of loss of load probability (LOLP) was reduced with the penetration of various PHEV scenarios in the VACAR sub-region (Virginia and Carolinas) of the Southeast Electric Reliability Council (Hadley, 2006).

EVs and the efficient use of electricity: charging scenarios and economic incentives

When to recharge is not unimportant

Electricity demand coming from EV owners will be added to the usual demand coming from the different parts of the economy and society, resulting in a load profile. A load profile shows demand (average, actual or forecast) of electricity at different times of the day. It is, of course, different on working days and holidays and depends on the seasons, whereas the effect of the weather is fairly predictable and is included in average load profiles.

Figure 11.1 represents the load profiles on the third Wednesday of a winter month (January 2010) and of a summer month (July 2010) for Italy. The power withdrawal over a day is characterized by a power profile curve with minimum values during the night (off-peak hours) and maximum values during the day, usually one peak in the morning and one in the afternoon (peak hours). The two peaks are clearly evident in winter, but in summer air conditioning inflates demand and smooths hourly differences during the day. The curve is characterized by a load factor defined as the ratio between average and maximum power, which provides a rough characterization of the power profile shape. To satisfy the huge demand made during peak hours, it is necessary to exploit inefficient generating plants – inefficient both from an energy and an economic point of view. In the present scenario electricity cannot be stored and to keep frequency constant and the system functioning, the power injected by the generators needs to be matched, minute by minute, with that withdrawn by the load plus the losses ('just-in-time production'). For this reason, the right amount of power from the necessary number of power plants needs to be committed over the day. The more total demand approaches total internal capacity plus import capacity, the more

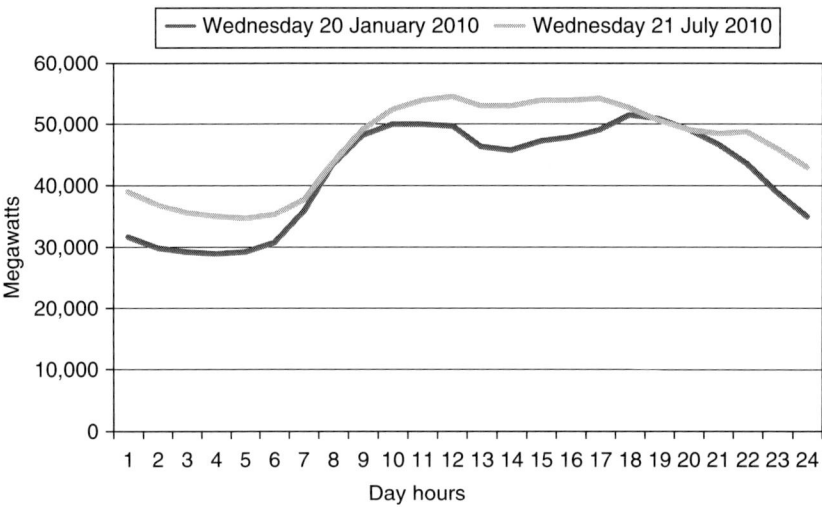

Figure 11.1 Examples of load profiles for Italy
Source: Terna (2011).

old, inefficient and polluting plants will be dispatched. This reflects on prices of supply. The clearing market price of the power exchange is high at those times, reflecting high marginal costs. Though few transactions occur on the power exchange, its price is taken as a reference point to fix bilateral contracts and small user tariffs.

It is important, then, to evaluate whether the daily charging profile, heavily influenced by individual habits, will balance or increment actual load fluctuations. For example, in one of the scenarios foreseen, large shopping centres will offer customers the possibility of recharging their battery while shopping. But this, unless the shopping mall or supermarket has its own generation or accumulation facilities, causes an increase in demand. The same would happen if drivers plug their cars in when returning home from work, if no delaying device is introduced.

It is therefore very important to plan regulation or governance through the prices of daily charging times. This is necessary to avoid uncontrolled recharging being carried out with inefficient or polluting sources, and to prevent major problems caused by local congestion.

Rational use of electricity: market mechanisms or top-down control?

A great deal of care is necessary to avoid the diffusion of EV becoming a problem for the electricity system, as well as nullifying the environmental benefit. Ways to encourage EV owners to avoid recharging in peak hours have to be found. Eventually, if the situation reaches a crisis because of local congestion, mechanisms to disconnect loads in certain situations are possible.

All this is related to electric system governance. As explained above regarding smart grids, the electricity supply used to be organized in a vertical matter with centralized control. The end of this monopoly and the introduction of competition implies a shift towards less centralized governance in which prices play a major role in guaranteeing allocative efficiency.

A similar choice now faces the control of the charging profile. Centralized governance of the charging process could be based on central charging stations, basing their charging profile on information regarding past peak hours, but also on data coming in real time from electricity utilities. In cases of emergency, supplies to these specialized stations could be switched off. In China, where electric mobility is looked on with particular interest, there is ongoing research on a wireless information system, controlling real-time vehicle positions and enabling charging needs to be forecast, as well as planning the rational use of electricity.

In the reverse of this top-down approach, market mechanisms base themselves on prices as signals of scarcity. In any given moment, EV owners, being assumed to act in a rational way, may decide whether to recharge or not after considering the price of this recharge and their battery level. If the recharge cannot be delayed because it is urgently needed, the consumer will incur extra costs, but in general s/he will postpone the charge until the low-cost hours.

Experiments exploring the attitudes of owners of BEVs in this early introduction phase have revealed behaviour that is too conservative and not rational, led by the fear that the car will not be available in a situation of unforeseen need. But this is linked to very little past accumulated experience. It is likely that, in a more mature phase, rational consumption patterns would be consolidated.

Ex-ante or real-time mechanisms?

The situation at the time of writing is characterized by scarce intelligence embedded in the distribution network, or, to be more precise, in the management systems of the distribution network. In fact, in some country smart meters have completely replaced old ones, and in other countries this renovation process is ongoing. But no distribution system can as yet manage real-time information on consumption flows. Even so, prices could be used to prevent the dangerous impact of EV diffusion on the electric network. Studying load profiles relative to different days (work days, Saturdays and holidays) and different seasons, a tariff system may be derived, with very high prices in peak hours. Charging the battery during those hours implies an extra cost, so customers would choose to do it only in cases of emergency. In this scenario, communication between suppliers and consumers happens only off-line, and the tariff is based only on historical information, the relevance of which might be reduced progressively by the diffusion of EVs themselves.

In a smart context, meters may receive real-time information on the level of global consumption and/or on the price of electricity. Charging devices in turn may also be smart, and decide whether to charge, to wait or even to sell accumulated electricity, considering both the market price and the battery level. Consumers may decide the relevant thresholds (or the parameters in the algorithms to calculate them) or exclude the automatic shift in case of necessity.

Uncontrolled and continuous charging may add considerable loads in periods of high demand, with consequent capacity requirements. Delayed charging practised through smart meters and chargers dramatically improves the situation by avoiding charging during the peak demand time of the late afternoon and early evening, while the optimal charging situation meets the overnight demand minimum. In 2007, PNNL conducted research on the impact of PHEV deployment on 12 main regions of the US power grid. The laboratory concluded that, with the existing infrastructure, up to 73 per cent of the estimated number of PHEV vehicles could be charged if vehicles were able to be charged at any time of the day. If charging was restricted to a 12-hour period, the study estimated that only 43 per cent of the vehicles could be charged (Kintner-Meyer et al., 2007). Research at Oak Ridge National Laboratory indicates the clear increase in electricity demand caused by PHEVs in 2020 and 2030 in 13 regions, as specified by the North American Electric Reliability Corporation (NERC), the US Department of Energy (DOE) and the Energy Information Administration (EIA), and upon which the data and analysis in EIA's *Annual Energy Outlook 2007* are based (Hadley and Tsvetkova, 2009).

The price system would be even more effective in the case of a network of charging operators (offering pay-per-mile contracts or leasing-maintenance contracts), who could plan the charging profile following price signals and also undertake long-term contracts with electricity producers or wholesalers, or even perhaps become electricity self-producers.

V2G: EV supporting the distribution grid

The availability of a rechargeable battery on board an EV provides a source of distributed storage that may be exploited in the efficient operation of smart distribution systems. How would this work?

The battery charger is based on power electronics converters that rectify the AC waveform into a DC form controlling the current and the voltage values applied to the battery. This well-established technology adopts a diode bridge (single-phase or three-phase) on the line side; in some cases, a power factor corrector is added to the diode bridge with the aim of meeting the power quality standards for the absorbed energy.

This circuit topology is intrinsically unidirectional. That is, the energy can flow only in one direction, from the grid to the vehicle. Recently, some authors (see Kempton and Tomić, 2005) have proposed the V2G approach, where the energy stored in the EVs connected to the grid is used to stabilize the grid and improve the power quality; this aspect is particularly interesting in the presence of an increasing number of distributed generators from renewable energy sources characterized by stochastic production. In this case, the power electronic converter must provide a bidirectional energy flow; both from the grid to the vehicle and vice versa; as a consequence the circuit topology must be different, for example including active rectifiers, which are more expensive.

The basic scenario for the exploitation of V2G is characterized by the deployment of technologies that allow the transfer of electricity to and from the power grid and the vehicle. EVs with V2G technology have these characteristics: the vehicle can connect to the electrical power grid to be fed by an electricity flow and, at the same time, is able to connect logically to the electrical power grid to communicate with the grid operator. This bilateral communication between the EV and the distribution grid allows the matching, in real time, of the needs of both, given their technological constraints. In addition, from an economic point of view, this communication allows for both the control of charging and the billing for the electricity provided to and from the grid.

From the point of view of the EV, the V2G strategy provides the possibility of charging the vehicle and being billed anywhere convenient at a reasonable cost while being able to 'sell' some services to the grid when the electricity need is not urgent, thus reducing the EV's electricity bill.

From the point of view of the grid, V2G represents a distributed source of stored energy for ancillary services, in terms of services needed to keep the network functioning while meeting demand and receiving power from the distributed generators. V2G may be used to supply power during peak periods, discharging batteries during the peaks, and contributing to smoothing the power profile and increasing the load factor. In an islanded grid where the frequency needs to be regulated, the battery may be charged (or discharged) when frequency deviates from the rated value.

The mismatch between generation and load may be matched by using V2G instead of acting on generators, giving V2G a 'load following' task. The voltage regulation in MV (and in particular in LV) distribution systems, in which voltage drops are related precisely to real power flows, may be compensated by V2G; the availability of storage may favour the accommodation of higher levels of penetration of intermittent distributed (and not distributed) power sources, such as wind power. New York ISO (2009) showed, for the city of New York, the benefits associated with V2G for PHEVs and BEVs, because charging profiles can follow power produced by wind turbines very well.

This role of EVs in reducing the mismatch between demand and supply is also important from an environmental point of view. Research has shown that the aggregated EVs as generators potentially reduce the required number of expensive peaking power plants (that is, plants working only when there is a peak in demand), which are usually inefficient and polluting (Yang and McCarthy, 2009). For the same reason, the GHG associated with peaking power plants also reduces, and this positive effect adds to the GHG reduction provided by the EVs as a kind of green vehicle (Scott et al., 2007).

Finally, the V2G approach is interesting in the management of emergencies; the storage from V2G may represent a source of distributed back-up power that may be made available in the case of a failure of the HV feeder of the distribution grid, and assure the supplying of loads, in co-operation with local distributed generators, for a time.

Some concluding remarks

EVs are considered to be a credible option for short/medium-range transportation in the coming decades; and their widespread use will have an impact on the world and on society in various ways. From an electrical energy system (EES) perspective, a massive diffusion of EVs may represent either a serious problem or new and challenging opportunity.

The electricity needed to feed the EVs has to be made available at the distribution level and there are huge differences, in terms of global impact, if it is produced locally by renewable sources or centrally by traditional oil-fired power plants.

In the present paradigm, for power distribution grids, the impacts of EVs would probably be considerable, and the centrally operated production/transmission system would be affected. But in EES we are facing a change in the paradigm from centralized power generation and operation to a distributed one, and EVs will be one of the key ingredients in the new paradigm, characterized at the distribution level by the so-called 'smart grids' (distributed decision-making and control, communication channels over the distribution grid, and electricity accumulation to face renewable generation variability over time).

The possibility of exploiting the positive impacts of EVs and reducing or countering the negative ones is closely related to the structure and operating strategies of the power distribution grid. Only in a future scenario of 'smart grids' will it be possible to exploit the advantages of EVs and to limit the problems they may cause.

A smart charging system, linked to price mechanisms or to a centralized charging administration, may avoid battery charging occurring during peak hours, which is linked to the use of inefficient and polluting plants in the short term and investment needs over the longer term. In the reverse of this scenario, EVs may represent added loads during off-peak hours, when there is a surplus of power on the transmission grid because of the impossibility of reducing the power output from thermal and nuclear-thermal power plants, which will contribute towards smoothing the daily national power profile and increasing the load factor.

In addition, a massive deployment of EVs may provide distributed storage making a remarkable amount of electricity available. This might decouple production from consumption and so allow for a remarkable change in the electricity industry, which up to now has been forced into a real-time balance between production and consumption. This would be a substantial advantage in the case of less predictable renewable power sources such as wind power.

Distributed storage may also play a role in supporting the distribution grid when the connection with HV distribution and transmission fails. The possibility of resorting to EVs as backup power to secure parts of the distribution system in islanding mode may enhance the reliability and reduce LOLP.

From an environmental and economic point of view, the impact of distributed storage from EVs may also be positive. Research has shown that the aggregated EVs as generators potentially reduce the required number of expensive peaking plants, thus decreasing both the cost of electricity and GHG emissions. These effects, which might happen in a smart grid scenario, would add to the positive impacts on the environment of the EV as a relatively environmentally friendly means of transport.

The impacts we can expect from a massive expansion of EVs in electric energy systems may be positive in so far as it will be possible to create a bidirectional and real-time exchange of information between the EVs and the grid that would allow for the best fit between the needs and requirements of both grids and EVs in real time. It is time for joint research projects and, above all, the development necessary to integrate the two paradigms so as to exploit their complementarities and avoid points of disagreement.

References

Boulanger, A. G., Chu, A. C., Maxx, S. and Waltz, D. L. (2011) 'Vehicle Electrification: Status and Issues', *Proceedings of the IEEE*, 99(6): 1116–1138.

Clement, K., Haesen, E. and Driesen, J. (2008) 'The Impact of Charging Plug-in Hybrid Electric Vehicles on the Distribution Grid', *Proceedings-4th IEEE BeNeLux Young Researchers Symposium in Electrical Power Engineering*, pp. 1–6.

Hadley, S. W. (2006) *Impact of Plug-in Hybrid Vehicles on the Electric Grid*. Oak Ridge, TN, USA: Oak Ridge National Laboratory.

Hadley, S. W. and Tsvetkova, A. A. (2009) 'Potential Impacts of Plug-in Hybrid Electric Vehicles on Regional Power Generation', *The Electricity Journal*, 22(10): 56–68.

Kempton, W. and Tomić, J. (2005) 'Vehicle-to-Grid Power Fundamentals: Calculating Capacity and Net Revenue', *Journal of Power Sources*; 144(1): 268–79.

King, J. and Ritchey, I., (2002) 'Marine Propulsion: The Transport Technology of the 21st Century?', *Ingenia online*: 12.

Kintner-Meyer, M., Schneider, K. and Pratt, R. (2007) 'Impacts Assessment of Plug-in Hybrid Vehicles on Electric Utilities and Regional US Power Grids, Part 1: Technical Analysis', Pacific Northwest National Laboratory, Richland, WA, USA, Paper No. 4, Online journal of EUEC 1.

Kintner-Meyer, M. et al. (8 contributors) (2008) 'Grid Analysis for Transforming the Transportation Sector from Fossil Fuel to Electricity' RDSI Program Review, October.

Lipman, T. E., Edwards, J. L. and Kammen, D. M. (2004) 'Fuel Cell System Economics: Comparing the Costs of Generating Power with Stationary and Motor Vehicle PEM Fuel Cell Systems', Energy Policy, 32(1): 101–25.

Lombardi, P., Heuer, M. and Styczynski, Z. (2010) 'Battery Switch Station as Storage System in an Autonomous Power System: Optimization Issue', Power and Energy Society General Meeting, IEEE, July: 1–6.

Maitra, A, Kook, K. S., Giumento, A., Taylor, J., Brooks, D. and Alexander, M. (2009) 'Evaluation of PEV Loading Characteristics on Hydro-Quebec's Distribution System Operations', Electric Vehicle Symposium, 24:1–11.

New York ISO (2009) 'Alternate Route: Electrifying the Transportation Sector', Technical report. New York: New York Independent System Operator.

Parks, K., Denholm, P. and Markel, T. (2007) 'Costs and Emissions Associated with Plug-in Hybrid Electric Vehicle Charging in the Xcel Energy Colorado Service Territory', Technical Report No. NREL/TP 640-41410, National Renewable Energy Laboratory, Golden, CO, USA.

Ragazzi, E. (ed.) (2011). 'Lo stato dell'arte sulle smart grid: orientamenti, attori, prospettive', Rapporti Tecnici Ceris, 36.

Rosero, J. A., Ortega, J. A., Aldabas, E. and Romeral, L. (2007) 'Moving Towards a More Electric Aircraft', Aerospace and Electronic Systems Magazine, IEEE, 22(3): 3–9.

Scott, M. J., Kintner-Meyer, M., Elliott, D. B. and Warwick, W. M. (2007) 'Impacts Assessment of Plug-in Hybrid Vehicles on Electric Utilities and Regional US Power Grids, Part 2: Economic Assessment', Pacific Northwest National Laboratory, Richland, WA, USA, Paper No. 5, Online journal of EUEC 1.

Yang, C. and McCarthy, R. (2009) 'Impacts of Plug-in Electric Vehicle Charging', EM Magazine, Air & Waste Management Association, June: 16–20.Zell, C. E. and Bolger, J. G. (1982) 'Development of an Engineering Prototype of a Roadway Powered Electric Transit Vehicle System: A Public/Private Sector Program', IEEE Vehicular Technology Conference, pp. 435–8.

References to internet sources

Becker, T. A., Sidhu, I. and Tenderich, B. (2009) 'Electric Vehicles in the United States. A New Model with Forecasts to 2030', Centre for Entrepreneurship & Technology (CET) Technical Brief, No. 2009.1.v.20. Available at: http://cet.berkeley.edu/dl/CET_Technical%20Brief_EconomicModel2030_f.pdf.

Boston Consulting Group (2009) 'The Comeback of the Electric Car? How Real, How Soon, and What Must Happen Next'. Available at: http://www.bcg.com/documents/file15404.pdf.

Deutsche Bank (2008) 'Electric Cars: Plugged In. Batteries Must Be Included', FITT Research, June 2008. Available at: http://www.d-incert.nl/wp-content/uploads/2011/05/deutch_bank_electric_cars.pdf.

Inage, S. (2010) 'Modelling Load Shifting Using Electric Vehicles in a Smart Grid Environment', International Energy Agency, Paris, France. Available at: http://www.iea.org/papers/2010/load_shifting.pdf.

International Energy Agency (2011) 'Electric and Plug-in Hybrid Electrical Vehicles'. Available at: www.iea.org/papers/2011/EV_PHEV_Roadmap.pdf.

McKinsey (2009) 'A Portfolio of Power-trains for Europe: A Fact-Based Analysis'. Available at: http://www.iphe.net/docs/Resources/Power_trains_for_Europe.pdf.

Roadmap 2050 (2010) 'A Practical Guide to a Prosperous, Low-Carbon Europe', Technical Analysis. Available at: http://www.roadmap2050.eu/attachments/files/Volume1_fullreport_PressPack.pdf.

Terna (2010a). 'Dati Statistici sull'Energia Elettrica in Italia 2009 – Introduzione'. Available at: http://www.terna.it/LinkClick.aspx?fileticket=MLAtrIiFDTg%3d&tabid=418&mid=2501.

Terna (2010b). 'Dati Statistici sull'Energia Elettrica in Italia 2009 – Carichi': http://www.terna.it/LinkClick.aspx?fileticket=nyNERWCnSFY%3d&tabid=418&mid=2501.

Terna (2010c) 'Dati Statistici sull'Energia Elettrica in Italia 2009 – Confronti Internazionali': http://www.terna.it/LinkClick.aspx?fileticket=PgmBzvheblE%3d&tabid=418&mid=2501.

Terna (2011) 'Dati statistici sull'energia elettrica in Italia, anno 2010': http://www.terna.it/default/Home/SISTEMA_ELETTRICO/statistiche/dati_statistici.aspx.

12
Agreements and Joint Ventures in the Electric Vehicle Industry

Giampaolo Vitali

Within the 'green innovation approach', the electric vehicle (EV) is one of the most important products as far as the reduction of pollution is concerned (OECD, 2010, 2011; Volpato and Zirpoli, 2011).

The EV is an emerging sector taking the first steps in its life-cycle, and there are many variables that are going to affect both the success of EV firms and the structure of the EV industry overall (Aggeri *et al.*, 2009; The Royal Academy of Engineering, 2010; Enrietti and Patrucco, 2011; Freyssenet, 2011).

First, the technology is facing a process of rapid evolution in many fields, such as the chemicals used in the batteries (lithium-ion and so on); the efficiency of the electric motors (using neodymium and other rare earths); the efficiency of the cooling system; the characteristics of the batteries (weight, size, cost, energy density, power density); the infrastructure of the recharging system (plug-in at home or on the road); the management of the batteries (leasing versus ownership); the billing of the entire system and so on. Even the configuration of the EV is a work in progress, with a variety of alternatives ranging between pure EV and hybrid EV, as well as PHEV (Chanaron and Teske, 2007).

Second, public institutions play a major part in defining the future success of the industry. On the one hand, many industrial standards need to be selected by international public institutions to prevent market failures that could reduce demand (such as the standard of the recharging system, for example). On the other hand, the diffusion of EVs is at the time of writing secured by public subsidies that cannot continue over a long period. In addition, the higher consumption of electricity resulting from the charging infrastructures will create some constraints in the energy market, which will be solved by the creation of smart grids, new green energy and so on. The solution is, again, public intervention into the market.

Third, the structure of the industry is not shaped by the big carmakers alone, as was the case within the traditional automobile sector, but there are new players coming from sectors that are apparently very far from the traditional car industry, such as chemicals for the batteries, plastics and new materials for the car bodies, electronics for motors and inverters, ICT for software and communication protocols, and finance and logistics for the recharging system. For example, Better Place, Tesla, Fisker and Juice Technologies are some of the new players that have entered the EV industry.

To deal with this challenging scenario, firms need to be very flexible as far as their growth strategies are concerned. They cannot afford to follow a high-risk strategy, or a high sunk-costs strategy, because the technological scenario could change in a few years according to the rapid changes in the technology. As firms try to avoid a strategy with a point of no return, they look for flexibility in the EV market: this is why the growth of the big carmakers is based mainly on joint ventures, agreements and other non-equity operations that are a good tool for sharing risks and financial investments in managing the new EV technologies.

This chapter describes the growth strategies of the companies involved in the emerging sector of the EV industry, focusing on agreements and joint ventures entered into by the main players in order to cope with the new technological paradigm. As the technology of EV depends critically on the performance of the batteries, motors, inverters and control systems, since the mid-2000s close relationships have been established between car manufacturers and component producers to develop new technologies and new products.

The next section concentrates on the new technologies that are shaping the structure of the EV industry and affecting the success of the traditional large carmakers. The third section is a survey of the theory regarding agreements and the multiple choices for firm growth. The fourth section describes some agreements and joint ventures established by firms in the EV industry, shedding some light on the importance of growth strategies based on these.

The concluding remarks describe the significant implications for company growth strategies and for policy-makers intending to promote the newly emerging industry at the local level.

Agreements, joint ventures and disruptive innovations in the EV industry

Within the EV sector, innovation has a radical nature and it could be defined as a new technological paradigm (Dosi, 1982; Teece, 2008), where the new EV components are based on innovation and technology that are localized at a distance from the traditional car industry. EV needs new components that are developed mainly by companies that are not the usual suppliers of the traditional automotive sector. The most important components for EVs are:

- *A power control unit*, which consists of an inverter, a voltage-boosting converter and an AC/DC converter to run the car on an electric motor.

- *An inverter*, which converts the DC supplied by the battery to AC to turn the electric motor and for use in the generator. Conversely, it converts AC generated by the electric motor and the generator into DC to recharge the battery. There is direct cooling of the switching device for cooling efficiency and to enable inverter downsizing and weight reduction.
- *A voltage-boosting converter*, which increases the low DC supply voltage to a maximum of 650 volts to feed the electric motor as required. This means that more power can be generated from a small current to ensure high performance from the motor. It also means that the inverter could be made smaller and lighter.
- *An AC/DC converter*, which reduces the high voltage from the battery to 12 V, to be used by ancillary systems and electronic devices such as the electronic control unit.
- *A synchronous AC generator*, which is capable of high-speed axial rotation, producing electrical power while the vehicle is running in the mid-speed range.
- *Regenerative braking*, to recycle kinetic energy by using the electric motor to regenerate electricity. Normally, an electric motor is turned by passing an electric current through it. However, if some outside force is used to turn the electric motor, it functions as a generator itself and produces electricity. This makes it possible to employ the rotational force of the driving axle to turn the electric motor, thus regenerating electric energy for storage in the battery and simultaneously slowing the car with the regenerative resistance of the electric motor. The system co-ordinates regenerative braking and the braking operation of conventional hydraulic brakes so that kinetic energy, which is normally lost as heat when braking, can be collected for later reuse in normal driving mode.
- *A reduction gear*, which amplifies torque from the electric motor. This reduction gear is designed to reduce the high rpm of the front electric motor so that the power produced can be transferred to the wheels, with the added benefit of torque amplification; that is, with greater power.
- *An electric motor*, a synchronous A/C motor of compact shape, light weight and high efficiency.
- *A battery charging system*, which allows the battery to be recharged from a household electrical outlet. At the time of writing, charging time is about 100 minutes at 220 volts, and about 180 minutes at 110 volts.
- *A secondary battery*, capable of storing electrical energy regenerated by the motor under deceleration.
- In hybrid vehicles, the gas/petrol engine is more energy-efficient, producing a higher output than conventional gas/petrol engines. Regarding the battery, lithium is the current preferred material for battery chemistry, but it is not the only option for a viable battery, since other materials can be used, such as sodium nickel chloride, lithium-sulphur, zinc-air, and bi-polar lead-acid. Within the Li-Ion batteries, new technology considers

the use of lithium cobalt oxide (LCO), lithium manganese oxide (LMO), lithium iron phosphate (LFP), lithium nickel manganese cobalt oxide (NMC), lithium nickel cobalt aluminium oxide (NCA) and lithium titanate (LTO). The system should maintain the battery charge at a constant level at all times by monitoring and computing the cumulative amount of discharge under acceleration and recharging by regenerative braking.

All the new components for EVs have to be developed to the maximum degree of efficiency, by investing huge financial resources in R&D projects. Individual carmakers cannot afford to invest such a huge amount of resources in such a wide range of different innovations, from new chemicals (for batteries) to new materials (for magnets), new software (to control or manage the system as a whole, or parts of it), new plastics (to reduce the car's weight), and new infrastructures (the recharging system).

This need boosts agreements and joint ventures between companies involved in the EV industry, mainly to reduce risks and to increase technological innovations, as described below.

Finally, it is important to underline that the role of public policy is important for the development of the new technologies within the EV industry. First, we have to consider that the present growth of the EV market is heavily dependent on the penalties for the use of ICEs (internal combustion engines) and the subsidies for EV use. Many governments have provided a welcome incentive for the introduction of urban EVs. However, it is difficult to scale up from the present trials of a few hundred vehicles around the world to the millions of EVs that will be needed in the future mass market. Many trials and plans for limited EV introductions have involved small numbers of vehicles operating in restricted geographical areas. For example, Renault–Nissan made ninety agreements with small municipalities around the world, even in China, to support sales of its 'Leaf'.

Second, the public support the EV industry derives from its contribution towards improving the environment by reducing pollution. The contribution of EVs to the environment justifies a public policy to support the demand for EVs, and such a policy has been applied in the 2000s in many countries. In China, the USA and the EU there are some schemes reducing the cost of purchasing the vehicles, energy costs, and direct and indirect taxes (VAT, registration tax and so on) on EVs, but these subsidies are still not sufficient to reduce the cost of EVs to the level of the cost of traditional ICEs, and it is necessary to apply a regulation scheme that has a negative effect on traditional ICE vehicles (Kley et al., 2010).

In addition, public policy should support not only EV demand, but also the EV industry, in order to increase innovations and cost reductions along with the EV sector. In this regard, it is important to stress the public role of innovation, where innovation networks could increase the efficiency of private R&D investments. As innovation is not only the result of the investment of each individual firm, but it is also the result of the R&D investments made by other firms and institutions in the local area, such as universities, R&D centres and public agencies, it could be important to set up a public policy to support R&D networks between EV companies and local institutions (Foray, 2006; Patrucco, 2008).

Economic theory, agreements and joint ventures in the EV industry

Different growth strategies

The main question of this section is about the kinds of company growth within the EV industry. As EV production is a newly emerging industry that is facing difficult challenges concerning technology, demand and public policy, we can find some answers to the multiple choices of growth strategies within the economic theories dealing with agreements, mergers and acquisitions (M&As), and outsourcing strategies.

Regarding the characteristics of the EV industry, economic theory suggests that agreements are the best solution for the growth strategy of traditional carmakers that deal with EVs, mainly because of the technological goals of those agreements. It is worth underlining that company growth can follow different approaches: hierarchical strategies (acquisitions or internal investments), market transactions strategies (sub-contracting and outsourcing) or partnerships strategies (both equity and non-equity agreements). Economic theory compares the different strategies according to the specific characteristics of the firms (goals, available resources, types of market and so on) and of the industry (an emerging or declining sector, traditional or high-tech and so on).

In this section, I shall consider four different groups of theories.

The first group is concerned with market imperfections (Williamson, 1985; Hill, 1990): a firm avoids the market transactions when the internal costs of co-ordination are lower than the transactional costs of the market itself, and agreements and joint ventures are defined as an 'intermediate form' between market and hierarchy, and are not optimal with respect to the hierarchical choice (first best). The latter is the best choice, because it reduces the high transactional costs of the market ('internalization approach'). As the EV industry is an emerging sector, there are very few opportunities to buy innovations 'on the market', because they are developed by specialized R&D companies generally located far away from the traditional automotive sector. If firms cannot buy innovations on the market (high transaction costs because of the lack of innovative products), they try to invest internally in innovations or to make an R&D agreement. For example, Toyota invested internally into another company in the Toyota Group (Panasonic/Sanyo), which is a leader within the battery sector; together they are pursuing R&D projects on new batteries that fit the characteristics of the EV very well. And in another case, General Motors (GM) made an agreement with LG Chem, producer of Li-Ion batteries, to develop a new EV battery.

The next group of theories considers the difficulty of creating such a huge number of innovations within a reasonably short time-frame. According to the complementary asset theory (Teece, 1986), firms need a great deal of financial, managerial, technological, manufacturing and commercial assets in order to overcome worldwide competition. Global companies obtain a wide range of products, to be traded at the worldwide level, thanks to a broad range of innovations. If a company does not have the financial resources to follow M&A strategies, it cannot have direct control over all the resources and it tries to have indirect control over them through agreements and joint ventures, especially at the international level.

The same choice of method applies to the alternatives between use of the market and agreements: since the complementary assets (innovations) are very complex and do not exist as 'standard goods' on the market, they can be obtained through an agreement with another company that holds such resources. In this case, the firm can obtain the exact investment it needs. Thanks to agreements, the internal capabilities of the company can quickly be improved, instead of pursuing internal growth at a slower pace (Blois, 1990).

This is even more important for companies in emerging sectors, where the risks are higher and innovation capabilities are not diffused. By forging agreements in new, emerging industries, the company can obtain all the complementary assets it needs for the new business without investing a large amount of its financial resources or bearing a high technological risk.

There is a long list of examples that confirms this strategy among EV companies. Traditional carmakers do not usually have many opportunities to develop EV innovations without the partnership of a specialized firm. For example, Renault–Nissan signed an agreement with NEC to develop a new battery. Daimler too tried to gain new ideas for its electric motors from its agreement with Bosch.

The third group is about growth strategies at the international level, because of the importance of globalization. Here too the choices are among internal investments (such as greenfield investments), external investments (M&As), use of the market (import–export flows) and international agreements. For example, theories dealing with foreign direct investments (FDIs) can be useful for focusing on the choice between acquisitions and joint ventures at the international level (Dunning, 1993). FDIs and cross-border M&As are mainly undertaken to control assets that are necessary for gaining a competitive advantage at the international level (ownership advantages), or to exploit external economies at the local level (location advantages), or to avoid market imperfections and transaction costs (internalization advantages) (Buckley and Casson, 1985). Some theories are based on an evolutionary approach, where experience gained from past growth strategies and the learning process affect future moves. The experience is important in order to overcome the organizational difficulties stemming from external growth, as a controlling acquisition is more difficult to manage than a non-controlling one, and the latter is more difficult than a non-equity agreement. To minimize growth uncertainty, the firm's preferred entry mode can move slowly from a non-controlling to a controlling acquisition, thanks to the accumulation of managerial resources gained from the learning process. This linear view represents the evolution of the organizational skills of the company, where the learning process allows it to search for and acquire missing resources and assets. On the basis of the Dunning's (1993) theory, the learning experiences represent an ownership advantage specifically related to the firm that encourages external growth at the national or international level.

As traditional carmakers need to participate at the global level, even when they produce EVs, they need to develop an EV project at the international level by choosing which strategy to implement in order to obtain all the complementary assets at the international level. This is very important in the EV industry, where we find many cross-border agreements designed to penetrate foreign markets with new EVs. For example, in 2010, Daimler made an agreement with the Chinese company BYD for the development of an EV for the Chinese market. China will be the most important market for the EV, and all the big carmakers are going to attempt to exploit this opportunity by making agreements with local Chinese firms (Burgelman and Grove, 2010). Volkswagen has an agreement with FAW, General Motors with SAIC, and Renault–Nissan with the municipality of Wuhan.

Finally, the fourth group of theories underlines that the structure of the industry affects company strategies. Within highly concentrated industries, agreements can be useful to exploit economies of scale without affecting the oligopolistic equilibria of such a sector.

Within the high-tech industries, agreements allow companies to produce and trade innovative products during the 'imitation lag' before they are copied, or become obsolete (Teece, 1986). This view is quite different from the traditional one, which focuses on market failure in the technology field: because of information asymmetry concerning technological transfer, firms prefer to have complete control of the acquired firm, and not to share the technology with a potential free-riding partner. These days, however, because of the high importance of competitive technological factors, agreements play a strategic role in creating inter-firm networks focused mainly on managing new technologies.

Within industries where trade barriers are low, it is possible to find a lesser degree of market imperfections, and so a lower level of transaction costs: this could reduce the advantages of internalization and foster the strategy of developing agreements with other companies.

Agreements, joint ventures and innovation

Companies can create innovations not only through internal R&D investments, but also by taking advantage of the positive externalities (spillovers) from the R&D investments made by other firms and local institutions, such as universities, R&D centres and public agencies. This means that a company fostering competitive advantage based on innovation can organize R&D networks with other EV companies and/or public R&D centres.

Within the EV industry, this approach could be very effective, as there are many problems concerning industrial standards and basic R&D, where sharing the risk of R&D failures is very useful. Economic theory focuses on the technological goal of agreements and joint ventures, mainly within the high-tech industries. Since the 1950s, scholars (West, 1959) have been emphasizing the 'pooling of know-how' in some sectors, such as chemicals and steel, and at the beginning of the 1980s they discovered a strong relationships between R&D investment and the number of R&D agreements that had been established (Berg et al., 1982).

Many studies have stressed that the most dynamic companies in the emerging sectors, mainly if these are high-tech sectors, are more likely to be involved in technological agreements. The characteristics of emerging sectors are strong determinants for fostering R&D agreements made by firms in order to reduce costs and risks of R&D investment, and to increase the number of projects in hand for the same amount of R&D effort (Rosenberg, 1982; Porter and Fuller, 1986).

The main characteristics of emerging sectors , supporting the creation of technological agreements, are:

- high levels of R&D costs related to the new technologies which increase the entry barriers and reduce the numbers of firms that can afford to invest in these innovations; even large car manufacturers do not have all the financial resources necessary to overcome the entry barriers in EV technology;
- high levels of risk in R&D projects in state-of-the-art technology increases the number of project failures and the costs of project success: as in the pharmaceuticals industry, for example, for every success story there are hundreds of failures, also in the EV industry the risk of not achieving any results from R&D investment is very high; EV companies try to share the risk with other companies in the EV sector, sometimes even with direct competitors;
- high levels of complexity in the integration of processes and products within the EV industry requires new forms of organization that enable carmakers to deal with such a complex business. The new organization is based on the best innovation supplied by each part of the EV sector: this is why an 'open innovation approach' could make carmakers more efficient than the traditional growth based on internal investments (Chesborough

and Teece, 1996). Open innovation defined by R&D networks and R&D agreements is one of the opportunities for large carmakers to co-ordinate highly complex technologies;
- the pervasiveness of the EV technology derives from the different levels of expertise related to EVs: new materials for electric motors, new chemicals for the battery, new software for the control units, new forms of organization for the recharging systems and so on. If a major carmaker needs to be updated on all the scientific fields of the new technology, this is supported by R&D agreements with small, high-tech companies that are leaders within their own narrow area of specialization; and
- the globalization of the business calls for international partners used to developing new markets, new products and new technologies. As a result of globalization, R&D agreements are made at the international more often than at the national level.

Agreements and joint ventures in the EV industry: some case studies

We have selected some agreements made by companies within the EV sector, to shed light on the different kinds of growth strategies that companies undertake to develop new technologies or new EV products.

We can divide agreements according to the types of operators involved in the partnership and according to their location within the EV sector.

Agreements, joint ventures and types of partners

According to the types of operators involved, we can define three kinds of agreements:

- partnerships between firms;
- partnerships between firms and public institutions; and
- partnerships between EV firms and electricity producers.

Partnerships between firms

The first group of partnerships is the most common, involving major carmakers, large automotive suppliers and small firms in non-automotive sectors. Most of these partnerships are set up by traditional carmakers who are trying to obtain the new EV components from other firms while avoiding large amounts of internal investment. The goal of such agreements could be R&D, joint production or joint trade. For example:

- PSA Group and Bosch agreed to develop a new split-axle hybrid which features an ICE on the front axle and an electrical drive on the rear axle for the new Peugeot '3008 Hybrid 4'; and

- Daimler and Bosch set up the EM-Motive joint venture to produce electric motors for EVs.

Most of the agreements focus on new batteries: Daimler and Evonik, Mitsubishi and GS Yuasa, GM and LG Chem, Toyota and Tesla, Renault–Nissan and NEC, Daimler and Tesla. One of the main goals of the agreements over developing new batteries concerns energy and power density. Energy density (Wh/kg) measures how much energy a battery can hold, and the companies signing agreements to work together are trying to increase the energy density in order to reduce the weight of the battery. Power density is a crucial item within the PHEV, as it indicates how much power a battery can deliver on demand (W/kg): agreements are made between relevant companies in an attempt to improve the performance of electric motors by increasing the power density of the battery.

Some agreements are managed by firms that are not within the traditional automotive sector. For example:

- General Electric and PPG formed the Azdel joint venture to develop new plastics for EVs to reduce the weight of the vehicles;
- Cobasys and A123System signed an agreement to develop a new battery (A123System) to complement EV characteristics;
- Dow Chemical and TK Advanced Battery agreed to develop new chemicals for batteries; and
- Schneider Electric (engineering and electricity) and Parkeon (parking systems) signed an agreement to develop recharging points.

Partnerships between firms and public institutions

The second group of partnerships is between firms and public institutions, in order to achieve a solution to standard problems, to obtain subsidies for R&D programmes, or for the diffusion of EV demand.

As far as standards are concerned, new products must be developed to meet international standards to provide the best diffusion at the worldwide level. The economic literature describes cases of both success and failure among industrial standards, such as the European GSM (a success story worldwide), or the case of the VHS video system (which was the victor in its dispute with Betamax). Globalization increases the difficulty of finding a common standard within the EV industry, because the standard has been defined not only at the EU level, as with an ordinary industrial standard, but also at the international level, where BRIC countries are one of the most advanced producers in the EV industry.

Since market failure would generate a lock-in effect in defining the best solution for new technology, a public policy is necessary to facilitate standard agreements between carmakers. Agreements between firms and public institutions are important for managing the recharging infrastructures, where there are two different types of recharging systems: en-route charging and charging at home. Both have pros and cons, and they are not alternatives to each other.

The en-route charging system can be managed by:

- *rapid-charging stations* (with the same performance as current fuel stations), but there are three big obstacles to these: the ability of the battery to absorb a charge in a short time, the ability of the local supply system to cope with the high instantaneous load and the difficulty of ensuring an efficient and user-friendly connection between the grid and the battery;
- *battery exchange stations*, where a discharged battery can be exchanged for a full one: this is a complicated system as there are problems deriving from logistics, different sizes of the battery for different EV types, battery ownership and so on. Better Place worked on this project and created a system for the fast exchange of non-standard batteries within a battery leasing scheme; and
- *recharging at the parking area*, but this would need investment in public infrastructures.

The second option, charging at home, would be comparatively easy to organize, but is restricted to the small proportion of consumers who can access an outlet at home.

Within the existing partnerships between firms and institutions, there are some interesting agreements, such as:

- Renault–Nissan has set up a programme with a regional development agency in the UK (One North East) to develop the EV industry and encourage demand in the UK;
- Better Place has organized agreements with local councils in California, Israel, Australia, Denmark and other countries.
- Fiat–Chrysler has been given some subsidies by the Piedmont region of Italy to develop a prototype of the Phylla solar car; and
- Renault–Nissan have made around 90 agreements with local municipalities worldwide (China, Europe, the USA) to test new infrastructures for recharging EVs.

Partnerships between EV firms and electricity producers

The third group of partnerships is between automotive firms and electricity producers, to find ways to manage the increase in electricity demand that EVs will foster, and to explore a new billing system for the recharging infrastructures. Agreements between firms and public institutions will be necessary to manage the infrastructural problems of electricity consumption. The future development of smart grids is based on partnerships between carmakers, energy producers and public regulation institutions.

Smart grids are necessary to increase the efficiency of the electricity market, as these supply dynamic, real-time flows of information about the network to deliver electricity more efficiently and reliably. Smart grids manage electricity loading on a street-by-street basis, by enabling electronic communication between consumers and electricity suppliers, so that load and generation can be scheduled as efficiently as possible.

Agreements are being made between carmakers and local producers of energy, such as Daimler and RWE in Germany; Daimler and Enel in Italy; PSA Group and Sorgenia in Italy; Toyota and EDF in France; Elektromotive and EDF in France; and Better Place and PowerStream in Canada.

Agreements, joint ventures and the EV sector

According to their location within the EV sector, we find the following groups of partnerships:

- *agreements in the upstream area*: these are usually made to foster the diffusion of industrial standards or R&D investments, by large consortia or through public R&D projects; sometimes the open innovation approach is involved in these types of agreement;
- *agreements in the core business of the EV industry*: the goals of these agreements is usually to produce or develop new components for the EV, such as batteries, electric motors, electronic control units, software, cooling systems and so on; and
- *agreements in the downstream area*: these are not related to the internal components of the EV but rather to the other services and infrastructures EVs need in use, such as new recharging infrastructures and standards, financial services for billing the use of an EV, smart grids to increase the supply of electric power and so on.

Agreements in the upstream area

As far as the agreements in the upstream area are concerned, there is not a long list of case studies: a few subsidized R&D projects by the EU or other national institutions, and a few large consortia composed of firms and institutions.

In the first group there are the cases of Better Place, which has set up three EU R&D projects with companies and public institutions (the EASYBAT, Green eMotion and Greening European transportation infrastructure projects); one with Fiat–Chrysler; one with the European Union seventh Framework Programme (EU FP7 2007–2013) in co-operation with Dublin University and Solar Print; and A*CAR, an R&D network of 11 partners backed by the Singapore government (regarding the latter, see http://acar.i2r.a-star.edu.sg).

The second group includes two Japanese consortia: the CHAdeMO consortium (with dozens of partners), working towards disseminating a quick-recharging standard; and the FUPET consortium (11 partners), pursuing R&D in electronics for EVs.

Agreements in the core business of the EV industry

As far as the agreements in the main core business of the EV sector are concerned, the long list of case studies is mainly composed of large carmakers that have made numerous agreements with OEM suppliers or with small, high-tech firms dealing with new products and technologies for EVs.

For example, in the list of major carmakers there are Renault–Nissan and its agreement with NEC for new batteries; Volkswagen and the FAW group, for EVs for the Chinese market; PSA Group with Mitsubishi (for the European i-MiEV model) and with Bosch (for a split axle hybrid); General Motors (GM) and SAIC, for an EV for the Chinese market; and Daimler with Bosch (for an electric motor), with BYD Automobile Co. (for a Chinese EV), with Evonik Industries (to develop a new battery), and with Tesla (for a new battery pack).

Agreements in the downstream area

The agreements in the downstream area are linked mainly to infrastructures for the recharging system or for smart grids. Within the first group of agreements are BMW and Siemens; Daimler and Enel or RWE; General Motors and Coulomb Technologies; Renault–Nissan and 90 local municipalities; and Toyota and EDF.

Some specialized firms are involved in partnerships working on the development of smart grids, mainly in the USA, such as Grid Point and Coulomb Technologies.

Within the downstream area, some agreements with other goals are included. For example: the financial services to support the demand for the i-MiEV, resulting from the agreement between Mitsubishi and Banque National de Paris; the development of a recycling scheme for exhausted batteries, from the joint venture (4R Energy Corporation) between Renault–Nissan and Sumitomo.

Some concluding remarks

The most likely scenario for the development of EVs is probably a mixture of PHEVs and pure EVs, in which the former are dedicated to long-distance journeys, and the latter to use in urban traffic. Both PHEVs and EVs stand at a crossroads in terms of becoming viable on the mass market. Technical change is proceeding as a result of many R&D agreements made by major carmakers and small high-tech firms, but the success of the EV industry relies on a number of infrastructural improvements, on agreements regarding standards and protocols, and on public policies to support the opening up of a huge mass market (Kley et al., 2010). If we consider the economic theory about company growth, we can find that agreements are a good tool of growth when the sector is in the early part of its life-cycle, as the EV industry is at the time of writing.

Every month new agreements are made by traditional carmakers to obtain complementary assets from specialized companies while avoiding being forced to invest huge financial resources into a wide range of new technologies. This chapter has provided some information about the wide range of agreements that can be explored by scholars with an interest in the EV industry.

Of course, there are significant implications for policymakers intending to promote the newly emerging industry, as they need to invest in subsidies to promote open innovation and innovation networks with local companies. The transfer of know-how that innovation networks usually provide could have a positive effect on the start of a local EV cluster of firms, as has been done within other high-tech sectors at the regional level (Vitali and Finardi, 2011).

According to some scholars, 'open innovation' could play a major role in defining the future of the EV industry. In the fields where intellectual protection is not so important, such as the standards or the basic R&D, case studies of Tesla and Better Place (Oliver Wyman, 2007; Enrietti and Patrucco, 2010; Ili et al., 2010) show that the open innovation approach can overcome some of the typical market failures of the R&D markets, not only at the company level but also as far as industrial policy is concerned (MacNeill and Bailey, 2010).

Finally, there are numerous local industrial policies that attempt to stimulate the EV market by providing a recharging infrastructure (OECD, 2011). This requires partnerships among national and local governments, car park operators, electricity distribution companies, bank card issuers and many other companies not usually concerned with carmakers or automotive suppliers. These kinds of partnerships are difficult to organize, and are quite unstable, because of the different cultures/countries of origin of the partners, the different sizes of the companies, and the different models of business of each industry, but they are necessary in order to overcome the technological problems that the emerging EV sector is facing.

References

Aggeri, F., Elmquist, M. and Pohl, H. (2009) 'Managing Learning in the Automotive Industry: The Innovation Race for Electric Vehicles', *International Journal of Automotive Technology and Management*, 9(2): 123–47.

Berg, S., Duncan, J. and Friedman, P. (1982) *Joint Venture Strategy and Corporate Innovation*. Cambridge: MA: Oelgeschlager, Gunn & Hain.

Blois, K. J. (1990) 'Transaction Costs and Networks', *Strategic Management Journal*, 11(6): 493–6.

Buckley, P. J. and Casson, M. C. (1985) *Economic Theory of the Multinational Enterprise: Selected Papers*. London: Macmillan.

Burgelman, R. and Grove, A. (2010) 'Toward Electric Cars and Clean Coal: A Comparative Analysis of Strategies and Strategy-Making in the U.S. and China', Stanford Research Paper Series, No. 2048, February.

Chanaron, J. J. and Teske, J. (2007) 'Hybrid Vehicles: A Temporary Step', *International Journal of Automotive Technology and Management*, 7(4): 268–88.

Chesborough, H. W. and Teece, D. J. (1996) 'When Is Virtual Virtuous? Organising for Innovation', *Harvard Business Review*, 74(1): 65–73.
Dosi, G. (1982) 'Technological Paradigms and Technological Trajectories: A Suggested Interpretation of the Determinants and Directions of Technical Change', *Research Policy*, 11(3): 147–62.
Dunning, J. H. (1993) *Multinational Enterprises and the Global Economy*. Wokingham, UK: Addison-Wesley.
Enrietti, A. and Patrucco, P. P. (2011) 'Systemic Innovation and Organisational Change in the Car Industry: Electric Vehicle Innovation Platforms', *Economia e Politica Industriale*, 38(2): 85–106.
Foray, D. (2006) *The Economics of Knowledge*. Cambridge, MA: MIT Press.
Freyssenet, M. (2011) 'Three Possible Scenarios for Cleaner Automobiles', *International Journal of Automotive Technology and Management*, 11(4): 300–11.
Kley, F., Wietschel, M. and Dallinger, D. (2010) 'Evaluation of European Electric Vehicle Support Schemes', Working paper, Sustainability and Innovation, 7/2010, Fraunhofer ISI, Karlsruhe, Germany.
Hill, C. W. L. (1990) 'Cooperation, Opportunism, and the Invisible Hand: Implications for Transaction Cost Theory', *The Academy of Management Review*, 15(3): 500–13.
Ili, S., Albers, A. and Miller, S. (2010) 'Open Innovation in the Automotive Industry', *R&D Management*, 40(3): 246–55.
MacNeill, S. and Bailey, D. (2010) 'Changing Policies for the Automotive Industry in an "Old" Industrial Region: An Open Innovation Model for the UK West Midlands?', *International Journal of Automotive Technology and Management*, 10(2/3): 128–44.
OECD (2010) *Interim Report of the Green Growth Strategy: Implementing Our Commitment For A Sustainable Future*. Paris: OECD.
OECD (2011) *Better Policies to Support Eco-Innovation*. Paris: OECD.
Oliver Wyman Consulting (2007) *Car Innovation 2015*. New York: Oliver Wyman Consulting.
Patrucco, P. P. (2008) 'The Economics of Collective Knowledge and Technological Communication', *The Journal of Technology Transfer*, 33(6): 579–99.
Porter, M. E. and Fuller, M. B. (1986) 'Coalitions and Global Strategies', in Porter, M. E. (eds), *Competition in Global Industries*, Cambridge, MA: Harvard Business School Press.
Royal Academy of Engineering (2010) *Electric Vehicles: Charged with Potential*. London: Royal Academy of Engineering.
Rosenberg, N. (1982) *Inside the Black Box: Technology and Economics*. Cambridge, UK: Cambridge University Press.
Teece, D. J. (1986) 'Profiting from Technological Innovation: Implications for Integration, Collaboration, Licensing and Public Policy', *Research Policy*, 15(6): 285–305.
Teece, D. J. (2008) 'Dosi's Technological Paradigms and Trajectories: Insights for Economics and Management', *Industrial and Corporate Change*, 17(3): 507–12.
Vitali, G. and Finardi, U. (2011) 'Nanotech Cluster and Regional Policy: The Case of Piedmont', Mimeo, Turin, Italy.
Volpato, G. and Zirpoli, F. (2011) 'The Auto Industry: From Unfettered Expansion to Sustainable Development. Challenges and Opportunities', *Economia e Politica Industriale*, 38(2): 5–24.
West, M. W., Jr. (1959) 'Thinking Ahead: The Jointly Owned Subsidiary', *Harvard Business Review*, 37(4): 165–72.
Williamson, O. E. (1985) *The Economic Institutions of Capitalism*. New York: Free Press.

13
Business Model Innovation and the Development of the Electric Vehicle Industry in China

Hua Wang and Chris Kimble

It is argued by many in the field of sustainable development that the need to use cleaner, more energy efficient and less environmentally damaging technologies has never been more urgent (Goldemberg, 1998), particularly in relation to transport and mobility (Romm, 2006). The problems of global warming and pollution, as well as issues of energy security, have acted as a spur to a global search for alternatives to the internal combustion engine (ICE) (Jacobson, 2009).

Authors such as Zhao (2006) have argued that China's industrial policy is aimed at addressing these challenges, and at positioning the country to take advantage of developments in alternative fuel vehicles in general and electric vehicles in particular. China's strategy of building a base of industrial competitiveness, founded in part on 'new-energy vehicles' (a classification that includes purely electric, electric hybrid and other forms of alternative energy vehicles) means that the country has now become a laboratory for the development of such vehicles, with several different types currently being evaluated by China's central and regional governments.

In previous works (Wang and Kimble, 2010a, 2010b), we have focused on China's capacity for technological innovation; however, as Chesbrough observes, 'Technology by itself has no single objective value. The economic value of a technology remains latent until it is commercialized in some way via a business model' (Chesbrough, 2010, p. 354). While it is not disputed that the development of new energy vehicles is a significant technological challenge, there are numerous examples to show that the successful adoption of a technology involves more than producing a technologically elegant solution (Leavy, 2007; Nair and Boulton, 2008). Consequently, in this chapter we turn to an examination of China's capacity for business model innovation rather than the development of technology.

Business models

The term 'business model' is relatively new and has yet to establish a solid grounding in economic or management theory. Some trace the use of the term to Chesbrough's analysis of technological innovation at Xerox (Chesbrough and Rosenbloom, 2002); others (Morris et al., 2005), trace it back to the mid-1990s; while Magretta (2002) claims that the concept existed as long ago as the 1890s. Whatever the origin, most agree that the term first entered popular use during the so-called 'dot com bubble' around the end of the 1990s (Osterwalder et al., 2005). Although the term can be defined in a variety of ways (Amit and Zott, 2001; Magretta, 2002; Osterwalder et al., 2005; Teece, 2010), most agree that the organization of a firm's resources around the twin activities of value creation and value capture lie at the heart of a business model (Chesbrough, 2007).

According to Teece (2010) a business model describes the way in which a firm 'delivers value to customers, entices its customers to pay for value, and converts those payments to profit' (Teece, 2010, p. 172). A good business model means that a product offers value propositions that are compelling for customers, provide an advantageous infrastructure for the enterprise and generate a substantial revenue stream. An inappropriate business model means that a product will either fail to deliver value or fail to capture it. In this chapter, we do not enter into the debate about what constitutes a business model, but simply make use of the term to highlight the fact that, in order to extract value from a technology, some way to exploit it must be found.

Business model innovation

While a good business model is required for short-term commercial success, it does not guarantee long-term competitive advantage. As Teece notes, 'in practice, successful business models very often become, to some degree, "shared" by multiple competitors' (Teece, 2010, p. 179). For a business model to continue to provide an advantage it must be differentiated clearly from others, and difficult to imitate. The process by which such models are created has become known as business model innovation. This term was first popularized by Mitchell and Coles (2003, 2004) and is built on earlier notions of 'disruptive technologies' (Bower and Christensen, 1995) and 'disruptive innovations' (Christensen, 1997).

A disruptive technology (Bower and Christensen, 1995) is a technology that disrupts an existing market by introducing a novel, and sometimes unlooked-for, value proposition. At first, such technologies appear limited and only able to satisfy the needs of a particular niche market in which the dimensions of performance at which they excel are valued. For the companies that serve mainstream markets, such technologies are perceived to be irrelevant and, at least initially, are ignored. However, over time, advances are made and the performance of the new technology improves until it reaches a point where it can satisfy the requirements of the mainstream market. At this point, the incumbent firms find they are unable to catch up with the conceptual and technological lead built up by those who focused on the disruptive technology and consequently lose their positions as market leaders.

Later Christensen broadened this idea to encompass innovation in general rather than in relation to specific technologies. He notes that:

> Generally, disruptive innovations were technologically straightforward, consisting of off-the-shelf components put together in a product architecture that was often simpler than prior approaches ... They offered a different package of attributes valued only in emerging markets remote from, and unimportant to, the mainstream (Christensen, 1997, p. 15).

Mitchell and Coles (2003, 2004) applied these ideas to business models and argued that similar advantages could be achieved by replacing an old business model with a new one that would leave competitors 'out of position and unable to respond effectively' (Mitchell and Coles, 2003, p. 15).

Markides (2006), however, notes that while business model innovation and technological innovation are similar notions, they have one fundamental difference. Bower and Christensen's technological innovations first disrupt and then dominate a market, but the effects of innovation in business models are less clear-cut; a new business model does not replace the existing business model completely but simply

> enlarges the existing economic pie ... business model innovators do not discover new products or services; they simply redefine what an existing product or service is and how it is provided to the customer (Markides, 2006, p. 20).

Thus the advantage gained by business model innovation is the way in which it allows the same basic product to be offered in a new way that yields more profit than a competitor could achieve using the current business model.

Business model innovation and emerging markets

London and Hart (2004) note that the approach to emerging markets taken by most multinational companies is one of *'waiting for Westernization'*:

> This perspective assumes that over time the local business environment will evolve into an economic setting that is familiar to Western managers: legal contracts will supersede social ones and competitive advantage will be grounded in the ability to protect resources and knowledge from unintended leakage outside firm boundaries (London and Hart, 2004, p. 354).

They argue that, by adopting this strategy, companies miss out on the potentially huge returns from the poorer *'base of the pyramid'* market. Christensen et al. (2001) make a similar point. Commenting on General Motors' attempts to develop a competitively priced electric car, they note,

> Globalization's real market opportunity lies with the billions of poor who are joining the market economy for the first time ... The crowded, polluted streets of Shanghai, Jakarta, and Bangkok could constitute a much more hospitable market for electric vehicles than do the expansive freeways of California (Christensen *et al.*, 2001, p. 92).

Most studies that look specifically at business model innovation in emerging markets focus on how firms from established economies adjust their business models to work in emerging markets. There appear to be very few studies that examine directly business models that have been developed within emerging markets. By ignoring developments in 'base of the pyramid' markets, we argue, companies also cut themselves off from a potential source of innovation and new ideas.

For example, Anderson and Markides' (2007) study of companies serving such markets in Africa, South Asia, East and Southeast Asia found innovation taking place on a number of levels. However, the companies they studied were those that had already been identified as having 'succeeded in serving customers living in poverty' (Anderson and Markides, 2007, p. 28). In practice, most were local subsidiaries of multinational companies that had overcome the problems of the affordability, acceptability, availability and awareness of their products in these markets.

Hart and Christensen (2002) is one of the first papers to draw attention explicitly to the value of business models that originate in emerging markets. Citing examples such as the Japanese firm Honda's success in selling low-cost motorcycles to the USA in the 1960's, and the Chinese firm Calanz's success in selling low-cost microwave cookers to Chinese consumers, they note,

> business models that are forged in low-income markets travel well; that is, they can be profitably applied in more places than models defined in high income markets (Hart and Christensen, 2002, p. 52).

Sánchez and Ricart (2010) have conducted one of the few recent studies directed specifically at the business models used in 'base of the pyramid' markets. They analysed the business models used by seven companies operating in low-income markets and evaluated their success. However, once again, most of the firms were companies that had their origins elsewhere; only a few of the firms were based in the countries in which such markets exist.

Summary

To summarize, environmental and geopolitical pressures have provided the driving force behind the search for alternative sources of motive power. In technological terms, firms from Europe and North America are probably leading the race, but firms from emerging economies such as Brazil, Russia, India and China (BRIC) are actively pursuing the same goal. However, while mastery of the technology is important, it is not the sole criterion of success, because 'The economic value of a technology remains latent until it is commercialized in some way via a business model' (Chesbrough, 2010, p. 354). Business models from emerging economies have not been the focus of many studies, but we believe there are two reasons why an improved understanding of these models might be of value.

First, the pressures that drive the search for alternative sources of motive power globally have been brought into particular focus by the growing pace of industrial development in the emerging economies. Thus, for a solution to be effective, it must be acceptable in, and applicable to, emerging as well as developed economies. As we have seen (Anderson and Markides, 2007; Sánchez and Ricart, 2010), a business model does not have to be developed in an emerging economy to be successful, but, as we have argued previously (Wang and Kimble, 2011), the likelihood of success will be greater if it is.

Second, the use of petrol-powered vehicles, at least in Europe and America, is well established, and the problems of breaking the Western 'dependency' on the car has long been a topic of debate (Newman et al., 1995). There is anecdotal evidence from studies (Hart and Christensen, 2002) that business models developed in emerging economies have the potential to be disruptive in a similar fashion to Bower and Christensen's original notion of disruptive technologies (Bower and Christensen, 1995). It is possible that a novel business model, from an emerging economy or elsewhere, could be sufficiently disruptive to provide a solution to Europe and America's fascination with the ICE.

The next section briefly reviews the move towards new-energy vehicles in China, and then looks at a case study of a company that has produced a specific form of new-energy vehicle, the low speed electric vehicle (LSEV). The case study is of particular interest as it contains many of the features associated with disruptive technologies and business models.

E-mobility in China

The EV industry, sometimes termed the e-mobility industry, began to develop in China in the early 1990s. In addition to electric cars, LSEVs include electric bicycles, scooters and motorcycles. E-bikes are simple bicycles with an electric motor attached, with an average speed of 20 km/h. E-scooters and e-motorcycles are equipped with heavier motors (1–5 kW) and have speeds of 40–80 km/h. Production of two-wheeled e-vehicles grew to 25 million units in 2010 and production volumes are expected to reach 35 million by 2015. However, despite the high volume of production, the industry is still in an early stage of growth. At the time of writing, the sector consists of around 2,700 licensed producers. The market share of the top 50 companies is only around 50 per cent, much lower than that of a mature industry.

With 140 million users of e-bicycles, e-scooters and e-motorcycles, EV solutions are widely accepted by Chinese consumers as the answer to their basic transportation needs: 90 per cent of the total production of e-bikes is for the domestic market. The acceptance of low-speed electric transport forms the base for a potential market for the LSEVs that will be described below. In addition to these consumers, there were 500 million users of standard bicycles in 2009. Over time, it is expected that a significant number of these will move to using e-bicycles, e-scooters or e-motorcycles and that some, together with a proportion of the 140 million current e-mobility users, will switch directly to LSEVs. Based on the modest assumption of 5 per cent of bicycle and current e-mobility users switching to LSEVs, the market for LSEVs in China would amount to around 32 million people.

Defining the low speed electric vehicle

There is currently no international consensus concerning the definition of a LSEV, and even in China, one of the leading producers of LSEVs, the LSEV is not officially recognized as a road vehicle. Below we describe the key features of a LSEV in China, together with a brief outline of how it is viewed in the USA and Europe.

China

The typical LSEV drive system comprises an accelerator, a set of brakes, a steering wheel and a lead acid battery pack. Gearsticks, air-conditioning and safety equipment are omitted to reduce the total cost. The electric motor is connected directly to the speed controller and most models do not have a sophisticated battery management or motor control system. A typical LSEV has a top speed of between 40 and 70 km/h, the dimensions of a compact car and weighs less than 1,100 kg. The lead acid battery can be recharged from a 220 V domestic electricity outlet and has a capacity of 120–250 Ah, giving a travel distance of 80 km, 100 km or 150 km, depending on the number of battery packs fitted.

As a rule, LSEVs may not be used on the roads in China. There are two main reasons for this: first, the companies that produce them are not listed in the *Announcement of Vehicle Producers and Vehicle Products*, published by the Ministry of Industry and Telecommunication; thus any products they produce are not recognized as road vehicles. Second, the Law of Road and Transportation Security of 2003, which applies to the whole of China, does not have any policies or regulations to cover the use of LSEVs; so consequently, in law, LSEVs have no right to use the roads.

However, while modifications to these regulations are not normally permitted, as we shall see, provincial governments in areas where LSEVs are produced have created 'temporary' local policies that include permission to use LSEVs on the road as well as road tax and road charge waivers for the owners of LSEVs.

The USA

In the USA, LSEVs are defined by Federal Motor Vehicle Security Standard No. 500 and Federal Crash Test Protocol TP-500-02. They are four-wheeled electric vehicles that can be driven on the road, are fitted with certain basic safety features, have their speed limited to 56 km/h and have an unladen weight of less than 1,362 kg.

Most LSEVs have a product architecture based on golf carts, are powered by lead-acid batteries and have a top speed of between 32 and 40 km/h. According to the National Highway Traffic Safety Administration, they are used mainly for short-distance transportation, shopping and recreation, by retired people and by golfers.

Currently, 46 states, with the exception of Connecticut, Mississippi, Montana and Pennsylvania, have legalized the use of LSEVs on roads. The speed limit in Texas and Alaska has been extended to 72 km/h. The leading American company, GEM (Global Electric Motorcars), established in 1998 and part of Chrysler between 2000 and 2011 (now owned by Polaris Industries), had sold more than 100,000 units (including three-wheelers) to 75 countries by the end of 2009.

Europe

At present, there is no agreed standard for LSEVs; the nearest there is to a definition of an LSEV is that of a quadricycle, or quadbike. Quadricycles are small, four-wheeled, motorized vehicles, powered either by internal combustion engines or electric motors, which have certain restrictions on weight, power and speed. France was the first country to define quadricycles in 1986; this was followed by European Union directive 92/61/EEC in 1992, and directive 2002/24/EEC in 2002.

Two categories of quadricycle correspond to the notion of an LSEV: L6e and L7e quadricycles. L6e (light) quadricycles, powered by electric motors with a power of less than 4 kW, have a maximum unladen weight of 350 kg and a top speed of 45 km/h. The technical requirements of L6e are broadly in line with three-wheel mopeds (category L2e).

Heavy quadricycles (L7e) have a maximum unladen weight of 400 kg (550 kg for models that carry goods), a motor with a maximum power of 15 kW, and a top speed of 60 km/h. No crash testing is required for either L6e or L7e models, and many European countries class these as category B vehicles that do not require a driving licence for their use.

Business model innovation: the Shifeng Group

We shall now present a case study of the Shifeng Group, one of the largest producers of LSEVs in China. This case study is of particular interest as an illustration of the role of the business model and business model innovation. First, in terms of business models, it is an example of a product that is still in the process of being defined: the technology for the product exists but an appropriate means of commercializing it does not. Second, in terms of business model innovation, it also illustrates a market for a product that corresponds closely to Christensen's notion of a disruptive technology and has grown, to date, without the support of central government and outside the boundaries of the mainstream automobile industry.

Background

The Shifeng (Group) Co. Ltd. is a state-owned enterprise, operating under the jurisdiction of the regional government of Shandong Province. It was established in 1993 and began production of low-speed, three-wheeled, diesel-powered light trucks. While such trucks are its core business, amounting to the production of 1.4 million vehicles annually, since 2008 Shifeng has become one of the key players in developing the market for LSEVs. The group's sales revenue in 2010 was ¥23.6 billion (US$3.6 billion) and the cumulative sales of their low-speed three- and four-wheeled vehicles has reached over 7 million units. Shifeng is still led by its founder, Mr Liu Yifa, and his son, Liu Cheng Qiang, became the firm's general manager in 1999. Further information on the history and background of this case study can be found in Wang and Kimble (2011).

Methodology

Broadly speaking, the methodology we employed is that of a descriptive case study (Yin, 2003). However, this methodology departs from the comparative or iterative approach described by Yin (2003) and Eisenhardt (1989) in that the analysis is in part a re-analysis of data collected in earlier case studies and in part the analysis of data that has been collected more recently. It also departs from the strict view of a descriptive case study, where the researcher sets out to explore cause/effect relationships using a set of propositions derived from existing theory, as some areas of existing theory, such as business models and business model innovation, were not sufficiently developed. Consequently, our approach to the case study is also exploratory, as the use of these concepts, particularly within emerging economies, is also not well developed.

The first contact with the company dates back to doctoral research in 2001 (Wang, 2002). In 2008, shortly after the first LSEVs had been produced, the president of the Shifeng Group, Mr Liu Yifa, was interviewed. Members of industrial associations, business competitors and journalists were interviewed during late 2010 and early 2011, to gain a broader view regarding the emerging market for LSEVs. These included the director of the Technology Service Center for Electric Vehicles, from the China Electric Engineering Technology Association; the president of Shanghai Kanleqiu Science and Technology Company; and a project general manager of Sina-Newchance New Energy Technology Co. Ltd.

Building a business model for LSEVs

At the time of writing, the main market for LSEVs in China is in rural areas. More than 70 per cent of the population of China, around 900 million people, live in such areas; however, their purchasing power is much lower than those who live in the cities. Consequently, the business model for LSEVs that has evolved so far has been aimed at servicing the needs of these consumers.

According to the *China Statistical Yearbook* (2010), the per capita annual income of rural households was ¥5,153 (US$790), compared to ¥17,175 (US$2,650) for urban households, giving rural consumers an income of less than a third of those who live in urban areas. Khan and Riskin (1998) found an increasing disparity between urban and rural areas between 1988 and 1995, while Yang and Hao (1999) show that this disparity has been growing since the opening up of the Chinese economy in the 1980s. Ranis, (1988, p. 74) describes this as a dual economy where 'two sectors which are basically asymmetrical – and thus dualistic – in terms of both product and organisational characteristics' co-exist.

The dual structure of the Chinese economy is a key factor in the development of the business model for LSEVs. At ¥25,000 (US$3,900), the purchase price of an LSEV is much more affordable than a small, traditionally-powered car. The attraction of an LSEV lies not only in its low price, but also in its low running costs. The cost of the electricity needed to travel 100 km is around ¥6 (US$0.9); for a small petrol-powered car, the cost for the same distance would be ¥49 (US$7.5), some eight times higher.

The battery can be charged overnight from an ordinary 220 V outlet. Households in rural areas have private parking spaces where vehicles can be recharged, which is more convenient than owning a petrol-powered vehicle, as the filling station network in such areas is not well developed. The cost of the battery is reduced through a system of recycling. The cost of a battery (about ¥4000/US$600) is included in the initial purchase price of the LSEV. The battery can be used for one to two years, depending on the distances travelled, after which consumers pay around ¥2000 (US$300) for a replacement. The used battery is then processed, recycled and reused.

While the LSEV offers a clear value proposition to (low-income) consumers in rural areas, it also presents advantages to a segment of the more affluent urban market. Since most urban commuting distances are less than 20 km and the top speed of an LSEV corresponds to the standard urban speed limit of 50 km/h or less, it could also meet the basic daily transportation needs of the many urban consumers who currently use mopeds, scooters and motorcycles. It has also been suggested that the simplicity of LSEVs, where the driver needs only to learn how to accelerate, brake and to drive forward and in reverse, might prove attractive to other categories of drivers, such as urban housewives and senior citizens, who may feel more comfortable driving such vehicles at low speeds. However, while this urban market exists in theory, LSEVs are not recognized as road vehicles, and changes to China's legislative framework are needed before this potential can be realized.

Building a market for LSEVs

The first plans by the Shifeng Group to build LSEVs were made in 2004 but production did not start until 2007. By the end of 2007, 5,000 units had been sold, mainly to existing rural customers living in the immediate vicinity of the plant that produced them. This modest local success acted as a spur to the Shifeng Group to look for a way to obtain a licence to produce and sell LSEVs on the national market.

The first problem they faced was that LSEVs are not seen as road vehicles. Two options were open to Shifeng. The first was to apply for an EV licence from the Ministry of Industry and Telecommunications, which would put the LSEV on the same footing as an electric car. However, as LSEVs do not fit within the current definition of an EV, this ruled out the possibility of gaining approval without first obtaining a change to the legislation that defines EVs. The other option was to apply for a licence from AQSIQ (the General Administration of Quality Supervision, Inspection and Quarantine of the People's Republic of China), an authority under the direct jurisdiction of the State Council, to use LSEVs as sightseeing vehicles. While this option presented fewer practical problems, it meant that the vehicles could only be used in designated areas and still could not be used on public roads.

Comparing these two options, the entry barrier to obtaining a licence to produce electric sightseeing vehicles was clearly lower. In June 2009, the Shifeng Group was granted its licence and its potential market grew from exclusively off-road rural use to rural and limited urban use. But to expand the market further, LSEVs would need to be allowed to use the public roads.

Despite modifications to national laws that deal with car registration, transportation security and management not being permitted, provincial governments are able to create temporary local policies. The Shifeng Group is located in the Gaotang County of Liaocheng City. The group contributes around 76 per cent of the total taxes paid to Gaotang County and has worked closely with local government to mobilize support for the legalization of LSEV use on public roads. In July 2008, the authorities there made a special arrangement to allow Shifeng's LSEVs to use the public roads. By the end of 2009, sales of LSEVs had reached 10,000 – 2,000 of which were sold on overseas markets. While Shifeng and other LSEV producers have had a measure of success at the local level, if the market is to expand further, change is needed at the national level.

Shifeng has actively sought changes in legislation at the national level. Several key decision-makers have been invited to visit the Shifeng Group, including members of the Development Research Centre of the State Council, the Ministry of Technology and the Ministry of Industry and Telecommunications. The status of the Shifeng Group as a state-owned enterprise has also allowed Mr Liu Yifa, president of the group, in his capacity as a Deputy to the National People's Congress (the legislative house in China), to argue for measures by central government to support LSEVs. The redrafting and delay in the publication of the policy document *Energy Efficiency and New-Energy Automotive Industry Planning (2011–2020)* is partly attributed to the mobilization of an interest group for LSEVs.

Outside China, LSEVs have already begun to find markets with foreign institutional buyers such as governments, police departments, hospitals, post offices and airports buying Chinese LSEVs as a low-cost 'green solution'. Similarly, a small number of private consumers in the USA and Europe have bought LSEVs as a low-cost alternative to a second or third car. However, as we shall see below, there is currently a great deal of uncertainty about the future of LSEVs in China, with a several outcomes being possible.

The implications for mainstream Western carmakers

In an earlier paper (Wang and Kimble, 2011) we outlined three possible scenarios for the development of the LSEV market in China.

The first is based on the previous experience of e-bicycles and low-speed farm vehicles in China. In this scenario, central government does not encourage the development of LSEVs but waits for the market to overcome existing legal and institutional barriers. If the development of LSEVs follows this trajectory then, as a result of the low cost of entry, the industry will remain fragmented for perhaps another ten years, after which restructuring and concentration, driven mainly by competition, will reduce the number of companies. Depending on developments elsewhere, this may or may not be to the disadvantage of Chinese LSEV manufacturers.

The second scenario offers a more pessimistic forecast. Here, the central government deliberately limits the growth of LSEVs, preferring instead to favour the development of designs that try to recreate the key features of Western passenger vehicles. If transport regulations are not changed to allow the use of LSEVs on the road and no standards for LSEVs are established, this will hinder the development of LSEVs and place Chinese manufacturers in direct competition with European and American car giants. While Chinese automobile manufacturers have shown themselves to be capable of remarkable innovations, this scenario will undoubtedly prove a challenge.

The final scenario sees the LSEV industry flourishing thanks to appropriate interventions by central government combined with the active engagement of individual companies and local governments. In this scenario, the expansion into international markets acts as a boost to LSEV production in China and the commercialization of low-speed electric vehicles elsewhere. While the outlook for the LSEV industry in China might be optimistic, it will almost certainly have a negative impact elsewhere.

Clearly, the direct competitive impact that the development of the LSEV in China would have elsewhere will depend on which of these scenarios is adopted, but regardless of this, the development of a successful 'LSEV business model', in China or elsewhere, could have profound implications for the existing automobile giants.

Looking first at the LSEV in China, because of its low price, low running costs and the ease of charging from a domestic 220 V electric outlet, the

LSEV offers a clear value proposition to low-income consumers, particularly those living in rural areas. It offers the basic utility of a car; the relatively short range is not a problem as most commuting distances are small; and it can be charged overnight. In addition, the use of LSEVs does not require the construction of the specialist charging stations needed for standard EVs, which has acted as an inhibitor to their spread in both rural and urban areas. In addition, as we have noted, most Chinese consumers do not have the fixed notions of 'a vehicle' (that is, a passenger car) or 'an electric vehicle' that mainstream Western consumers have, which may make this type of simple technology more easily accepted.

Outside China, the potential of LSEVs appears to be more limited. The utility of a low-speed, short-distance EV might seem obvious, but could it have the significant, or even revolutionary, implications alluded to in the earlier sections of this chapter? Will the LSEV ever be more than a technically inferior product that serves the specific needs of a particular group of consumers in a niche market? The key question here is whether the LSEV can offer a similar value proposition to people outside China.

Without entering into an analysis of the history of the automobile in Europe and America such as that offered by Calkins (2009), it is safe to say that, for most Western consumers, a passenger car is thought of as a long-distance cruising vehicle that has the capacity to reach relatively high speeds. The size of the fuel tank, the power of the engine and a long history of use has led us to expect our cars to carry us for long distances at speeds in excess of 100 km/h. However, the reality is that most of people live in urban areas, are subjected to speed limits of 50 km/h or less and they travel less than 50 km/day. Viewed as a simple matter of economics, this approach is a clear waste of resources for most urban users.

For LSEVs to offer an alternative there would have to be a change in the way we think about personal transport, perhaps using an LSEV for urban travel and public transport or some other form of fuel-efficient vehicle for long distances; however, this would raise a series of complex questions. These include questions of politics (Calkins, 2009), the social utility of car ownership (Steg *et al.*, 2001), the legacy of urban planning based around car use (Newman *et al.*, 1995) and a growing environmental concern among consumers regarding dwindling natural resources and pollution (Goldemberg, 1998).

While the LSEV is unlikely to provide the answer to all these issues, it is clear that it offers many of the same advantages to Western consumers as it does to those in China: simplicity, low cost and the removal of one of the main hurdles to the spread of EVs, namely charging stations. Thus, if the right business model can be found to commercialize this potential then the LSEV could prove to be disruptive in the sense that Mitchell and Coles (2003) use the term, placing competitors 'out of position and unable to respond effectively' (p. 15).

Currently the LSEV is something of a curiosity: a product that in global terms is clearly inferior but that serves the needs of a specific geographical and social niche. However, if the right business model can be found, then the LSEV also has the potential to be 'disruptive' in Bower and Christensen's (1995) sense of the term, currently only suitable for the needs of a niche market but with the capability of improving to the extent that it could meet the (changed) needs of a wider market.

References

Amit, R. and Zott, C. (2001) 'Value Creation in E-business', *Strategic Management Journal*, 22(6–7): 493–520.
Anderson, J. and Markides, C. (2007) 'Strategic Innovation at the Base of the Pyramid', *MIT Sloan Management Review*, 49(1): 83–8.
Bower, J. L. and Christensen, C. M. (1995) 'Disruptive Technologies: Catching the Wave', *Harvard Business Review*, 73(1): 43–53.
Calkins, M. (2009) 'King Car and the Ethics of Automobile Proponents' Strategies in China', *Journal of Business Ethics*, 85(supplement 1): 157–72.
Chesbrough, H. W. (2007) 'Business Model Innovation: It's Not Just About Technology Anymore', *Strategy & Leadership*, 35(6): 12–17.
Chesbrough, H. W. (2010) 'Business Model Innovation: Opportunities and Barriers', *Long Range Planning*, 43(2–3): 354–63.
Chesbrough, H. W. and Rosenbloom, R. S. (2002) 'The Role of the Business Model in Capturing Value from Innovation: Evidence from Xerox Corporation's Technology Spin-off Companies', *Industrial and Corporate Change*, 11(3): 529–55.
China Statistics Press (2010) *China Statistical Yearbook*.
Christensen, C. M. (1997) *The Innovator's Dilemma: When New Technologies Cause Great Firms to Fail*. Boston, MA: Harvard Business Press.
Christensen, C. M., Craig, T. and Hart, S. (2001) 'The Great Disruption', *Foreign Affairs*, 80(2): 80–95.
Eisenhardt, K. M. (1989) 'Building Theories from Case Study Research', *Academy of Management Review*, 14(4): 532–50.
Hart, S. L. and Christensen, C. M. (2002) 'The Great Leap: Driving Innovation from the Base of the Pyramid,' *MIT Sloan Management Review*, 44(1): 51–5.
Jacobson, M. Z. (2009) 'Review of Solutions to Global Warming, Air Pollution, and Energy Security', *Energy & Environmental Science*, 2(2): 148–73.
Khan, A. R. and Riskin, C. (1998) 'Income and Inequality in China: Composition, Distribution and Growth of Household Income, 1988 to 1995', *The China Quarterly*, (154): 221–53.
Leavy, B. (2007) 'Managing the Risks That Go With High-Impact Strategies in Uncertain Markets', *Strategy & Leadership*, 35(4): 43–6.
London, T. and Hart, S. L. (2004) 'Reinventing Strategies for Emerging Markets: Beyond the Transnational Model', *Journal of International Business Studies*, 35(5): 350–70.
Magretta, J. (2002) 'Why Business Models Matter', *Harvard Business Review*, 80(5): 86–92.
Markides, C. (2006) 'Disruptive Innovation: In Need of Better Theory', *The Journal of Product Innovation Management*, 23(1): 19–25.
Mitchell, D. W. and Coles, C. B. (2003) 'The Ultimate Competitive Advantage of Continuing Business Model Innovation', *Journal of Business Strategy*, 24(5): 15–21.

Mitchell, D. W. and Coles, C. B. (2004) 'Business Model Innovation Breakthrough Moves', *The Journal of Business Strategy*, 25(1): 16–26.

Morris, M., Schindehutte, M. and Allen, J. (2005) 'The Entrepreneur's Business Model: Toward a Unified Perspective', *Journal of Business Research*, 58(6): 726–35.

Nair, A. and Boulton, W. R. (2008) 'Innovation-oriented Operations Strategy Typology and Stage-based Model', *International Journal of Operations & Production Management*, 28(8): 748–771.

Newman, P., Kenworthy, J. and Vintila, P. (1995) 'Can We Overcome Automobile Dependence? Physical Planning in an Age of Urban Cynicism', *Cities*, 12(1): 53–65.

Osterwalder, A., Pigneur, Y. and Tucci, C. L. (2005) 'Clarifying Business Models: Origins, Present, and Future of the Concept', *Communications of the Association for Information Systems*, 15: 1–25.

Ranis, G. (1988) 'Analytics of Development: Dualism', in H. Chenery and T. N. Srinivasan, (eds) *The Handbook of Development Economics*, Vol. 1. Amsterdam: North-Holland.

Romm, J. (2006) 'The Car and Fuel of the Future', *Energy Policy*, 34(17): 2609–14.

Sánchez, P. and Ricart, J. (2010) 'Business Model Innovation and Sources of Value Creation in Low-Income Markets', *European Management Review*, 7(3): 138–54.

Steg, L., Vlek, C. and Slotegraaf, G. (2001) 'Instrumental-reasoned and Symbolic-Affective Motives for Using a Motor Car', *Transportation Research Part F: Traffic Psychology and Behaviour*, 4(3): 151–69.

Teece, D. J. (2010) 'Business Models, Business Strategy and Innovation', *Long Range Planning*, 43(2–3): 172–94.

Wang, H. (2002) *The Reconstruction of the Chinese Automobile Industry: What Is the Trajectory Towards Globalization?* Grenoble, France: University of Pierre Mendès.

Wang, H. and Kimble, C. (2010a) 'Betting on Chinese Electric Cars? Analysing BYD's Capacity for Innovation', *International Journal of Automotive Technology and Management*, 10(1): 77–92.

Wang, H. and Kimble, C. (2010b) 'Low-cost Strategy Through Product Architecture: Lessons from China', *Journal of Business Strategy*, 31(3): 12–20.

Wang, H. and Kimble, C. (2011, June) 'Business Model Innovation in the Chinese Electric Vehicle Industry', Paper presented at the 19th International GERPISA Colloquium, 'Is the Second Automobile Revolution on the Way?', Paris, 8–10 June.

Yang, D. T. and Hao, Z. (1999) 'Rural–Urban Disparity and Sectoral Labour Allocation in China', *The Journal of Development Studies*, 35(3): 105–33.

Yin, R. K. (2003) *Case Study Research: Design and Methods*, Vol. 5 (3rd edn). Thousand Oaks, CA: Sage.

Zhao, J. (2006) 'Whither the Car? China's Automobile Industry and Cleaner Vehicle Technologies', *Development and Change*, 37(1): 121–44.

Zuev, V. V., Burlakov, V. D., El'nikov, A. V. and Goldemberg, J. (1998) 'Leapfrog Energy Technologies', *Energy Policy*, 26(10): 729–41.

14
Urban Mobility as a Product of Systemic Change and the Greening of the Automotive Industry

Francesco Garibaldo

The focus of the automotive industry is shifting and must shift from merely delivering cars to delivering mobility with cars being by-products of this.

One of the profound traits of our condition as modern humans is the conquest of personal mobility as a social good, progressively widened to include all social classes. The years during which the modern world was being shaped, according to the historians – that is, between 1815 and 1830 – saw the beginning of the transport revolution. At the end of the decade between 1820 and 1830, both Paris and London disposed of efficiently organized public transport systems based on horses: the omnibuses. A quotation from Paul Johnson (1994, p. 882) aptly introduces the first mobility pattern change:

> With the invention of the omnibus, everyone has his own carriage, and how cheap it is! The bane of the domestic servants has come to an end, along with the cost of their liveries, the bills of the coach-builders and the veterinary surgeons, the need to keep our fellow creatures waiting for us in the cold of the night or in the rain, while in the hall there is a bustle of dancing or another hour is being spent taking one's leave in a leisurely fashion before the glowing hearth. We needn't concern ourselves over the means of transport until we need to, and at that point the 'bus arrives at our doorstep, otherwise we go out and a few minutes later we see it appearing at the bottom of the street.

The transition from carriages to the omnibus is a transition from the private ownership of the means of locomotion to a solution designed for a specifically urban situation; a similar transition is the one we have to cope with today.

The crisis of car-focused mobility

Mobility, or should I say the absence of mobility because of congested city traffic, is undermining one of the historical advantages of cities: the extreme ease with which different peoples and cultures come into contact with one another – what economists term a 'reduction of transaction costs'. In the interpretation I am proposing, then, what is relevant is not the technical means of assuring mobility, but mobility as a social asset and an individual right to be safeguarded.

At the beginning of the twentieth century, the way of looking at mobility changed radically – both a democratization and an urbanization of the problem occurred – and so did the technical mode for its realization; the previous omnibus pattern was de-structured and a new culture came to the fore, able to come to terms both with a different articulation of society and individual rights, and with the changing role of cities that became industrial locations and thus dense and large enough for collective, reticulated solutions. This shift in perspectives has some preconditions as well as social and economic consequences, but this is not the place to analyse them thoroughly. I just want to mention some themes: the extramural space as a public space needing infrastructure; the great cities seen as spaces to be covered by the construction of inter-nodal transport networks, with the consequent enhancement of the radial connections between city centres and their outskirts, rather than those links that are merely used for transit; therefore, there is a need for planning and for an increase in public action as the only way to provide an answer to the problem of mobility. This was why the tramcar system was such a success.

These changes led to new economic–industrial developments: substantial fixed investments, the birth of new productive sectors and of new services, and a continual search for the lowest service prices to increase the number of users and/or clients.

Shaping the network

There is a 'network dimension' of each system of mobility that must be separated from the specific means of locomotion. Historically, there have been two different transport patterns: (i) a highly flexible system, such as carriages and cars, based on private ownership of individually oriented means of locomotion; and (ii) a lowly flexible or reticulate solution, such as the omnibus, trolley bus, motorbus or underground railway, based on planning and an increase in public action, but not necessarily on public ownership. This was evident in the tramcar system but is still true in the case of the motor vehicle system.

The car is perceived as being 'private', but it cannot be used in isolation, only within the scope of metropolitan and social space inside which the physical and information technology networks are structured, such as the traffic light systems, the car park layouts, the spatial typologies of the roads; that is, a specific *forma urbis* – the shape of the city.

The network will therefore become the basic structure for mobility, and the publicly made agreements on the nature of the network will become the basic determinant of the process of production of mobility, via laws, rules, standards and so on.

Mobility as the actual product

Looking at mobility today interweaves different issues:

- mobility as the actual product: a split between what Marx called the utility value and a good's exchange value;
- overproduction and market segmentation: a strategic crisis of the car industry and the growing differentiation between urban and non-urban environments;
- energy supply and the time horizon: new powertrains; and
- the search for a new sustainable social urban model.

In other words, if cars are a by-product of the real product – that is, mobility – then the focus is on designing the city and the infrastructure for mobility at the same time; cars are just part of a broader system. In addition, each new design implies the search for new products to support mobility, and of new industrial processes to build them.

It implies the necessity for a vertical integration between the classical automobile sector and a newly emerging urban design and management sector.

Utility value as a vector for change

The levels of technological and economic lock-in of some investments have never to date prevented the radical reorientation of technological and economic investments when the level of social and cultural lock-in that sustained the previous model has been de-structured. This has not always happened in the same way and at the same pace – the horse-drawn omnibus has disappeared, for example, but the tramway is enjoying a second lease of life and the underground railway has never been toppled from its position as a key solution in the metropolis – in each phase, what has changed is the reciprocal relationship between the different technological strands. The fundamental vector of change we think we can isolate – the utility value of the different goods for mobility – is the problem of mobility as a collective as well as a personal right.

At a certain point an insoluble conflict has arisen between the specific modality for realizing mobility in a given period and its actual use, either collectively or individually. In short, there has been a split between what Karl Marx called the utility value and the exchange value of the good. At this point, the conditions are determined that allow for a transformation, and when it takes place, as a general rule the drive is so powerful that the process does not stop, even before the destruction of the non-amortized fixed capital or before the vested interests of the losing industries, however powerful they might be; a case in point is the destruction of the tramways or the urban sprawl to make way for the car. This was the case with the success of cars on street railways (cable cars) in the USA, as this 1933 report to U S president, Herbert Hoover, witnesses (D'Eramo, 1995, p. 96):

> The rapid popular consensus for the new vehicle is mainly due to the fact that it has given its owner the control over his own movements that he had been denied by the previous means [this was only possible in the previous aristocratic and upper middle-class model of the carriage]. Within easy reach and ready for instant use, it takes its owner from his doorstep to his destination according to itineraries that he himself has chosen and within a time schedule and an itinerary that he himself has decided.

The process of change has hitherto come about blindly – that is, without any prior awareness of the medium- or long-term consequences of a given choice; which is another way of saying how much the drive towards mobility is a 'primary drive' that needs to be 'elaborated'.

The car industry still represents best example of industrial rationality. It is the industry with the highest manufacturing employment rate for both developed and developing countries; in short, it is a strategic industry. The amount of direct and indirect investment by individual states, households and private capital in the car industry is enormous. What is increasingly crisis-ridden is the car understood as the prime and most rational means for assuring mobility both inside and between cities. To cope with this strategic crisis means the development of a transition to a new equilibrium regarding the means of assuring personal mobility and the role of cars. It implies a new industrial perspective:

- the search for a new business model in the automotive industry;
- the development of the hybrid manufacturing model (Bryson, 2011), also in the automotive industry;
- a closer integration between the strategic goals pursued in the different social and economic activities affecting the mobility system, and the nature of the mobility system (Chiao, 2011). For example, deciding to stop the extension of the urban sprawl implies the choice of 'vertical shaped' cities; to cope with clean energy production and use leads to a portfolio of solutions for people mobility and goods delivery, based on clean, smart vehicles and new infrastructures such as elevators, escalators and cableways, but also to recover old means, such as bicycles (Brillembourg, 2011);
- the development of an urban mobility platform using ICT and communicating with smart vehicles (Sassen, 2011); and
- publicly available clean, smart vehicles, based on a web-shaped network.

How can these different dimensions be managed?

Overproduction and market segmentation

At the time of writing, the automotive industry is in structural crisis because of the convergence of many different dynamics. On the one hand there are the demographic trends: the majority of the world's population is now living in urban areas and these areas are growing in size. The cities of the future will have up to 3–5 million inhabitants, not including the already existing megacities. This trend will generate and/or exacerbate many interrelated negative consequences: first, traffic congestion in these areas, because of the prevailing model of the individual use of cars for short runs; second, a growing inadequacy of cars to assure efficient and effective mobility, as a result of traffic congestion and the very slow average speed of the traffic; third, the pollution caused by traditional petrol and diesel engines; and finally, an energy delivery problem resulting from the foreseeable end of the availability of affordable oil supplies and the need for energy supplies for any kind of alternative engine.

The competitive model of recent decades has generated a situation of structural overproduction in Europe. The European market, still very relevant with around a quarter of all the registered light vehicles in the world,[1] is indeed a mature market, mainly of substitution for old cars, and European producers are competing to conquer substantial market shares outside Europe, namely in the booming BRIC market – in Brazil, Russia, India and China. This race to export is based mainly on the development of new facilities in these countries and on the flow of new models, generates a structural overcapacity, though distributed unevenly.

Europe therefore finds itself in a classical Marxian overproduction crisis, the foundations of which lie in the stagnation since the early 1980s. It means that, in a capitalistic regime, the excess of supply over demand – a paradox in itself, because of the unbelievable amount of existing and unfulfilled individual and social demand – is relative. It depends on the impossibility of selling goods and services at a profit; to be more precise, at an acceptable profit. To be 'acceptable' is a social and not an absolute measure (Garibaldo, 2011, p. 165).

Overcapacity and the stagnation of working-class incomes compelled countries to find other markets for their output, and to choose between neo-mercantilism and an economy based on debt. This, in turn, has created enormous room for manoeuvre for financial capital. This has been one factor, among many others, contributing to the political implosion of the European Union (EU), unable to find common industrial and labour policies to face up to the crisis, and giving way, instead, to nationalistic attempts to defend each country's status quo, as seen in the Opel–General Motors quarrel.

This situation is leading to a new industrial concentration of the car industry, but not as deep as the Fiat CEO, Sergio Marchionne, has forecast

(*Financial Times*, 2009). The reason is the birth of global niches, the still extreme segmentation of the market, and the shift of the focus of the car market to the East. Up to now, the main demand from the East to European producers has been for cars in the premium sector, but the creation of a new segment, the city car, is now of the utmost importance for the future of the European automotive industry. The way to assure urban mobility in this new scenario is the key strategic choice for the automotive industry all over the world.

Energy supply and the time horizon: a portfolio of solutions

Vehicles need energy made available by an endothermic engine, using oil or gas, or by an electric engine. Oil and gas are non-renewable sources of energy, and according to some statistical studies they will be unavailable or extremely expensive to find by 2050 to 2060.[2] Electric power can be produced through so-called 'fuel cells', or through nuclear fusion. Fuel-cell technology, according to a joint report issued by all the European carmakers (with the exception of Fiat), the European Climate Foundation and the European Fuel Cells and Hydrogen Joint Undertaking, is already beginning the industrialization process[3] but it is still at an experimental stage. Nuclear fusion is still at the research phase and there are no realistic forecasts on its availability in the near future. The only existing industrial alternative is electric batteries, or new urban infrastructure such as an electromagnetic grid placed under the road network, as hypothesized for ecological new towns in China. Electric batteries are still in a very intense phase of industrial R&D to overcome the main obstacles to the possibility of fully substituting for endothermic engines: the low energy density. However, they are already suitable for use in urban environments.

What is relevant is that, according to the joint report quoted above, the EU's goal of an 80 per cent reduction in CO_2 emissions in Europe implies a decarbonization of the car emissions of up to 95 per cent. To reach this level of decarbonization would be impossible with endothermic engines, so the authors of the joint report support the idea of a portfolio of electrical solutions as the only realistic way forward.

A systemic change

The required change is very complex and cannot rely on market dynamics alone. Public planning and public investments are required, both for urban planning and renewal, and for basic infrastructure – for example, a publicly available system of recharging for electric cars and an ICT-based urban mobility platform; for industrial policies – for example, how to accommodate the industrial transition; and for environmental policies; and finally for R&D policies on new technologies.

Forma urbis and infrastructures

The specific features of the infrastructure depend on many different issues – for example:

- the degree of physical embeddedness of the inherited *forma urbis*: for example, in Italian cities with a medieval legacy of small and very dense central areas, such as Perugia and Ferrara, escalators have proved very effective in the former, and bicycles in the latter;
- the size of the city: the central area of middle-sized Italian towns is often smaller than the passenger area of Frankfurt Airport. This means that escalators, and/or moving walkways can be effective solutions.
- the interaction of GPS, electronic mapping and smart vehicles allow the setting up of an integrated platform to manage traffic.

This means that on the user side there is the possibility of a portfolio of mobility modalities within as well as outside the urban environment, which leads to a portfolio of solutions that need a new infrastructure of services, new vehicles based on new powertrains that must be clean; that is, designed for zero emissions. It implies a deep restructuring of the automotive industry and of business models, as well as a new set of employee skills and competencies.

Industrial policies

A set of European industrial policies should be a co-ordinated effort oriented ecologically in different fields, such as energy, urban planning, industrial renewal and technologies, to guarantee an efficient and effective mobility of people and goods. These policies should rely on and to enrich the industrial inheritance of the automotive industry, both as industrial and technological knowledge as well as of people skills and competencies. It is possible to use part of this inheritance for the renewal of other industrial sectors – for example, energy production. It means managing a long industrial and social transition. The time scales are 40 to 50 years for a brand new and widespread mobility system; 20 to 30 years for a portfolio of new and old solutions; and 10 years for a relevant restructuring of the European automotive industry, avoiding intolerable social costs.

The idea that the future will be without cars is totally unrealistic; the use of cars will remain an effective and efficient solution in non-urban areas and, in a certain proportion and with new business models, in urban areas – for example, the possibility of renting a car to use by the hour instead of owning one. The main big change will be the ongoing new segmentation that will overcome the idea of all-purpose cars in favour of specific-purposes vehicles; that is, cars designed for a specific environment, such as cities. This is the other side of the concept of a portfolio of solutions.

To accomplish a socially responsible industrial transition means also changing the proportions between exports and the internal market in Europe. If Europe is prepared to gamble on a complete renewal of the transport system in its many large conurbations there will be a new operative space for the automotive industry, but it means deploying new products.

Integrated systems and the role of R&D

Integrated urban mobility systems are made up of public transport (buses, underground trains, cable cars and so on) and publicly available pedestrian systems – (escalators, moving walkways, for example) as well as private systems –traditional private ownership, but also new leasing systems. They may be accessed as part of a centralized governance system such as a mobility platform, or by individuals buying a 'mobility carnet', made up of a portfolio of means such as the possibility of using a car when needed (or a bicycle or a motorbike), or perhaps even an electric vehicle.

This perspective requires huge public investment in education, to support the process of industrial restructuring and renewal, and in R&D in the fields of batteries, fuel cells and nuclear fusion as well as developing new urban infrastructures.

Conclusions

The multifarious features of mobility make difficult to organize a specific set of policies to maintain its social and personal value.

Some dimensions or features have been selected and highlighted in this chapter. They are deeply interconnected in a dynamic way, but two drivers for change have been illustrated: the networked social dimension of all kinds of mobility system, on the one hand, and the role of utility value on the other. These drivers are merely logical design tools, but what is needed is a social coalition and social actors both willing and with enough power to do it, as well as public powers thinking in a strategic way in order to achieve full employment, a sustainable environment and a satisfactory social life for their citizens. This agenda is not one that is emerging naturally from the free movement of the market and the invisible hand of the different strategies of industrialists – or from out of thin air. Government polices and the strategic behaviour of actors in civil society, such as trade unions, non-governmental organizations and civil movements are required. The transition from the car-based system of mobility to a new system based on a portfolio of demands and solutions can be achieved, and it is possible to organize the transition in a socially acceptable way.

Notes

1. http://www.acea.be/collection/statistics, accessed 27 January 2012. In addition, according to the Boston Consulting Group, 'Winning the BRIC Auto Markets',

January 2010, BRIC countries will represent 30 per cent of all registered cars in 2014; see http://www.bcg.com/expertise_impact/publications/AllPublications.aspx?page=2&practiceArea=Automotive&sortBy=Date, page 4, accessed 18 August 2011.
2. According to the BP Statistical Review of World Energy, June 2010 – see http://www.bp.com/liveassets/bp_internet/globalbp/globalbp_uk_english/reports_and_publications/statistical_energy_review_2008/STAGING/local_assets/2010_downloads/statistical_review_of_world_energy_full_report_2010.pdf, accessed 31 August 2011. The time horizon forecast is 40–50 years for oil and 60–70 years for natural gases.
3. The report is available at http://www.europeanclimate.org/documents/Power_trains_for_Europe.pdf, accessed 18 August 2011.

References

D'Eramo, M. (1995) *Il maiale e il grattacielo. Chicago una storia del nostro futuro*. Milan: Feltrinelli.
Garibaldo, F. (2011) 'Verso una nuova politica industriale', in G. Volpato and F. Zirpoli, (eds), *L'industria dell'auto dopo la crisi: player, politiche e nuove tecnologie sostenibili*. Milan: Brioschi Editore.
Johnson, P. (1994) *La nascita del moderno, 1815–1830*. Milan: Il Corbaccio. English edn (1991) *The Birth of the Modern: World Society 1815–1830*. New York: HarperCollins.

References to internet sources

Brillembourg, A. (2011) 'Empowering the Bottom of the Pyramid with Good Design', McKinsey and Company. Available at: http://whatmatters.mckinseydigital.com/the_debate_zone/as-the-world-urbanizes-will-the-most-successful-cities-result-from-top-down-planning-or-bottom-up-innovation#a; accessed 31 August 2011.
Bryson, J. R. (2010) 'Hybrid Manufacturing Systems and Hybrid Products: Services, Production and Industrialisation'. Available at: http://www.internationalmonitoring.com/downloads/trend-studies/hybrid-products.html;accessed 31 August 2011.
Chiao, S. C. S. (2011) 'Planning China's Megacities', McKinsey and Company. Available at: http://whatmatters.mckinseydigital.com/cities/planning-china-s-megacities; accessed 31 August 2011.
Financial Times (2009) 'Driving on Regardless', Interview with Sergio Marchionne, 4 May. Available at: http://digital.olivesoftware.com/Repository/ml.asp?Ref=RlRFLzIwMDkvMDUvMDQjQXIwMDcwMA%3D%3D&Mode=Gif&Locale=english-skin-europe.
Sassen, S. (2011) 'Talking Back to Your Intelligent City', McKinsey and Company; available at: http://whatmatters.mckinseydigital.com/cities/talking-back-to-your-intelligent-city; accessed 31 August 2011.

15
A Forecasting Framework for Evaluating Alternative Vehicle Fuels, Using the Analytic Hierarchy Process Model and Scenarios

Tugrul Unsal Daim and Jubin Dilip Upadhyay

'Green' is the magic word in today's world. Many different fuels and fuel motors are being developed, assuming they will be adopted and used in the future as an alternative to fossil-fuel-based technologies. But many questions still remain unanswered. Is there going to be a supply problem? Is the 'Running out of fossil fuels' scenario merely a hype? Do we have any green alternative that can be adopted on a mass scale? Is there any alternative fuel capable of giving all the benefits of current fossil fuels and simultaneously eliminating pollution and price problems? And many more such questions! Alternative fuel development research arose from the perennial uncertainty in oil prices. This chapter attempts to study the reasons behind oil price fluctuations, believing them to be the root of the problem. Then the Analytic Hierarchy Process (AHP) model is presented. The AHP model has different levels, like a tree, the lowest level displaying all the alternatives to the problem. Each type of alternative motor fuel is briefly explained. Issues such as cost and the pollution impact, are assessed for each. Then, in the next level of the AHP model, the criteria are presented. Criteria include the different factors that have an impact on the adoption of each alternative. Economic, cultural, environmental, sustainability and development time criteria are considered. Finally, in the uppermost level of the AHP tree model is the goal: to find the best motor fuel for the future. To prove the robustness of each alternative fuel, four different scenarios are used, and quantitative results for each scenario are presented. It is hoped that the quantified values for each alternative might give us an idea of where the solution lies.

The challenge

In these times when a range of factors are affecting oil prices, carmakers are pursuing the quest for alternative vehicle fuels. Different technologies are being developed, but to try to forecast which of these will have a promising future we need to evaluate the different underlying factors that could determine their performance on the market. One of the most important underlying factors is the price of oil. This is determined by a delicate balance between supply and demand.

The 'oil peak', which could be defined as a depleted oil supply caused by a growing demand, is the dominant factor, and this has worried forecasters and economists alike. The thought that one day oil might run out and that the world will not be prepared for that when it happens is scary. This thought is what encouraged the researchers, government and industry to think about what might replace petroleum.

The meaning of the oil price

A low oil price indicates enough supply to satisfy demand, whereas a high price suggests a depleted supply which cannot satisfy demand. Supply and demand can be affected by a variety of factors. If we extrapolate the history of the oil price curve we should be able to see when the oil price peak of the 1970s will repeat itself again. As of August 2011, the price per barrel reached US$88.74 (Oilprice.net, 2011). The question is not *whether* the price peak will happen, but *when*. Let us try to identify which factors could trigger an oil price peak. It is obvious that a balance between demand and supply is necessary, but what are the factors that might produce imbalances between the two? An analysis of the fluctuation of the price of oil plus some additional research lead us towards understanding the factors that give rise to supply and demand imbalances, and these imbalances play a major role in determining the price of oil. These factors are listed below:

- *Consumption patterns*. Prices may fall as consumers respond to higher prices by adopting more energy-efficient practices.
- *Emerging economies*. Large increases in crude oil prices and petrol prices is partly a result of a rapid growth in demand from China and India:
  - Potential threats to American oil interests are India, China and Russia.
  - The International Energy Agency (IEA) estimates that oil imports by China and India will increase exponentially, and we can already see that this is happening. By 2020, India is expected to have to import 80 per cent of its energy needs.
  - India already has around more than 30 times more cars than it did in 1990; and China about 90 times – by 2030 China will have more cars than the USA, according to the Energy Research Institute of Beijing.
  - What worries Western powers most are China's and India's growing ties with Iran (Pocha, 2005; Kronstadt and Katzman, 2006). India recently cleared the entire US$5 billion oil debt of Iran and signed various energy co-operation agreements, promising more investment in Iran (Financial Express, 2011).

- *Environmental issues*. Greater conservation or increased fuel efficiency could decrease demand, thus decreasing oil prices.
- *International efforts* to reduce the generation of greenhouse gas (GHG) emissions could cause a reduction in the demand for crude oil through the development and use of more fuel-efficient processes.
- *Alternative sources of energy improvement*. Innovations that reduce the costs of alternative sources of energy could reduce the demand for crude oil and thereby ease price pressures (biodiesel, hydrogen fuel cells and so on).
- *Investment in production capacity*. Prices may fall as oil companies invest in more crude oil producing and refining capacity.
- *A lower level of inventories* may cause prices to be more volatile when a supply disruption occurs.
- *Unexpected refinery outages or accidents* may cause prices to increase (for example, the effects of Hurricane Katrina in 2005 on oil wells in the Gulf of Mexico).
- *The difficult location* of new sources of oil which may be more expensive to develop (for example, tar sands, beneath deep water in the ocean and so on).
- *Technological innovations* improve the ability to extract and process oil, which will increase the available future supply; for example:
  - The increased use of 3D seismic data reduces drilling risk.
  - Directional and horizontal drilling leads to improved production in many reservoirs.
  - Financial instruments are used to limit exposure to price movements.
  - Increased use of CO_2 flooding and improved recovery methods improve production in existing wells.
- *OPEC behaviour*. OPEC members have collectively agreed to restrict the production of crude oil in order to increase world prices. Analysts warn that oil prices – which have doubled since 2009 – are unlikely to fall in the near future. One reason is that the diplomatic showdown over Iran's nuclear ambitions is escalating, raising the prospect of sanctions against the nation, the second-largest oil producer in the Organization of the Petroleum Exporting Countries (OPEC). Analysts and traders fear that international penalties against Iran might cramp its oil industry, slow energy investments or remove sorely needed barrels from the market.

Another source of concern is Nigeria, also an OPEC producer, where armed rebels have threatened more attacks against the country's energy sector – protesting, among other things, President Olusegun Obasanjo's plan to seek a third term in last year's elections. Since the beginning of 2006, attacks by militants in the Niger Delta have forced oil companies to cut production by about 500,000 barrels a day, or more than 20 per cent of the country's output.

'Without question, this is the worst political-risk year we've seen for energy supplies since 1973,' said Ian Bremmer, president of the Eurasia

Group, a consultancy in New York, referring to the year of the first Arab oil embargo. 'The likelihood of escalation from the two biggest threats out there – Iran and Nigeria – remains very strong. In both cases, the worst is still ahead of us.' (*The New York Times*, 18 April 2006). Nigeria remains the fifth-largest supplier of oil to the USA, with a little over 10 per cent of the nation's imports coming from the Nigerian Delta (Upi.com, 2009).

- *Regulations* that increase the cost or otherwise limit the building of refineries and the storage capacity for imported crude oil may lead to supply disruptions and price volatility.
- *Instability in the Middle East, and particularly the Persian Gulf region* has caused major disruption to oil supplies. (The US government spent more than US$120 billion in 2010–11 alone through the Department of Defense to secure seaways and ports from terrorists, pirates and hostile nations.)
- *Value of the US dollar.* Oil-producing countries may wish to increase prices for their crude oil in order to maintain their purchasing power in the face of a weakening US dollar.

The instability in the Middle East, considered to be the main focus of oil supply, the rising number of terrorist activities, the organized network of Somalian sea pirates active in the channels of the Gulf of Aden and the Yemen, and the rise in oil consumption by developing countries, have all contributed the spike in the oil price at the time of writing.

Such peaks and troughs in the international price of oil are neither disastrous nor surprising. As may be true in most businesses, ups and downs are cyclic in nature, but the underlying truth is that oil, while renewable over a very long time scale, fossil fuels are being exploited at rates that may deplete reserves in the near future, and therefore oil is not considered to be a sustainable source of energy.

New fuels are certain to be widely adopted for future vehicles. Regardless of various assessments of the abundance of petroleum, deposits are finite and vehicle energy supplies must ultimately come from another source. According to Riley and Chee (2003), even if economically recoverable petroleum reserves are three times greater than today's known reserves, we shall probably have to move away from oil as a primary source of transportation energy by the year 2020. So alternative motor fuels or energy sources are inevitable. It is also possible that alternative fuels may ultimately be used more as fuels for the generation of electricity, rather than as fuels consumed by individual vehicles. Electric cars might then complete the loop (Riley and Chee, 2003).

In this introductory section we have presented the problem: a growing oil demand and a shrinking oil supply, which have forced researchers, industry and governments to move on to find alternative motor fuels. The purpose of this chapter is to use the Analytic Hierarchy Process (AHP) as a forecasting tool to try to determine which underlying factors will play an important role in determining the successful implementation of different alternative motor fuels for transportation. Once we have determined which factors are important, we go on to do a sensitivity analysis by using different scenarios, to see if our key factors remain consistent if the scenario changes. In the next section we present the definition of AHP, how it has been used in the past, its benefits as well as it disadvantages, and the steps needed to carry out the AHP analysis. The method is based on the earlier work of Winebrake and Creswick (2003). They integrated AHP and three scenarios for hydrogen fuelling systems.

The AHP model

The Analytic Hierarchy Process (AHP) is a theory of relative measurement on absolute scales of both tangible and intangible criteria based on a made paired comparison judgement by knowledgeable experts (Ozdemir and Saaty, 2006). AHP is a flexible multi-criteria decision-making tool, which helps to set priorities to make the best decision when both qualitative and quantitative aspects of a decision need to be considered (Ahsan and Bartlema, 2005). It includes ranking and comparison models. It is based on the innate human ability to make sound judgements; it is natural to human intuition and general thinking.

AHP in essence is the use of ratio scales in elaborate structures to assess complex problems (Saaty, 1994). The output of a Delphi method can be used as an input to AHP to prioritize different performance criteria (Ahsan and Bartelma, 2005).

AHP is used in over 20 countries (Saaty, 1994), in diverse disciplines from environment to health, manufacturing to economic planning, computer selection to portfolio selection, marketing to database system selection, and more. Below is a list of advantages and disadvantages of AHP:

- It has fewer restrictive assumptions than other decision-modelling techniques, and the ability to incorporate and trade off values and influences with a greater accuracy of understanding.
- It has the ability to include judgements resulting from intuition and emotion as well as from logic.
- It allows for a combination of conclusions from different people (Saaty, 1994).
- There are limitations in the way that pairwise comparisons are used, and the way that AHP evaluates alternatives.
- It may reverse the ranking of alternatives when an identical alternative is introduced within the same tier of the AHP model (Saaty, 1994; Ahsan and Bartelma, 2005); however, the distribute mode of AHP overcomes this deficiency (Ahsan and Bartelma, 2005).

Alternatives

Petrol

The main source for the manufacture of petrol is crude oil. All types of petrol-driven vehicles are available today, and petrol is available at all filling stations (afdc.energy.gov, 2011).

Pollution and safety issues

Petrol produces harmful emissions; however, petrol itself and petrol-driven vehicles are rapidly improving and emissions are being reduced. Petrol is not an energy-secure option because it is manufactured using mainly imported oil. It is a relatively safe fuel since people have learnt to use it safely; however, it is not biodegradable, so a spill could pollute soil or water (afdc.energy.gov, 2011).

Electricity

Electricity is the only alternative fuel source that provides direct mechanical power rather than providing power through chemical energy and combustion.

Pollution and safety issues

Since the process takes place without combustion, there is no pollution emitted, therefore electrical vehicles (EVs) have zero tailpipe emissions; however, some quantity of emissions can be attributed to power generation. Since batteries for electric vehicles have a limited storage capacity, they can be recharged from the existing power grid or through solar or wind energy (news.stanford.edu, 2008). Electricity is generated mainly through coal-fired power plants (for example, only 2.9 per cent of the USA's electricity comes from renewable resources; about half is generated in coal-burning plants). Coal is the USA's most plentiful fossil energy resource, and is the most economical and price-stable fossil fuel (mapawatt.com, 2010).

Biodiesel

Unlike spark-ignition engines, diesels rely solely on high compression in the cylinder to raise the temperature of the air enough to ignite the fuel. Consequently, diesel engines are tolerant of fuels of varying quality, and the high compression results in high efficiency. Diesels extract more energy from each gallon of fuel than do petrol engines, and less energy is lost as heat leaving via the exhaust pipe than with a petrol engine (Baker, 2011).

Pollution and safety issues

According to the US Department of Energy (DOE), pure biodiesel reduces carbon monoxide (CO) emissions by more than 75 per cent over petroleum diesel. A blend of 20 per cent biodiesel and 80 per cent petrodiesel, sold as B20, reduces carbon dioxide (CO_2) emissions by around 15 per cent. Therefore, biodiesel reduces particulate matter and global-warming gas emissions compared to conventional diesel; however, oxides of nitrogen (NOx) emissions may be increased. Biodiesel is manufactured using imported oil, which is not an energy-secure option. Diesel is also a relatively safe fuel, since people have learned to use it safely. In the same way as petrol, diesel is not biodegradable, so a spill could pollute soil or water (afdc.energy.gov, 2011).

Ethanol

Ethanol is an excellent, clean-burning fuel, potentially providing more horsepower than petrol. Alcohol-based ethanol is made from distilled starch crops such as sugar cane, corn, barley and wheat. Combined with gasoline it makes E85, which can fuel cars, which are known as flexible fuel vehicles (FFVs; USDOE, 2008). Ethanol is ethyl alcohol, often referred to as grain alcohol; E85 is a blend of 85 per cent ethanol and 15 per cent petrol. The performance of E85 vehicles is potentially higher than of petrol vehicles because E85's high octane rating allows a much higher compression ratio, which translates into greater thermodynamic efficiency. However, FFVs that retain the capacity to run on petrol alone cannot really take advantage of this octane boost, since they also need to be able to run on pump-grade petrol. Also, growing corn is an intensive process that requires pesticides, fertilizer, heavy equipment and transport. When considering the viability of ethanol, the total impact of all this activity needs to be taken into account.

One acre (c.4,046 sq. metres) of corn can produce 300 gallons (c.1,136 litres) of ethanol per growing season. So, in order to replace 200 billion gallons (757 billion litres) of petroleum products, American farmers would need to dedicate 675 million acres (c.2,732 million sq. km), or 71 per cent of the nation's 938 million acres (c.3,796 million sq. km) of farmland, to growing feedstock. Clearly, ethanol alone won't kick the present-day fossil fuel dependence in the USA – unless the country wants to replace its oil imports with food imports (Naylor, 2007).

Pollution and safety issues

According to the DOE, the growing, fermenting and distillation chain actually results in a surplus of energy ranging from 34 per cent to 66 per cent. Moreover, the CO_2 that an engine produces started out as the atmospheric CO_2 that the cornstalk captured during growth, thus making ethanol GHG neutral. Recent DOE studies note that using ethanol in blends lowers CO and CO_2 emissions substantially. In 2005, burning such blends had the same effect on GHG emissions as removing 1 million cars from American roads (cbo.gov, 2009). Ethanol's benefits are that its source can be farmed, making it domestically renewable, and vehicles fuelled by ethanol usually produce fewer injurious emissions. Some environmentalists point out, however, that some companies producing ethanol are releasing volatile organic compounds and CO into the air during production (cbsnews.com, 2010). E85 vehicles can demonstrate a 25 per cent reduction in ozone-forming emissions compared to reformulated petrol. However, ethanol can form an explosive vapour in fuel tanks. In accidents, however, ethanol is less dangerous than petrol because its low evaporation speed keeps the alcohol concentration in the air low and non-explosive (afdc.energy.gov, 2011).

Liquefied natural gas (LNG)

Natural gas or liquefied natural gas (LNG) is composed mainly of methane with a mix of other hydrocarbons. Natural gas is attractive as an alternative fuel source because it burns cleanly and can be produced domestically. The fuel can be stored as compressed gas or in liquid form, and can also be mixed with hydrogen to create a fuel blend (web.mit.edu, 2011). In general, natural gas supplies are abundant, and pipelines for fuel transport and distribution are extensive and adequate. Even under conservative conditions, it is estimated that the recoverable gas resources in the lower 48 states of the USA are sufficient to serve the current demand for gas for another 60–70 years. However, temporary natural gas supply shortages may occur in states that experience extremely cold temperatures in winter.

Pollution and safety issues

LNG vehicles can demonstrate a reduction in ozone-forming emissions compared to some conventional fuels; however, volatile hydrocarbon (HC) emissions may be increased. LNG is produced domestically and typically costs less than petrol and diesel fuels. Cryogenic fuels require special handling procedures and equipment to properly store and dispense them (afdc. energy.gov, 2011).

Propane (LPG)

Liquefied petroleum gas (LPG), or propane, produces fewer emissions as a fuel source than petrol and is accessible to the public because of an already-existing infrastructure of storage, processing and transportation methods. A by-product of natural gas processing and crude oil refining, propane also tends to be a cheaper fuel source than petrol (eia.gov, 2011).

Pollution and safety issues

LPG vehicles demonstrate a 60 per cent reduction in ozone-forming emissions compared to reformulated petrol. LPG is the most widely available alternative fuel with an estimated 3,400 refuelling sites nationwide in the USA. The disadvantage of LPG, however, is that 45 per cent of this fuel in the USA is derived from oil.

Adequate ventilation is important for fuelling an LPG vehicle because of the increased flammability of LPG. LPG tanks are 20 times more puncture resistant than petrol tanks and can withstand a high impact (afdc.energy.gov, 2011).

Hydrogen

Hydrogen is the lightest alternative fuel source, allowing for greater ease of transportation and storage. The gas is being used experimentally for fuel-cell vehicles and combustion engines. The benefits of hydrogen fuel are that it is an easily renewable fuel source which can be produced by using a variety of methods, and its use may improve air quality and reduce GHG emissions. Though hydrogen can be used to drive a modified internal combustion engine (ICE), most developers see hydrogen as a way of powering fuel cells to move vehicles electrically. The only by-product of a hydrogen fuel cell is water.

Pollution and safety issues

The environmental impacts of burning this fuel are zero regulated emissions for fuel-cell-powered vehicles, and only NOx emissions possible for ICEs operating on hydrogen. Hydrogen can help to reduce US dependence on foreign oil by being produced from renewable resources. It has an excellent industrial safety record, and codes and standards for consumer vehicle use are under development (afdc.energy.gov, 2011).

In this section we presented each alternative solution to the problem: petrol, hybrid/electric, biodiesel, ethanol, natural gas, propane and hydrogen. We emphasized, among other issues, infrastructure cost and pollution problems. In the next section, we apply the AHP to the problem presented in the introductory section, by following each of the steps to apply the AHP to the problem:

- construct the hierarchies;
- perform pairwise comparisons;
- synthesize the results; and
- conduct a sensitivity analysis.

Applying AHP to the problem

Constructing the hierarchies

The primary goal of the decision analysis is always placed at the top of the hierarchical tree. To evaluate this goal, specific criteria that have a direct impact on the goal are then established and organized into layers on the tree. At the bottom of the tree, the decision alternatives are listed. In this chapter, the goal is to find the 'best motor vehicle fuel'. Nineteen criteria were identified to evaluate this problem, grouped into five major categories: (i) economics; (ii) culture; (iii) environment; (iv) long-term sustainability;

and (v) development time. These five major categories cover the different external influences with which technology developers have to deal. Broadly speaking, the economics criterion refers to the question 'Is the motor vehicle fuel economically feasible?' The culture criterion looks at sociological and political issues and asks 'Is the motor vehicle fuel culturally and politically accepted? The environment criterion looks at ecological issues, and asks 'Is the motor vehicle fuel environmental friendly?' Time is also an external influence, therefore the long-term sustainability criterion asks 'Is the motor vehicle fuel a long-term solution to the problem?'; and finally, the development time criterion asks 'Does the motor vehicle fuel have a short development time?'

These criteria, each with a short description, are presented in Table 15.1.

Having defined in the previous sections the alternatives, and the criteria and sub-criteria, we have now all the levels of the tree, which are depicted in the Figure 15.1.

Scenarios

In this section, we present a number of scenarios. These are used to make a sensitivity analysis of the AHP model shown in Figure 15.1. The purpose of using scenarios is to test the 'robustness' of each motor fuel by depicting a variety of environments, from favourable to unfavourable. Scenarios are a forecasting tool that can be used to produce a macroscopic technology forecast, with tangible and intangible measures, as we explain in more detail below. Scenarios are defined as follows:

> hypothetical sequences of events constructed for the purpose of focusing attention on causal processes and decision-points. They answer two types of questions: (a) precisely how might some hypothetical situation come about, step by step? And (b) what alternatives exist, for each actor, at each step, for preventing, and diverting, or facilitating the process? (Martino, 1983)

Unlike trend analysis and expert judgement, macroscopic factors beyond the scope of quantified variables can be used in scenarios. Macroscopic factors could be non-technical factors, such as politics, economics, and social and technological factors. Scenarios are less useful when precise, micro answers are needed, such as those produced by statistical models.

Scenarios are an outline or synopsis of some aspect in the future. Scenarios are one tool belonging to multi-option analyses (Millet and Honton, 1991). Its conceptual foundation differs from trend analyses and expert judgements in the sense that 'not only is the future uncertain, but it is also influenced in part by what we ourselves do to make it', whereas trend analysis and expert judgment implicitly assume that the future will be the future

Table 15.1 Criteria to find the best motor vehicle fuel

Economics	Infrastructure B/C ratio	• The B/C ratio of the refuelling infrastructure building. Benefits such as revenue and costs such as initial investment and periodic maintenance should be taken into account. Alternatives with a B/C ratio greater than 1 are best. • Key question: *Which alternative's infrastructure has a B/C ratio greater than 1?*
	Fuel cost	• The fuel cost seen by consumers measured in dollars/gal or dollars/gge (gge = energy content in gallon of gas (petrol) equivalent). Alternatives with low costs fare the best, especially if the fuel cost is cheaper than petrol. • Key question: *Which fuel is cheaper at the pump? Is the alternative cheaper than petrol?*
	Vehicle cost	• The capital cost of the vehicle, including processor and on-board storage systems. Alternatives with low capital costs are best, especially if a new vehicle buyout is not required, but only adaptations to the current vehicle are necessary. • Key question: *Which alternative requires the least initial investment?*
	Vehicle maintenance cost	• The cost of vehicle maintenance. Alternatives with low capital costs are best. • Key question: *Which alternative's maintenance is cheapest?*
	Efficiency	• Measured in miles per gallon (mpg) or miles per gallon of gas (petrol) equivalent (gge). Alternatives with greater efficiency are best. • Key question: *Which alternative would give more miles per gallon (mpg) or miles per gallon of gas (petrol) equivalent (mgge)?*
Culture	Lifestyle changes	• A measure of the implicit lifestyle change posed by an alternative. Consumers resist change; therefore, if an alternative is to be successful, it has to imply the least lifestyle change possible. Alternatives with fewer implicit lifestyle changes are best. • Key question: *Which alternative would allow consumers to update their vehicles, rather than to buy new ones? Which alternative would preserve the consumer's city image?*
	Government support	• A measure of the level of government support that the particular alternative has. A change towards a 'greener vehicle' would gather momentum if the government supported it. Alternatives that have government support are perceived to be better because of incumbent tax benefits. • Key question: *Which alternative has the most government support?*

(continued)

Table 15.1 Continued

	Refuelling convenience	- A measure of the alternative's convenience of use. Since petrol vehicles are ready to be used 24/7, options that cannot be used 24/7 because of refuelling issues won't be as appealing to consumers as petrol vehicles. Alternatives that are ready to use 24/7 are best. - Key question: *Which alternative does not pose refuelling hurdles, so the vehicle is ready to be used 24/7?*
	Personal safety	- A measure of personal risk posed by the particular alternative. These risks are posed to consumers while refuelling their vehicles. Alternatives with low personal risk are best. - Key question: *Which alternative would pose the least personal risk?*
	National safety	- A measure of national risk posed by the particular alternative. National risks refer to terrorism target risks posed to the nation while producing or storing the fuel. Alternatives that pose fewer terrorism risks are best, as well as those that help to decrease dependency on foreign oil. - Key question: *Which alternative would pose fewer terrorism risks for the country? Which alternative would make the country less dependent on foreign oil?*
Environment	Global warming	- Emissions of GHG along the entire fuel cycle. Alternatives with the low GHG emissions (on a per mile basis) are best. - Key question: *Which alternative has the lowest GHG emissions?*
	Air pollution	- A measure of the emission of pollutants. Alternatives with low emissions (on a per mile basis) are best. - Key question: *Which alternative produces the least toxic emissions?*
	Renewable energy use (wind, solar)	- A measure of the environment friendliness of the particular alternative. Environmental friendly fuel production will impact less negatively on the environment. Alternatives that can use renewable energy such as wind or solar power are best. - Key question: *Which alternative can use wind or solar power?*
	Land use	- A measure of the land use impacts associated with the distribution infrastructure needed for a particular alternative. Alternatives that pose the least impact on land use are best. - Key question: *Which alternative would pose the least risk of deforestation and damage to the land?*

	Animal friendly	• A measure of the environment friendliness of the particular alternative. Environmental friendly fuel production will impact less negatively on the environment. Alternatives that are animal friendly are best. • Key question: *Which alternative is the most animal friendly?*
Long-term sustainability	Fuel use lifetime	• The estimated number of years that the particular alternative can be used when fully developed. The alternative lifetime depends (among other factors) in the renewability of the fuel, fuel cost, and fuel and infrastructure availability. Alternatives that have a longer use lifetime are best. • Key question: *Which option could be thought of as a long-term solution, rather than a quick fix to alleviate the pressure on oil stocks?*
	Renewability	• A measure of how renewable is the alternative's fuel. Renewable energy sources, such as solar power, are those whose stocks are rapidly replenished by natural processes, and which are not expected to be depleted within the lifetime of the human species. On the other hand, sustainable energy sources that are not renewable are those whose stocks are not replenished, but for which the presently available stocks are expected to last as long as human civilization cares to use them. These energy sources will be derived from nuclear energy, as other forms of stored energy found on the Earth do not have sufficient energy density to supply humanity indefinitely. • Alternatives that can use wind or solar energy as renewable sources, or nuclear power to produce their fuels are best. • Key question: *Which alternative is renewable, rather than non-renewable?*
Development time	Fuelling infrastructure development time	• The number of years that are needed to put the particular alternative's infrastructure in place. Development time include research, design and testing of the infrastructure until it is ready to commercialize. Alternatives with shorter development times are best. • Key question: *Which alternative's infrastructure would take the least time to be developed?*
	Vehicle development time	• Years needed to have the particular alternative's vehicle ready to use. Development time include research, design and testing of the vehicle until ready to commercialize. Alternatives with shorter development time are best. • Key question: *Which alternative's vehicle would take the least time to be developed?*

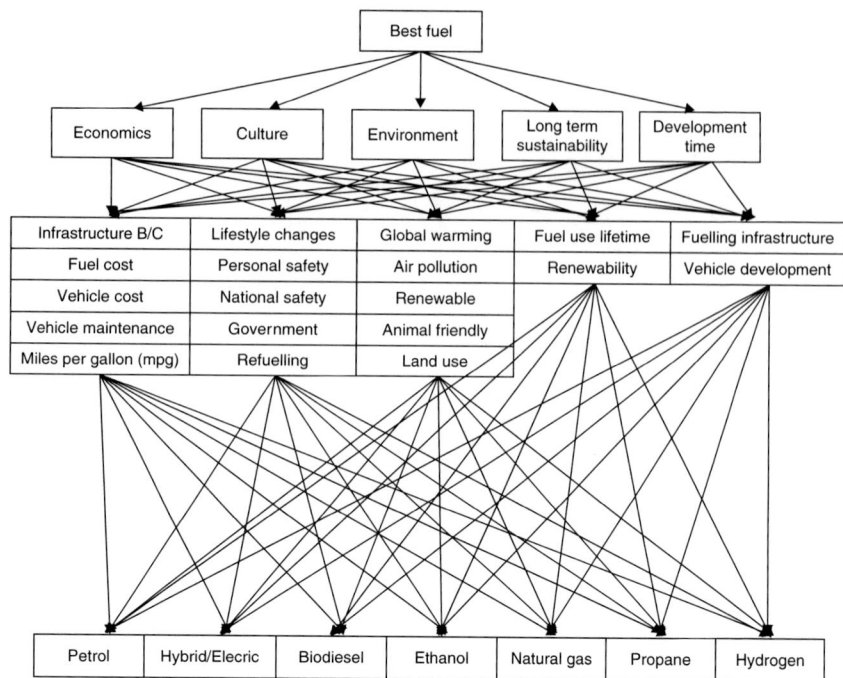

Figure 15.1 The AHP model

no matter what we do. It is beyond our control. Therefore, the basic idea behind scenarios is that we look at likely outcomes in the future, define the outcome we would like to see happen, and more important, are looking for strategies to make that outcome happen.

In this chapter we use scenarios to identify global conditions that will help the different alternative fuels to become the best motor fuel in the future. Once identified, these global conditions can be used to draw up strategies to make the chosen conditions happen. Scenarios are also used to help to prepare for unfavourable conditions that might arise from pessimistic scenarios, hence the value of this method. Overall, we propose four different scenarios, each stressing crucial factors that could have a direct impact on the adoption of the different motor fuels as the best alternative.

Scenarios offer flexibility and certain advantages, for example:

- 'Scenarios are excellent at integrating political, social, economic, and technological factors at the most aggregate level of analysis' (Millet and Honton, 1991). As presented above regarding the AHP model, factors such as 'economics', 'environment' and 'culture' are found to play an

important role in the adoption of the best vehicle fuel of the future; therefore, by using the scenarios tool in combination with the AHP model, we are able to include these factors, to stress their value or decrease their importance in different scenarios to discover to what extent they affect the choice for the best vehicle fuel.
- Another characteristic of this study is great uncertainty. Because we are looking at long-term time-frames, the nature of the factors surrounding the adoption of the best motor fuel of the future is unstable, in particular the oil price. Scenarios are estimated as general portrayals of future conditions, which are well suited for addressing situations of great uncertainty.
- Non-quantitative factors are included in our analysis. Factors such as culture cannot be quantified. Scenarios do not ignore unquantifiable factors; on the contrary, their value is considered.

In the next section we present a description of the different scenarios we have used to produce a sensitivity analysis of the motor fuel alternatives. A 40-year horizon is observed for all scenarios. Each scenario was planned to emphasize particular criteria that can play decisive roles in the different scenarios. Our goal is to find out how much the alternatives for the best motor fuel can vary. Each scenario is based on a different set of assumptions, but the issues addressed by these assumptions are interrelated, as we shall see below.

The principal *players* or *actors* of the following scenarios are: (a) consumers; (b) the automotive industry; (c) technology researchers; and (d) governments. Consumers represent the market pull force; in other words, the demand. The automotive industry invests in technology research, which represents the technology push force and provides the final product embodying the technology that meets market needs, and finally, governments provide a political factor, which can make the adoption of any alternative motor fuel easier or harder.

The two *external influences* that have an impact on the development of alternative motor fuels are *oil price* and *environmental concerns*, which are the consumers' environmental concerns regarding issues such as global warming, and air and water pollution. If there were two perpendicular axes, one representing oil price and the other representing environmental concerns, there would be four different scenarios: (i) Status Quo; (Ii) Environment Challenge; (Iii) Economic Challenge; and (Iv) Catastrophe. These scenarios are adopted from Winebrake and Creswick (2003) and expanded for this application. The aim of each scenario is different. The *Status Quo* scenario is set to focus on the current time. *Environment Challenge* scenario is set at a time when it is optimal to start looking for alternative motor fuels, before it's too late. The *Economic Challenge* scenario is set during times of economic hardship, when the best motor fuel has to be efficient, cheap, and sustainable. Finally, the *Catastrophe* scenario is set in time when oil has been depleted and global warming has taken its toll on the planet. In this scenario, the best motor fuel would have to have a short development time, because it does not have the luxury of time. We believe that these four scenarios would be the most useful to any alternative motor fuel technology developer; they will help them adapt to volatile factors, such as the oil price, and to prepare for unexpected environmental problems. The four scenarios are depicted in Figure 15.2.

Scenario I: 'Status Quo'

In the Status Quo scenario, we have IT that the oil price is low, and that environmental concerns are also low. In this scenario, economics is not an issue, since the oil price is low. People drive their petrol-fuelled cars happily, without feeling guilty for causing toxic emissions. There is neither interest nor support from the government to encourage or to carry out research into alternative motor fuels. In the same way, the automotive industry is not spending money on such research. Oil companies are making big profits, because extra reserves are being found, and therefore any oil shortage is pushed another twenty years away. Petrol prices are rising, but not dramatically. Emergent economies, such as China and India, are adding a strain to existing oil reserves. The general idea in the Status Quo scenario is: *Environmental problems such as global warming are far away. They might become*

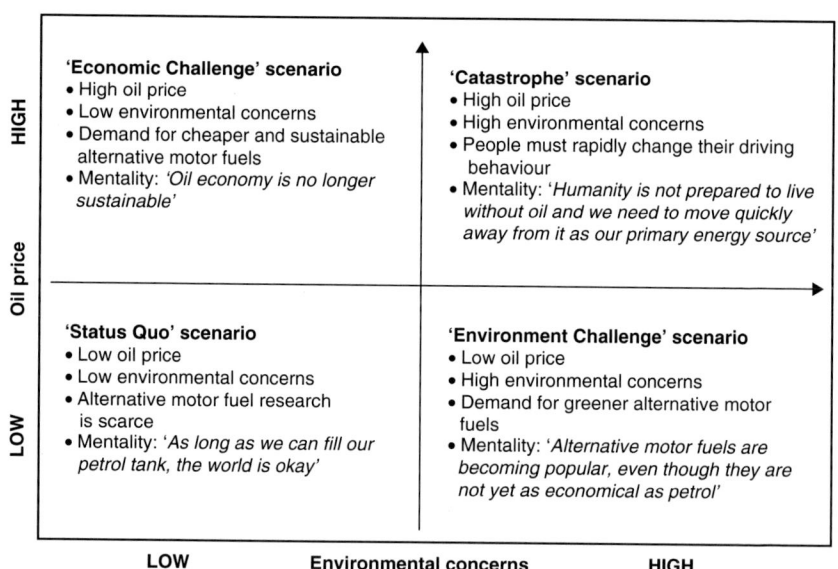

Figure 15.2 The four scenarios
Source: Adapted from Winebrake and Creswick (2003).

apparent in 100 or 200 years hence. Let's not worry about it now. As long as we can fill our petrol tanks, the world is okay.

Highlights:

- extra oil reserves are found, therefore any shortage is delayed for another 20 years;
- petrol prices rising, but not dramatically;
- environmental concerns are low;
- global warming and air quality are issues, but not very important; and
- emergent economies/world powers adding strain to existing oil reserves and may erase extra reserves.

Scenario II: 'Environmental Challenge'

In the Environmental Challenge scenario, we have it that oil prices are still low, but environmental concerns are high, which means that consumers are starting to realize that environmental problems such as global warming are real, and are happening, and that it is time to do something to remedy them. Global warming is worse than predicted. Too many petrol-powered vehicles have created problems of air pollution. Consumers demand that governments implement policies that encourage and promote research into and the use of alternative motor fuels. Carmakers begin to invest heavily in research into vehicles using alternative fuels. Technologies that rely on fossil fuels such as propane or natural gas begin to decrease in popularity. Also, bio-technologies, such as ethanol, which put a strain on the land, are decreasing in popularity. Renewable energy sources are becoming the answer for environmental problems, therefore carmakers are looking for motor fuels that can benefit from these, such as hybrid and electric cars that can use electricity generated by either solar or wind power. Industry is investing heavily in improving batteries to store electricity. At the same time, interest in hydrogen technology is increasing. The quest is for a long-term, sustainable motor fuel. The general idea in the Environmental Challenge scenario is: *There is a battle between economic and environmental interests, petrol prices are still low, petrol is still the best motor fuel option from an economic point of view, but alternative fuels are becoming popular, even though they are not yet as economical as petrol.*

Highlights:

- petrol is cheap;
- environmental concerns are high;
- global warming is worse than predicted;
- extra CO_2 is poisoning the environment;
- ozone is being destroyed in the lower stratosphere, and there is too much at ground level; and
- too many petrol-powered vehicles are creating air pollution problems.

Scenario III: 'Economic Challenge'

In the Economic Challenge scenario we have it that oil prices are high, but environmental concern is low. Consumers do not feel that global warming is a real threat. But because oil prices are extremely high as a result of long conflicts in the Middle East caused by a slow-growing demand that is restraining the supply of oil, and a probable imminent oil shortage, consumers are demanding carmakers to develop alternative motor fuels. The cost of living is increasing. The need to move on to cheaper motor fuels is making alternative motor fuels more efficient as technologies are improving. Batteries are becoming more efficient and safer. Hydrogen infrastructure is becoming more prevalent, but still not common. The general idea in the Economic Challenge scenario is: *We have reached the price limit for oil. Petrol is no longer acceptable, for economic reasons. It's very expensive but demand is still growing, which means that petrol prices can only go up. We are tired of this aggravation. We demand cheaper fuel motor alternatives. The oil economy is no longer sustainable.*

Highlights:

- environmental concerns are low;
- petrol is very expensive, but there is no imminent fuel shortage;
- cost of living is increasing;
- hybrid cars are becoming cheaper;
- petrol-driven cars becoming cheaper; and
- the alternative fuel infrastructure is becoming more prevalent, but still not common.

Scenario IV: 'Catastrophe'

In the Catastrophe scenario we have it that oil prices are high and environmental concerns are also high, meaning that consumers realize that global warming is an issue, and something must be done. Sea levels are rising, forcing coastal cities to be abandoned. People forced to move to inland cause an increased demand for petrol; this growing demand strains oil production and triggers an exorbitant increment in oil prices. Global warming causes hurricanes that severely damage the oil wells, and oil companies have to shut them down for repair. Oil production ceases. Stocks get low. Middle East conflicts and terrorism increase oil prices even more. Oil reserves are much smaller than previously thought. Time is an issue in this scenario. Since petrol vehicles are not a sustainable alternative, development time for alternative motor fuels becomes critical. People must change their driving behaviour in a very short time. The general idea in the catastrophic scenario is: *Global warming is taking its toll. It is a real problem. Petrol-driven cars are no longer an affordable option, neither from an economic nor an environmental point of view. We are not prepared to live without oil, but we have to move away from it as our primary energy source.*

Highlights:

- environmental concerns are high;
- there is time pressure;
- oil reserves much smaller than previously thought;
- Petrol prices rise to astronomical levels;
- terrorism erupts because of oil shortages;
- people must change their driving behaviour; and
- there is an imminent war over oil.

Results and discussion

The results of the AHP analysis undertaken using expert pairwise comparisons are tabulated in Figures 15.3 and 15.4.

Scenario I: 'Status Quo'

This scenario seems to be the most prevalent one for long periods of time, because sudden spikes in oil prices are only experienced occasionally. The results are therefore not too surprising, since the absence of any urgency

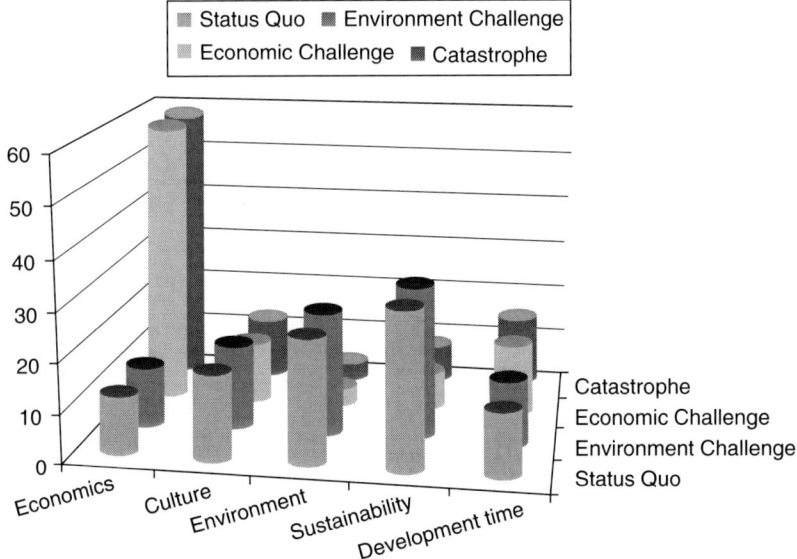

Figure 15.3 Scenario analysis and first-tier criteria comparison results
Source: Survey data and pairwise comparison calculations.

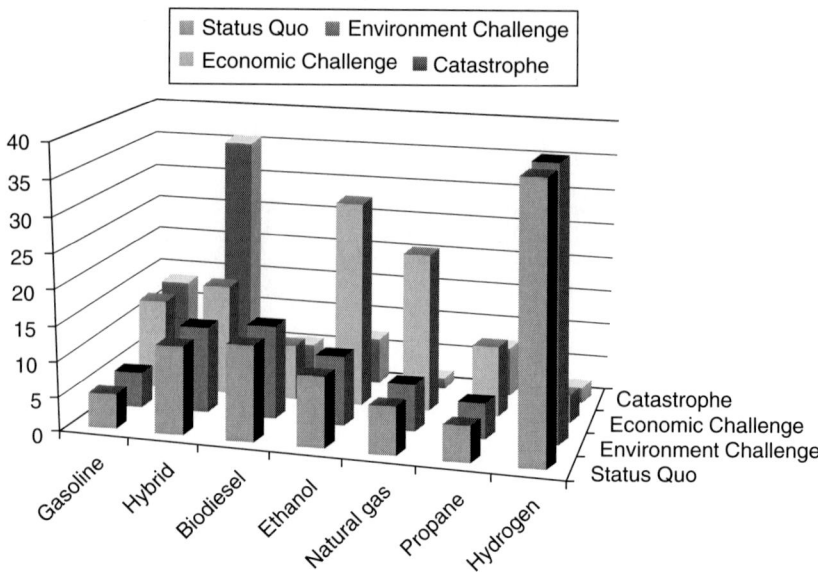

Figure 15.4 Scenario analysis and choice of fuel results
Source: Survey data and pairwise comparison calculations.

tends to develop a more relaxed atmosphere regarding the awareness of foreign dependencies and monetary consequences. In this kind of a situation, the policy-makers, business analysts, energy stakeholders and the 'common man' enjoy the luxury of understanding the environmental implications of fossil fuels and remain conscious that efforts must be made to bring pollution under control. The absence of urgency to curb oil prices and the need to move towards alternative fuels results in the promotion of a different set of priorities. 'Sustainability' tops the chart, followed by 'environment' and 'culture', respectively.

The choice of fuel also follows a similar trend, with hydrogen winning the race, followed by the biodiesel and hybrid options. Though being comparatively unsafe, and still under development, the hydrogen option is clearly suitable from the sustainability perspective as it is one of the most abundant naturally occurring elements. There is no fear of harmful emissions (SO_x, NO_x) as water is the only by-product hydrogen combustion produces.

Scenario II: 'Environmental Challenge'

The Environmental Challenge scenario almost replicates the results of the Status Quo scenario, since here too the assumptions are somewhat similar. The tipping point in the Environmental Challenge scenario is that awareness regarding the deteriorating environmental situation is high. People are aware that something must be done to keep pollution in check, and control emissions, but at the same time petrol prices are not putting a strain on pockets. Hence, having an option of sustainable fuel followed by 'environment' holds the winning positions. The choice of fuel in this case is also hydrogen as it is both sustainable and carbon-free.

Scenario III: 'Economic Challenge'

This is a very interesting scenario. It seems more relevant to today's situation. Economic problems, the deteriorating power of the US dollar, continued slow-paced economic recovery, dependency on foreign energy, and the stupendous domestic consumption of oil, all point towards an economic challenge. The rising price is the biggest concern among the population and policy-makers alike, as energy has a direct link to national security, peace and growth. In such a case it is natural to see that the economics criterion will win the priority race, followed by development time, as the urgency to replace conventional fossil fuels would be more like the Status Quo scenario.

The country would be looking towards a source of fuel that could replace fossil fuels as quickly as possible, but also be economical. The choice of fuel would therefore depend on what could be developed and adopted within a short time-span. Ethanol is the highest-scoring option, followed by the natural gas and hybrid options.

Scenario IV: 'Catastrophe'

This nightmarish scenario throws new light on the fear of what the world would look like with the sudden disappearance of fossil fuels. Chaos would ensue, assuming that coastal cities would have to be abandoned and people would be forced to move inland.

Most of the world's major cities are situated on or close to the coast. They depend on ports and sea mercantile access. In such a situation the financial burden on people's lives and assets would be enormous along with the chaotic political situation likely to erupt into a 'war'. Everybody will have personal safety and personal finance on their minds, followed by a concern to live life as they have always done. This situation will be combined with the unavailability of oil and stress on agricultural sources (for ethanol/biodiesel production) because of population migration. Hence, the economics criterion is the topmost choice followed by development time.

The choice of fuel in this scenario turns out to be hybrid, followed by more conventional fuels and propane. This appears to provide a logical rationale because there is a nationwide electrical grid system, and commercially available hybrid vehicles are already on the road, with a developing market.

Conclusion

It would not be entirely incorrect to say that it is a very human preference to not want to change one's existing lifestyle. Once human beings are settled comfortably, they become resistant to change, hence, to make significant changes in the future, very strong drivers must be in place. For us to conclude, from examining the graphical results, that if the present conditions were to prevail, the general idea would be to continue to research alternative technologies without any sense of urgency. On the other hand, if a disastrous economic downturn were to happen, or conditions were to move towards a catastrophe, a sense of urgency and desperation would lead to changes in lifestyles and a strong push towards faster development of alternative technologies. The mass of the population would move towards alternative technologies and fuels in such a scenario. In the case of the luxury of time and abundant supply, sustainability and continuing research remain among the preferred areas, while under economic and catastrophic pressure the population would tend to move towards newer fuels, choosing the one that would also satisfy the economic constraints. People's priorities change drastically with the changing dynamics of the world, and these have a strong influence on personal decision-making. This effect of the collective influence of an entire population will create a trend.

The results throw light on scientific research being affected, and sometimes even being led by changes in human behaviour. We can only research and propose the different scenarios that might materialize, but what will actually happen will remain unseen in the real world, until time takes its own course.

References

Ahsan, M. K. and Bartlema, J. (2005) 'Monitoring Healthcare Performance by Analytic Hierarchy Process: A Developing-Country Perspective', *Journal of Transactions in Operational Research*, 11(4): 465–478.

Martino, J. P. (1983) *Technological Forecasting for Decision Making*. New York: North-Holland.

Millet, S. M. and Honton, E. J. (1991) *A Manager's Guide to Technology Forecasting and Strategy Analysis Methods*. Columbus, OH: Battelle Press.

Ozdemir, M. S. and Saaty, T. L. (2006) 'The Unknown in Decision Making: What to Do About It', *European Journal of Operational Research*, 174(1): 349–59.

Riley, R. Q. and Chee, W. (2003) *Alternative Cars in the Twenty-First Century: A New Personal Transportation Paradigm*, Warrendale, PA: SAE International.

Saaty, T. L. (1994) 'How to Make a Decision: The Analytic Hierarchy Process', *Interfaces*, 24(6): 19–43.

Winebrake, J. J. and Creswick, B. P. (2003) 'The Future of Hydrogen Fueling Systems for Transportation: An Application of Perspective-Based Scenario Analysis Using the Analytic Hierarchy Process', *Technological Forecasting and Social Change*, 70(4): 359–84.

References to internet sources

Baker, C. J., 'Understanding Today's Diesel'. Available at: http://www.bankspower.com/techarticles/show/26-Understanding-Todays-Diesel; accessed 10 October 2011.

CBS Interactive Inc. (2010) 'Ethanol Pollution Surprise', 8 January. Available at: http://www.cbsnews.com/stories/2002/05/03/tech/main508006.shtml; accessed 10 October 2011.

Congressional Budget Office – United States (2009) 'The Impact on Food Prices and Greenhouse-Gas Emissions', April. Available at: http://www.cbo.gov/ftpdocs/100xx/doc10057/04-08-Ethanol.pdf; accessed October 2011.

Financial Express, The (2011) 'India Clears $5 Billion Iran Payment', 4 September. Available at: http://www.financialexpress.com/news/India-clears--5-billion-oil-debt-to-Iran--says-report/841533; accessed 5 September 2011.

Kronstadt, K. A. and Katzman, K. (2006) 'India–Iran Relations and U.S. Interests', State Government, – Department of Foreign Policy, August. Available at: http://fpc.state.gov/documents/organization/70294.pdf; accessed October 2011.

Map-a-Watt (2010) 'Where Does US Electricity Come From?', 29 November. Available at: http://mapawatt.com/2010/11/29/where-does-u-s-electricity-come-from; accessed 10 October 2011.

MIT (2009) 'The Future of Natural Gas: An Interdisciplinary MIT Study'. Available at: http://web.mit.edu/mitei/research/studies/documents/natural-gas-2011/NaturalGas_Report.pdf. Accessed October 2010.

Naylor, R., Burke, M. and Falcon, W. (2007) 'The Ripple Effect: Biofuels, Food Security, and the Environment', Science and Policy for Sustainable Development, November. Available at: http://www.environmentmagazine.org/Archives/Back%20Issues/November%202007/Naylor-Nov07-full.html; accessed October 2011.

Oil Price (2011) Available at: http://www.oil-price.net; accessed 5 September 2011.

Pocha, J. (2005) 'The Axis of Oil'. Available at: http://www.globalpolicy.org/empire/economy/2005/0131oilaxis.htm; accessed 23 August 2006.

Stober, D. (2007) 'Nanowire Battery Can Hold 10 times the Charge of Existing Lithium-ion Battery', 18 December. Available at: http://news.stanford.edu/news/2008/january9/nanowire-010908.html; accessed 11 October 2011.

UPI.com (2009) 'Hillary Clinton Improves Ties with Nigeria', 14 August. Available at: http://www.upi.com/Business_News/Energy-Resources/2009/08/14/Hillary-Clinton-improves-ties-with-Nigeria-one-of-Americas-largest-oil-suppliers/UPI-97141250280963; accessed 2 September 2011.

US Congressional Budget Office (2009) 'The Impact of Ethanol Use on Food Prices and Greenhouse-Gas Emissions', April. Available at: www.cbo.gov; accessed 10 October 2011.

USDOE (US Department of Energy) (2008) 'Fact Sheet April 2008'. Available at: http://www.afdc.energy.gov/afdc/pdfs/41597.pdf; accessed 10 October 2011.

USDOE (US Department of Energy) (2009) 'The Future of Natural Gas: An Interdisciplinary MIT Study'. Available at: http://web.mit.edu/mitei/research/studies/documents/natural-gas-2011/NaturalGas_Report.pdf; accessed October 2010.

USDOE (US Department of Energy) (2011) 'Clean Cities Alternative Fuel Price Report'. Available at: http://www.afdc.energy.gov/afdc/pdfs/afpr_jul_11.pdf; accessed 11 October 2011. US Energy Information Administration (2011) 'Weekly National Price Data'. Available at" www.eia.gov; accessed 10 October 2011.

16
Consumer Attitudes Towards Alternative Vehicles

Marc Dijk, Jorrit Nijhuis and Reinhard Madlener

A quick glance at the automobile market suggests steady progress in the eco-efficiency of cars in recent decades. The conventional internal combustion engines (ICE) in automobiles have improved in energy efficiency and reduced emissions. Moreover, a new market niche of alternative vehicles[1] has emerged since 2000. By the spring of 2011, more than 3 million hybrid electric vehicles (HEVs) had been sold worldwide and various car firms are currently introducing mass-produced electric car models. But do these trends really reflect a more fundamental change in demand for alternative car engines? Or are the sales of eco-efficient vehicles merely the result of stricter environmental regulations on car emissions and environmental subsidies being offered for eco-efficient vehicles? In this chapter, we assess this question by reviewing six recent studies on car consumers (TNS-Emnid/AutoScout24, 2004; Lane, 2005; Mytelka, 2008; Nijhuis and van den Burg, 2009; Dijk, 2011; Ozaki and Sevastyanova, 2011).

More specifically, we consider 'demand' to be more than simply sales levels. It also encompasses consumer framing; that is, a consumer perspective (Garud and Rappa, 1994; Porac *et al.*, 2001; Dijk, 2011). A *frame* is the way in which the innovation is described or interpreted by consumers. The framing metaphor can be understood as a window or some sort of spectacles (worn by the actor group) that filter the total amount of information in a first impression (what it is about and what is important for them), and focuses attention on key elements and aspects within. More precisely, it is the structure of (relevant) beliefs, knowledge, perceptions and appreciation that underlie consumer attitudes.

Several studies have analysed consumer perspectives on automobiles in general (Heath and Scott, 1998; Steg *et al.*, 2001), but there are hardly any that have examined perceptions of car *engines* specifically. The most notable is a study reporting that the majority of HEV owners saw their vehicles as projecting images linked to larger values such as social awareness, responsibility and concern for others (Heffner *et al.*, 2005), and another study found that 31 per cent of HEV buyers said they purchased an HEV because it 'makes a statement about me' (CNW, 2006). Steg *et al.* (2001) found that, for car buyers, symbolic value plays a significant role in addition to functional characteristics, but the question of how people frame their vehicles' propulsion technology remains unanswered. Another study analyses the relative importance of fuel economy, compared to other car buyer preferences (Mytelka, 2008). The author argues that the oil price shocks of the 1970s seem to have affected the early 1980s preference structure with its emphasis on fuel economy, whereas five years later, fuel economy was of less importance to consumers, and price and reliability had become the prime concerns.

A study by business consultants Maritz (2006) noted changes of consumer habits as a result of rising fuel prices (in France, Germany and the UK, in 2006, with responses from 1,240 new-vehicle owners). To the statement 'I am thinking about buying, or have bought, a vehicle with a more economical engine', 57 per cent agreed mildly or strongly, while only 23 per cent disagreed mildly or strongly. Based on these figures, the authors concluded that a major proportion of European drivers are changing their car purchase considerations because of rising fuel prices.

Achtnicht (2009) examined whether CO_2 emission per kilometre is a relevant attribute in car choices, based on a stated preference experiment among potential car buyers in Germany. The results suggest that the emissions performance of a car matters substantially, but the consideration of this varies strongly across the sampled population. More precisely, they find that women are willing to pay more (€87) for a reduction of one gram per kilometre of CO_2 than are men (€60). Further, people under 45 years are willing to pay more than people aged 45 and older, and people who have had a higher level of education are willing to pay more than those who have not.

While insightful, these studies have not clarified consumers' multifarious perspectives or attitudes towards the different types of engines (ICE, hybrid electric, fully electric and hydrogen-powered). This is our focus in this chapter. In the next four sections, we successively discuss the changing context of automotive purchasing since the mid-1990s through the increasing use of the internet as an information source, and the decreasing influence of salespeople in the showroom; trends in the stated preferences of car buyers; the changing mental perspectives regarding conventional and alternative engines; and car consumer segmentation.

Though most of the studies we have drawn upon are focused on the Netherlands, we feel that the results are indicative also of car buyers' behaviour in industrialized countries.

Trends in automotive consumption practice

In the economics and marketing literature, the purchase of a car is defined as a form of 'complex buying behavior'[2] (Reed *et al.*, 2004). Given that the car is a high-involvement product, the car-buying process is also seen as a high-involvement process, leading to active search and use of information, deliberate evaluation of alternatives, and a careful choice. More specifically, the prospective car buyer's search for information usually includes both 'internal search'– retrieval of information based on previous searches and personal experiences – and 'external search' – the accessing of different types of information sources (Klein and Ford, 2003). Furthermore, research has highlighted that the car purchase can be seen as a two-stage process; in the first stage, the vehicle class is decided, based on cost and car capabilities, while in the second stage consumers carry out a more profound review of vehicles (Lane, 2005; Teisl et al., 2007).

Since the late 1980s the automotive market has undergone a significant change in the way information is provided to consumers. The advent of the 'information age' has had an enormous influence on the purchasing process of complex products, including car purchase. Traditionally, automotive dealers were seen as the dominant source of information, resulting in a situation of buyer–salesperson interfaces in which the salespeople 'led' the potential purchaser through the buying process (Reed et al., 2004). Market research conducted in Germany by TNS-Emnid/AutoScout24 (2004) showed a number of interesting developments in the automotive sector. The practice of information seeking has shifted in such a way that the majority of the people purchasing a car make use of the internet as a source of information, thus making the internet one of the most dominant sources of information.

Interesting enough, another related development that can be witnessed is the decrease in customer ties, meaning that formerly fixed customer–supplier relations have become increasingly fluid (Capgemini, 2010). While important brand differences remain, in general emotional attachment to a specific brand has lessened as a consequence of increased similarity in styling and performance and decreased quality differences among automobiles (Capgemini, 2010). The result is that, in the 2000s, strong consumer empowerment has taken place, which may be summarized by the slogan: 'negotiate like a pro', according to the car vending site Edmunds (www.edmunds.com).[3] Not only have car salespeople noticed that consumers arrive at the showroom armed with a mass of background information about the automotive sector. And the purchase process itself has changed; instead of visiting a number of different showrooms before making their choice, most potential buyers make a pre-selection of approximately three car types, which they investigate intensively. As a consequence, the role of the salespeople has also shifted, from leading to guiding, and from salesperson to adviser.

Furthermore, there is a general trend in consumption towards more comfort and convenience (Shove, 2003). This trend, together with higher safety requirements, has led to an increased demand for size and luxury-level of cars (Van den Brink and Van Wee, 2001). As Shove (2003) indicates, attributes typically start out as an extra capacity or luxury, but can soon become accepted as the norm, thus shifting consumer preferences and expectations regarding automobile characteristics. As a consequence, the price of an average new car in the Netherlands has doubled from some €13,500 in 1970 to €26,000 in 2009;[4] but, more important, the fuel consumption of vehicles has not seen any substantial decrease since the 1980s (see Figure 16.1).

All these trends in the practice of automotive purchase and use that we observed are also reflected in what consumers acknowledge as their preferences, to which we turn now.

Stated preferences of car consumers

Looking at consumers' stated preferences, most studies find that environmental factors currently do not seem to play a major role in car buyers' choices. Mytelka (2008), for example, analysed the relative importance of fuel economy, compared to other car buyers' preferences, between 1980 and 2005[5]; see Table 16.1. Mytelka argues that the oil price shocks of the 1970s

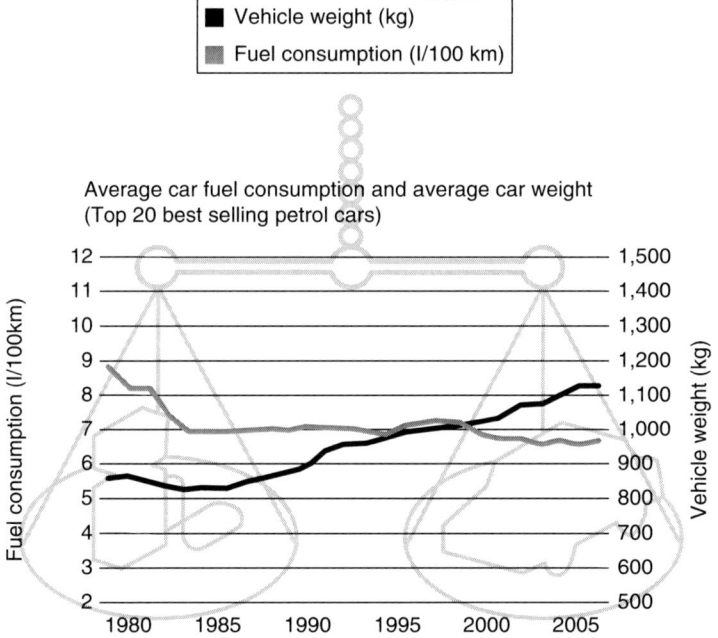

Figure 16.1 Trends of average car mass and fuel consumption
Source: Obtained directly and adapted from BOVAG/ACEA.

seem to have affected the early 1980s' preference structure, with its emphasis on fuel economy, whereas five years later, fuel economy was of the least importance to consumers, and price and reliability had become the prime concerns. *Dependability*[6] has always been a prominent attribute for (US) consumers. *Fuel economy* fluctuates in importance, most probably because of fluctuations in the oil price. Nevertheless, surprisingly, the importance of fuel economy decreased in 2004 and 2005, despite the fuel price rising in those years.

Another study by Lane (2005) found somewhat similar conclusions (see Table 16.2), though they are difficult to compare. Lane distinguishes fuel economy from environmental impact (see both highlighted in bold in Table 16.2), so that the former becomes an issue of price. He found that consumers do care about fuel consumption (operation cost), but very little about vehicle emissions. So, even though consumers mention sustainability issues as a major consumer concern (Florschütz and Bakker,, 2002), the infamous attitude–action gap reveals that consumers' concern for the environmental impact does not often translate into behavioural change in car purchases (Lane, 2005). In addition, even though fuel consumption is mentioned as an important factor, most car buyers make little effort to compare fuel consumption differences across vehicles during their decision-making process (Boardman *et al.*, 2000; Lane, 2005). One conclusion could be that people tend to be more concerned about status value and less about environmental performance than they would like to admit (Johansson-Stenman and Martinsson, 2006). This would fit well with the often-heard claim made by policy-makers and car producers, that 'consumers are just not interested in environmentally-friendly cars'.

However, desk research by Ben Lane (2005) sheds more light on this paradox. First, many buyers assume that there are no major differences in fuel efficiency within the same vehicle class. By buying a new car, consumers automatically assume they are buying a car with a high energy efficiency and one that complies with strict environmental regulations. It is also still widely believed that an environmental choice involves some degree of sacrifice: in comfort, performance, or financially. Furthermore, because con-

Table 16.1 Trends in vehicle attribute preferences in the USA

Attributes	1980	1987	1996	2001	2004	2005
Fuel economy	42	4	7	10	22	12
Dependability	31	44	34	29	26	33
Low price	14	31	11	8	10	6
Quality	4	8	19	22	19	20
Safety	9	14	29	29	23	26

Source: Adapted from Mytelka (2008) and Kubik (2006).

Table 16.2 Factors mentioned as being important in the purchasing decision

Most important (10%–30%)	Fairly important (5%–10%)	Least important (<5%)
Price	Performance/Power	Depreciation
Fuel consumption	Image/Style	Personal experience
Size/Practicality	Brand name	Sales package
Reliability	Insurance costs	Dealership
Comfort	Engine size	**Environment**
Safety	Equipment levels	**Vehicle emissions**
Running costs		Road tax
Style/Appearance		Recommendation
		Alternative fuel

Note: Bold type as in original; author distinguishes fuel economy from environmental impact.
Source: Adapted from Lane (2005).

sumers' knowledge is incomplete, they often find the environmental effects of car use confusing and complex. The relationship between fuel efficiency, CO_2 emissions and climate change is only very loosely understood. Finally, the differences between local and global emissions are often unclear (Lane, 2005).

Different environmental information tools and taxation schemes have been developed over the years to help consumers take the environmental aspects of car purchase into account. Indeed, these strategies were designed specifically to tackle the above-mentioned problems and misconceptions. But, as we have seen, not all of these points have been resolved (Nijhuis and van den Burg, 2009).

Finally, Ozaki and Sevastyanova (2011) have reviewed various studies on the motivation to buy a hybrid vehicle (Haan *et al.*, 2006; Heffner *et al.*, 2007a, 2007b; Klein, 2007) and their conclusions are therefore difficult to compare with the previous studies on cars or car engines in general. They find that five groups of motivations, broadly of equal importance, are key for HEV buyers: financial benefits and other regulation-related advantages; the symbolic meaning of hybrids (that is, 'environmentalism'); compliance with the social norms of peers; the high-tech character of hybrids; and the reduction of oil dependence. Through their own questionnaire survey among 1,250 UK Prius buyers, however, Ozaki and Sevastyanova find that, in particular, the environmental impact, the comfort of the driving experience of a Prius, and the social connotation or desirability are significant motivations. They devote little attention to their quite surprising results, however; they only argue that the comfort of driving may be a post-purchase effect. In fact, all their results may be signalling post-purchase justification rather pre-purchase motivation, since their respondents had all owned a Prius for up to 24 months.

Mental perspectives of conventional and alternative engines

Though many studies have been performed on the social perception of automobiles (Steg et al., 2001; Heffner et al., 2006), few studies have examined perspectives specifically on car *engines*, and it is not clear how people perceive, or *frame*, the propulsion technology of their vehicle. A frame, as introduced earlier, is the way in which the innovation is described or interpreted by consumers. Through an analysis of newspapers, and in particular their car magazine supplements, Dijk (2011) was able to determine which features attracted attention among potential users, and how such features were valued (as something positive or negative). The changes in frame-attribute appraisal reflect the technological innovations that have occurred, on the one hand, and changes in social perceptions on the other (emphasis on fuel economy, attention to climate change and so on). Here, we summarize the main findings for battery electric vehicles (BEVs), HEVs and diesel engines in turn. The analysis disclosed that for BEVs in 1996, *range* was perceived as a salient attribute (mentioned in 75 per cent of the accounts), as well as *price* (55 per cent). Both range and price were appraised negatively, which shows that market actors were dissatisfied with both functionality and price. *Environmental impact* (35 per cent) had dropped in frequency in comparison to 1990, though it was appraised positively (an assessment score of +5). The *social connotation* score of BEV was 0 by 1996, which means a neutral stance. By 2000 there was only one salient attribute: *range* (71 per cent). This attribute was evaluated negatively (assessment score –5). Hence, perceived functionality was still low, and attention paid to BEVs had decreased.

In 2000, Toyota's hybrid electric Prius was launched in the Netherlands. The analysis disclosed that, for HEVs, prominent attributes up until 2000 were *fuel use* (59 per cent), *price* (55 per cent) and *environmental impact* (55 per cent). By 2005, they were *fuel use* (63 per cent) and *environmental impact* (37 per cent). The higher price as a negative factor dropped in importance to 15 per cent. Fuel use, being a positive factor, was increasingly appraised: scoring +15 in 2000 and as high as +32 in 2005. The negative score for *price* improved from –14 in 2000 to –3 in 2005. The score for *social connotation* rose from +3 in 2000 to +9 by 2005. Our analysis of the articles also disclosed that hybrids were mainly associated with low fuel use, while remarks on performance were few and far between. *Acceleration*, an aspect of performance, was mentioned in only 17 per cent of the accounts, and appraised as +5 in 2005. This satisfaction was low in comparison to diesels.

For HEVs, there has been a shift in the frames between 2000 and 2005, which is shown in Figure 16.2. Most remarkable is the decrease in emphasis of *environmental impact* (from 55 per cent to 37 per cent), and simultaneously to a slight increase of emphasis on *fuel efficiency* (59 per cent to 63 per cent). This may well have to do with increasing fuel prices in the early 2000s, and with a tendency of consumers to focus more on individual benefits and less

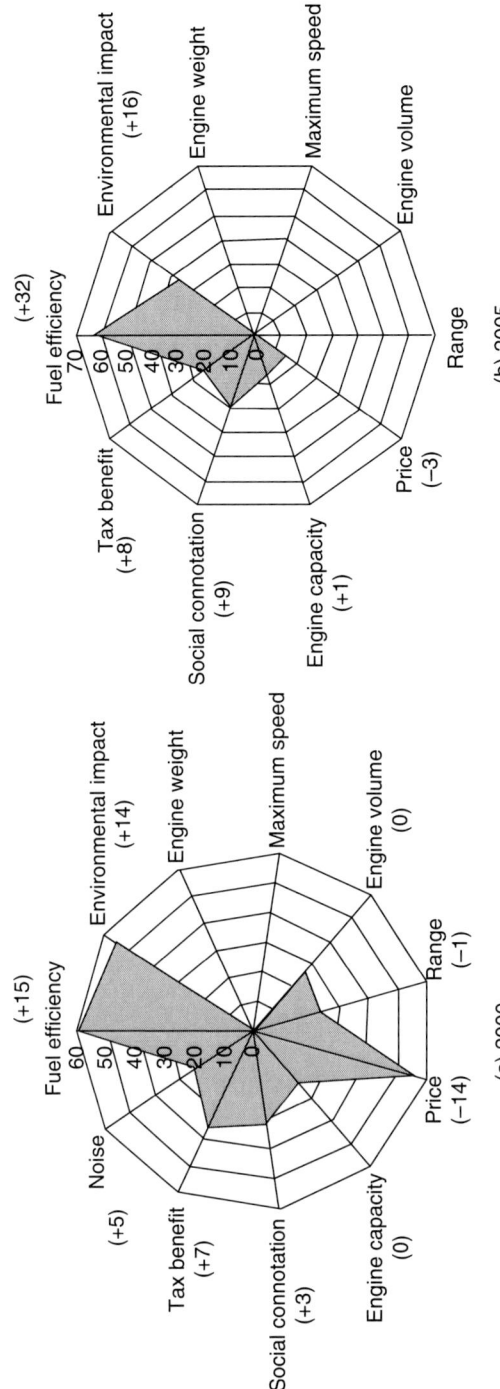

Figure 16.2 Frames for hybrid electric vehicles, 2000 and 2005
Notes: The shaded area indicates engine attributes; the outer numerical values indicate the appraisal scores for each feature.

on public goods (such as air quality).[7] The change in frame for HEVs was quite substantial, and took place mainly prior to the diffusion process: in 2005, the percentage share of HEVs in new car sales was only 0.5 per cent. Apparently the frame was fairly flexible at that time. For BEVs, in contrast, there were quite a number of changes in the frames (between 1996 and 2000), though it was stable in its negative appraisal of range and refuelling time.

We also examined frames for diesel engines, which were increasingly equipped with direct injection (DI) systems. Performance attributes, such as *engine capacity* ('horsepower'), *acceleration* and *torque*, attracted the most attention. Innovations in diesel engines were well appreciated, both in terms of (perceived) functionality, and the social connotation. *Overall performance* was appraised at +9 in 2000 and +12 in 2005. *Social connotation* scored +11 in 2000 and +8 in 2005. Only *price* was unsatisfying for most people: –2 in 2000 and –7 in 2005.

Between 1996 and 2005 the shift in framing was only minor (both in structure and appraisal), as shown in Figure 16.3. The drop in appraisal for fuel efficiency is the only notable change. It might be because users had become accustomed to the relatively high fuel efficiency of the DI diesel engine, and so found it to be the norm by 2005.

Consumer segments

Among car buyers there are likely to be various sub-frames relating to various consumer groups. Adopters of new vehicles can choose their preferred type (and size) of engine from a range of alternatives (usually between two and six). We analysed the outcomes of such consumer choices for the 30 top-selling car types in the Netherlands, in total more than 200,000 choices (per year), between 2003 and 2008 (see Appendix 1 and Dijk (2011), for more details of this analysis). We found three types of (engine) consumer groups. The first group, consisting of (on average) 30–35 per cent of consumers, chose the cheapest engine. The cheapest engine tends to be the lightest, which results in a relatively high fuel economy (for its class). A second group of buyers, the largest proportion, was willing to pay a few thousand euros more for a more powerful engine (but with lower fuel economy): this was the case for 60–70 per cent of the consumers. Car buyers in this group have different reasons for this choice: typically to drive more conveniently on motorways, or because they like sporty driving, for status, or to tow a caravan. Few buyers are willing to pay more (a few thousand euros) for a cleaner engine: if such an option was available, it was chosen by only 5 per cent of the buyers of vehicles; however, often such an option was not available.[8] Adding these 'green' ICE car consumers to hybrid electric vehicle drivers (and LPG drivers), makes the estimated share of 'green consumers' in the whole population around 2–3 per cent. We found no evidence that this group is growing: hybrid vehicle sales stabilized at 2,800 in

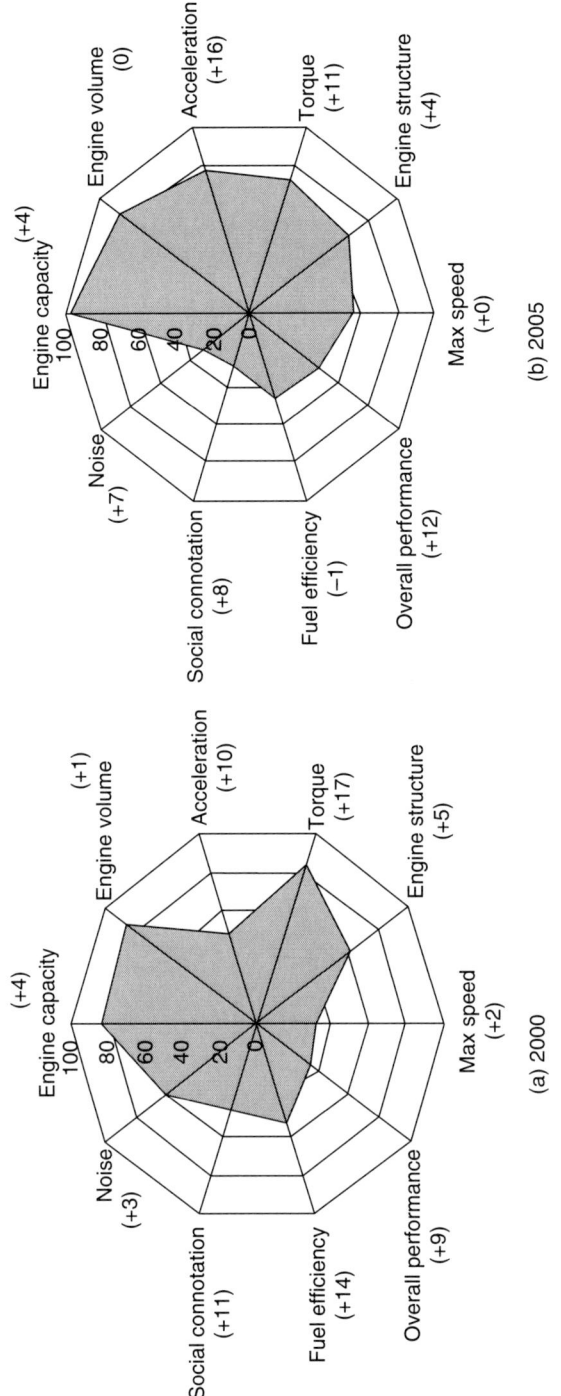

Figure 16.3 Frames for direct injection diesels, 2000 and 2005
Notes: As Figure 16.2.

the Netherlands in 2005–7, representing about a 0.5 per cent market share;[9] nor did LPG grow.

To get a flavour of the three sub-groups, we combine discourse analysis (from the section above, headed 'Mental perspectives of conventional and alternative engines') with the revealed preferences, and portray three sub-groups or sub-frames within the aggregated market frames depicted in Figures 16.2 and 16.3. Table 16.3 shows how we portray these three groups with their different levels of attention towards functionality, social connotation, price and environmental impact. Figure 16.4 visualizes the three groups shown in Table 16.3. Individual consumers will be inclined to belong to one of these three groups. The rationale here is that when the three user groups (frames) are merged, taking into account their sizes, they should deliver an (aggregate and averaged) socio-technical frame, as shown in Figures 16.2 and 16.3. Clearly, this is not a perfectly sound method, but valuable as a first estimate for consumer sub-groups.

Summary and conclusion

In this chapter we have assessed whether the introduction of hybrid electric and electric vehicles reflect a more fundamental change in demand structures for car engines. We find that, since the mid-1990s, the automotive market has gone through a significant change in the way information is available to consumers. The internet has become one of the most dominant sources of information, and in response the role of automobile salespeople has shifted from leading to guiding, and from a selling role to advising. Also, formerly fixed customer–dealer ties have become increasingly fluid.

An important outcome of this trend is that the information age has disclosed information about the environmental impact of (alternative) vehicles that was until recently inaccessible to consumers. Environmental information and taxation schemes put a spotlight on the issue of environmental performance of car engines. In particular, information about the fuel

Table 16.3 Sizes and characteristics of consumer sub-frames

Consumer sub-frame	Size (%)	Weights			
		Functionality*	Social connotation	Price**	Environmental impact
1 'penny-pinchers'	35	0.30	0.20	0.45	0.05
2 'power for convenience'	60	0.50	0.20	0.30	0.00
3 'green'	5	0.05	0.20	0.30	0.45

Notes: *Functionality includes engine capacity, acceleration, maximum speed, etc. **Price includes purchase price and operational cost (which relates to fuel efficiency).

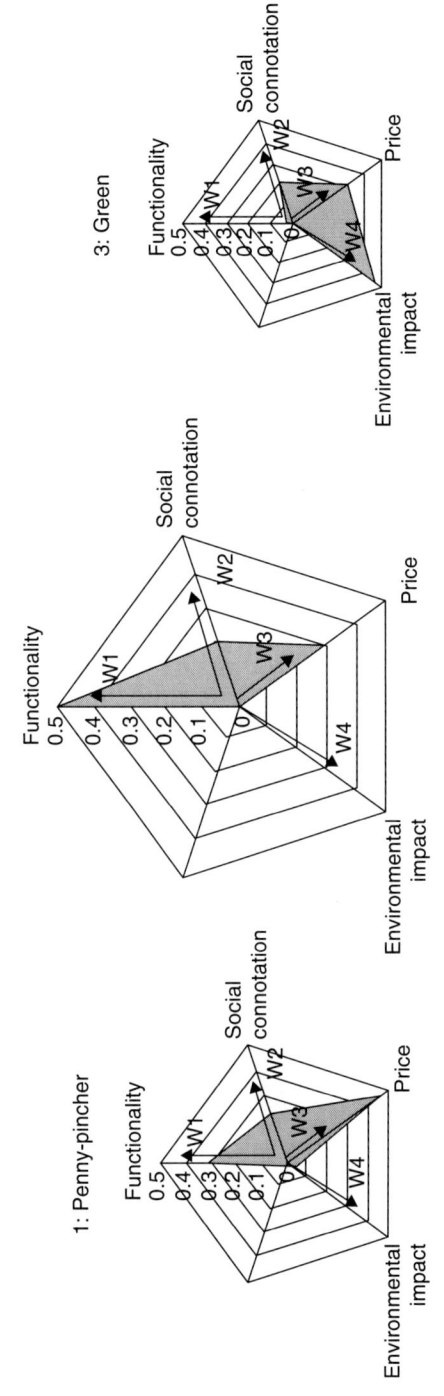

Figure 16.4 Consumer sub-frames: a visualization of Table 16.3
Notes: As Figure 16.2.

efficiency of vehicles is now much easier to find and compare than even a few years ago.

Nevertheless, regarding alternative engine types, we do not find any evidence that stated preferences are becoming more favourable towards these vehicles. Most studies come to the conclusion that environmental factors currently do not apparently play a major role in consumer car choices. Thus, whether the recent disclosure of the environmental performance of cars will lead to a significant shift in consumer preferences remains to be seen.

Regarding consumer perspectives, we find that conventional and alternative engines are framed rather differently. In frames of conventional diesel engines, the main attributes are *engine capacity* (number of kW), *engine volume* (number of litres) (both appearing in at least 83 per cent of accounts since 1996) and *torque* (amount of force, in Nm). Further, *environmental impact* was mentioned in 20 per cent of the accounts in 1996, 3 per cent in 2000, and 10 per cent in 2005. It was therefore a minor consideration, and a very minor one after the year 2000. Environmental concerns are apparently associated only loosely with discussions of new diesel engines throughout the whole period.

For hybrids, on the other hand, *fuel efficiency* is by far the most prominent attribute, and thus hybrids were associated mainly with low fuel use, while remarks on performance were few. *Acceleration*, an aspect of performance, was mentioned in only 17 per cent of the accounts, and appraised as +5 in 2005. This satisfaction score was low in comparison to diesels.

Further, we find that, over time, the framing of a conventional engine (diesel) is more stable than that of an unconventional engine (hybrid and fully electric). This corresponds with earlier studies of technological change (Bijker, 1995), which state that actor perspectives 'stabilize' in the course of market development, and thus that the framing of an established technology (conventional diesel engines) is more rigid than that of an unconventional engine.

The development of the social connotation of electric, and especially hybrid electric, vehicles shows some remarkable dynamics. In the mid-1990s, we found that attention paid to BEVs was fairly high: 20 accounts, which is double the amount for direct-injection diesels in that year. Social connotation, in contrast, was underdeveloped: we could identify it in only 15 per cent of the accounts, and in quite neutral statements. Most accounts were summing up technical characteristics. Around the year 2000, interest in BEVs had collapsed, and social connotation had not developed. Meanwhile, attention paid to HEVs had grown considerably, up to 22 accounts by the year 2000. Social connotative attributes appeared in a third of the accounts. Most were positive in tone (using adjectives such as 'high-tech', 'environmentally friendly', 'modern'), some were neutral, and one was negative. By 2005, attention had risen to 41 accounts, even more than for direct injection diesels in that year. Reference to social connotation appeared in about a third of the accounts, and its mainly positive appraisal had caught up with references to direct injection diesel. Since our findings are based on articles in national newspapers, they are difficult to compare to a survey of American HEV adopters, which found that 31 per cent of HEV buyers said they purchased an HEV because the vehicle 'makes a statement about me' (Heffner et al., 2005). There was a slight suggestion that social connotation was more important for US buyers compared to those in the Netherlands. Finally, we found some evidence for sub-frames among consumers. Supported by an analysis of actual sales in the Netherlands, our research indicates that three sub-groups can be distinguished in the total population of consumers of new cars. For the first group (about 35 per cent), price is the most salient attribute. These buyers are satisfied with the functionality of the cheapest engine. The second group (about 60 per cent) is willing to pay more for a more powerful engine. Adopters have different reasons for this, as noted earlier. Third, we find a small 'green car' segment; that is, a group of consumers willing to pay more for a cleaner engine, comprising only a small percentage of the consumer base (of those who buy a *new* vehicle). We conclude that alternative engines will become more widespread when they are embraced by the first group: those for whom price is salient. When oil prices become higher, the fuel efficiency benefits of alternative vehicles is likely to open up these models to the larger segment of private car buyers: those who favour the cheapest drive. Consumer-oriented policies, such as financial benefits for eco-efficient vehicles, may provide a further push in that direction. Recent studies are moderately positive about the influence of financial incentives on increasing sales of fuel-efficient cars (MMG Advies, 2008; Nijhuis and van den Burg, 2009; PBL, 2009).

Appendix 1: Analysis of engine choice

Buyers of new vehicles can choose their preferred type (and size) of engine from a range of alternatives. For example, as Figure A16.1 shows, Ford Focus consumers could choose between four petrol and three diesel engines in 2007. The smallest petrol engine (1400 cc) was chosen by 3.3 per cent of them, and 81.8 per cent chose the smallest diesel (1600 cc). Car salespeople (see Appendix 2) explained that these consumers typically buy the cheapest engine in the range. We carried out a similar analysis for the 30 top-selling vehicles in the Netherlands. For all these models we found that, on average, 34.8 per cent chose the cheapest and lightest engine in 2007. The average number differs significantly across brands, models and fuel. Typically, the larger models of the premium brands with petrol engines represent the lowest percentage of this group (for example, Audi A6, Mercedes S-class), whereas consumers of smaller vehicles or diesel engines have a greater preference for the least expensive version.

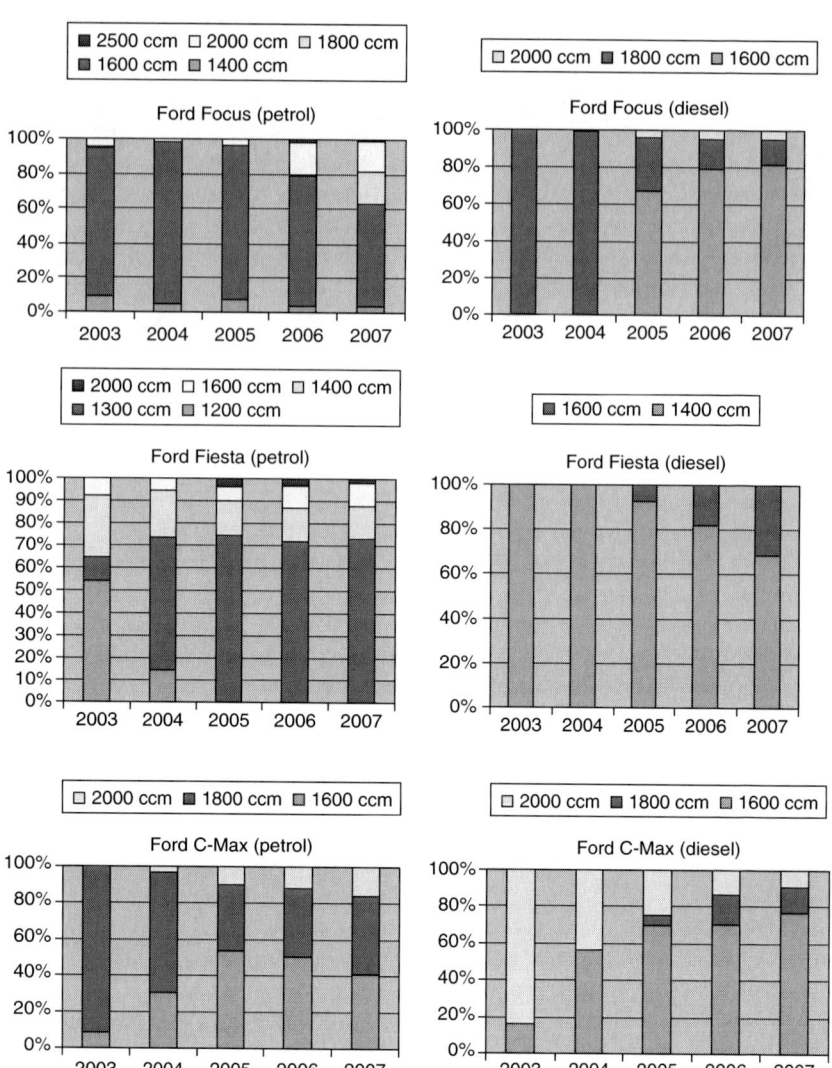

Figure A16.1 Sales figures for three Ford models in the Netherlands, 2003–7
Source: Ford Nederland/RDC.

Appendix 2: Stakeholder consultation

For the study mentioned in the text, car salespeople were interviewed at car dealers of three different brands (Peugeot, Audi and Opel – see Table A16.1). In these interviews we provided them with the data of national sales figures for different engine types (2003–7, see Appendix 1), and asked them whether they could explain why consumers choose the various types of engines.

Table A16.1 Interviews with motor companies/dealerships, 2005 and 2009

Organization	Interviewee	Position	Date
Peugeot dealer (Maastricht)	Mr Van der Cluijs	Salesman	Spring 2005 / Spring 2009
Opel dealer (Maastricht)	Frank Geurts	Salesman	Spring 2005 / Spring 2009
Audi dealer (Maastricht)	Jean Lucassen	Salesman	Spring 2005 / Spring 2009
Ford Nederland (Amsterdam)	Roland Roordink	Sales & Marketing Department Manager	Spring 2009

Notes

1. In this chapter we have omitted to mention compressed natural gas (CNG) and biofuels, since they are not alternative engines but alternative fuels. Hence, by the term 'alternative vehicles', we mean 'cars other than those with conventional, sole IC propulsion'.
2. Complex buying behaviour means that consumers are highly involved in the purchase; the product itself is expensive, is bought infrequently, is perceived to be risky and is highly self-expressive.
3. An important note here is that there might be a significant difference between purchasers of second-hand cars and new-car purchasers. Furthermore, Lambert-Pandraud et al. (2005) have pointed out that older adopters, who constitute an important market segment, repurchase a brand more frequently when they buy a new car.
4. These figures represent the catalogue price as paid by buyers from the private household sector in 2009 euros (adjusted for inflation). It ignores environmental subsidies but includes net catalogue price, private motor vehicle tax, and VAT.
5. Importantly, the study is on car preferences, not on car engines.
6. That is: reliability.
7. Also, consumers may have been influenced by deliberate marketing strategies. The Prius I (2000–4) was marketed specifically as an environmentally friendly vehicle. The Prius II (2005–9), however, was marketed as an innovative high-tech product with an energy-efficient motor, and less as an eco-product (information based on an interview with a Toyota Marketing Manager in 2006).
8. Here, engines that are specifically engineered for high fuel economy, without compromising on power. An example is the 3-litre engine from Audi/Volkswagen, which can travel 100 km on 3 litres of petrol. These engines are more expensive. Not all manufacturers offer these high-efficiency engines, but after 2007, more versions have appeared on the market (for example, Volkswagen BlueMotion).
9. These three years are convenient for comparison, since regulatory support for vehicles did not change during that time in the Netherlands. From 2008 onwards, a new regulation was introduced: taxation schemes for car leasing and car purchase. Subsequently, sales of HEVs grew enormously in the Netherlands: to 12,000 in 2008 and 16,000 in 2009. We see a clear link between the two, and therefore this does not change our consumer segmentation. At that time, eco-friendly cars became cheaper to drive, thus enlarging the potential market of HEVs to new consumer segments (that is, the 'penny-pinchers' in our typology).

References

Achtnicht, M. (2009) 'German Car Buyers' Willingness to Pay to Reduce CO_2 Emissions', ZEW Discussion Paper No. 09-058, Mannheim, Germany.

Bijker, W. B. (1995) *Of Bicycles, Bakelite, and Bulbs: Toward a Theory of Sociotechnical Change*. Cambridge, MA: MIT Press.

Boardman, B., Banks, N., Kirby, H. R., Keay-Bright, S., Hutton B. J. and Stradling, S. G. (2000) *Choosing Cleaner Cars: The Role of Labels and Guides. Final Report on Vehicle Environmental Rating Schemes*. Oxford/Edinburgh, UK: TRI (Transport Research Institute)/ECI (Environmental Change Institute).

Capgemini (2010) 'Listening to the Voice of the Consumer', Cars Online, 10/11.

CNW (2006) *Hybrid Motivators. Report #135Q*. Bandon, OR, USA: CNW Market Research.

Dijk, M. (2011) 'Technological Frames of Car Engines', *Technology in Society*, 33(1–2): 165–80.

Florschütz, C. C. and Bakker, M. H. N. (2002) *Informatie over duurzaamheid: een zoektocht (A Quest for Information about Sustainbility)*. Leeuwarden: NIDO.

Garud, R. and Rappa, M. (1994) 'A Socio-Cognitive Model of Technology Evolution', *Organization Science*, 5(3): 344–62.

Haan, P., De Peters, A. and Mueller, M. (2006) 'Comparison of Buyers of Hybrid and Conventional Internal Combustion Engine Automobiles: Characteristics, Preferences, and Previously Owned Vehicles', *Transportation Research Record: Journal of the Transportation Research Board*, 1983: 106–13.

Heath, A. P. and Scott, D. (1998) 'The Self-concept and Image Congruence Hypothesis', *European Journal of Marketing*, 32(11/12): 1110–23.

Heffner, R., Kurani, K. and Turrentine, T. (2005) *Effects of Vehicle Image in Gasoline-Hybrid Electric Vehicles*. Davis, CA: Institute of Transportation Studies, University of California, Davis.

Heffner, R., Kurani, K. and Turrentine, T. (2006) 'A Primer on Automobile Semiotics', ITS Working Paper 06-01. Davis CA: Institute of Transportation Studies, University of California, Davis.

Heffner, R., Kurani, K. and Turrentine, T. (2007a) 'Symbolism and the Adoption of Fuel Cell Vehicles', *World Electric Vehicle Association Journal*, 1: 24–31.

Heffner, R., Kurani, K. and Turrentine, T. (2007b) 'Symbolism in California's Early Market for Hybrid Electric Vehicles', *Transportation Research Part D*, 12(6): 396–413.

Johansson-Stenman, O. and Martinsson, P. (2006) 'Honestly, Why Are You Driving a BMW?', *Journal of Economic Behaviour and Organisation*, 60(2): 129–46.

Klein, J. (2007) *Why People Really Buy Hybrids*. Topline Strategy Group Report. Wellesley, MA, USA. Available at: www.cleanenergycouncil.org/files/Topline_Strategy_Report_Why_People_Really_Buy_Hybrids.pdf.

Klein, L. R. and Ford, G. T. (2003) 'Consumer Search for Information in the Digital Age: An Empirical Study of Prepurchase Search for Automobiles', *Journal of Interactive Marketing*, 17(3): 29–49.

Kubik, M. (2006) 'Consumer Views on Transportation and Energy (Third Edition)', Technical Report NREL/TP-620-39047, January. Golden, CO, USA: National Renewable Energy Laboratory.

Lambert-Pandraud, R., Laurent, G. and Lapersonne, E. (2005) 'Repeat Purchasing of New Automobiles by Older Consumers: Empirical Evidence and Interpretations', *Journal of Marketing*, 69(2): 97–113.

Lane, B. (2005) *Car Buyer Research Report: Consumer Attitudes to Low Carbon and Fuel-Efficient Cars*. Bristol, UK: Ecolane Transport Consultancy.
Maritz (2006) *Maritz Poll 2006: Automotive Sector in Europe*. Maritz Automotive Research Group, Hamburg. (Available via the author.) MMG Advies (2008) *Evaluatierapport: Werkgroep evaluatie energielabel en bonus-malusregeling BPM 2006* (Evaluation report, energy labels). Den Haag, Netherlands: MMG Advies BV.
Mytelka, L. K. (2008) 'Hydrogen Fuel Cells and Alternatives in the Transport Sector: A Framework for Analysis', in L. K. Mytelka and G. Boyle (eds), *Making Choices About Hydrogen: Transport Issues for Developing Countries*. Tokyo: UNU Press/Ottawa: IDRC Press.
Nijhuis, J. and van den Burg, S. (2009) 'Consumer-oriented Strategies for Car Purchases. An Analysis of Environmental Information Tools and Taxation Schemes in the Netherlands', in T. Geerken and M. Borup (eds), *System Innovation for Sustainability 2: Case Studies in Sustainable Consumption and Production – Mobility*. Sheffield, UK: Greenleaf.
Ozaki, R. and Sevastyanova, K. (2011) 'Going Hybrid: An Analysis of Consumer Purchase Motivations', *Energy Policy*, 39(5): 2217–27.
PBL (Planbureau voor de Leefomgeving) (2009) *Energielabels en autotypekeuze: Effect van het energielabel op de aanschaf van nieuwe personenauto's door consumenten* (Energy labels and choice of car types). Bilthoven, Netherlands: PBL.
Porac, J. F., Rosa, J. A., Spanjol, J. and Saxon, M. (2001) 'America's Family Vehicle: Path Creation in the U.S. Minivan Market', in R. Garud and P. Karnøe (eds), *Path Dependence and Creation*. Mahwah, NJ: Lawrence Erlbaum Associates.
Reed, G., Story, V. and Saker, J. (2004) 'Information Technology: Changing the Face of Automotive Retailing?', *International Journal of Retail & Distribution Management*, 32(1): 19–32.
Shove, E. (2003) *Comfort, Cleanliness and Convenience: The Social Organisation of Normality*. Oxford, UK: Berg.
Steg, L., Vlek, C. and Slotegraaf, G. (2001) 'Instrumental-Reasoned and Symbolic-Affective Motives for Using a Motor Car', *Transportation Research Part F – Traffic Psychology and Behavior*, 4(3): 151–69.
Teisl, M. F., Noblet, C. L. and Rubin, J. (2007) 'The Design of an Eco-Marketing and Labeling Program for Vehicles in Maine', in U. Grote, A. K. Basu and N. H. Chau (eds), *New Frontiers in Environmental and Social Labeling*. Berlin: Physica-Verlag (Springer), pp. 11–35.
TNS-Emnid/AutoScout24 (2004) *Internet gewinnt bei der Fahrzeugvermittlung weiter an Bedeutung* (The internet gains importance in vehicle placement), Report, 19 August.
Van den Brink, R. M. M. and Van Wee, B. (2001) 'Why Has Car-Fleet Specific Fuel Consumption Not Shown Any Decrease Since 1990? Quantitative Analysis of Dutch Passenger Car-Fleet Specific Fuel Consumption', *Transportation Research Part D*, 6(2): 75–93.

17
The Second Automotive Revolution Is Under Way: Scenarios in Confrontation

Michel Freyssenet

Since the publication of the previous GERPISA book, *The Second Automotive Revolution* (Freyssenet, 2009b), many signs seem to confirm that the transition to cleaner vehicles is actually beginning (Jullien 2010). The use of alternative fuels, natural gas or liquefied petroleum gas (LPG) is growing steadily, and there has been a significant increase in the numbers of hybrid vehicles, battery electric vehicles and plug-in hybrid cars being produced. Projects planning to launch more of these types of cars by 2015 are still more numerous (see Table 17.1). Car manufacturers are increasing the testing of plug-in and electric vehicles under real conditions. Plans for installing the infrastructure for battery charging or exchange are starting to be realized. Many governments and local authorities have confirmed, and sometimes extended, despite the current financial crisis, their support programme for new powertrains. Research into batteries with an autonomy similar to that of the internal combustion engine (ICE) is well established, and research on fuel cells continues. Associations, foundations, consortia, the media and 'initiatives' for the promotion of electric vehicles are proliferating. Finally, one can observe the evolution of the position and of the strategy of some governments and some carmakers, a sign that the battle for the various transition scenarios is well and truly under way.

Three of the four conditions for a new automobile revolution are about to be fulfilled

At the 17th International Conference of GERPISA (Freyssenet, 2009b), we highlighted the four major conditions that made possible the automobile revolution at the turn of nineteenth and twentieth centuries, and the choice of an oil-powered ICE to the detriment of steam boilers and electric motors (Bardou *et al.*, 1982; Mballa, 1998; Mom, 2004). At the time of writing, three of these four conditions have been met and make for a probable new car revolution (Freyssenet, 2009b).

The first of these conditions is the urgency of solving the crisis of the 'door-to-door' transport system, based solely on animal traction in the past,

Table 17.1 Number of cleaner car models on sale across the world, 2011, models to be launched by 2015, and prototypes (non-exhaustive census, June 2011)

Cleaner car models	Hybrid	Plug-in hybrid	Electric
On sale 2011	36	4	70 + 37 LS*
Launched by 2015	25	29	60 + 4 LS*
Prototypes	6	7	41 + 3 LS*

Note: *LS = low speed car model.

and today on the oil-powered engine. This urgency is affecting all 'automobilized' countries. Indeed, nothing now seems able to stop the trend of an increasing real price of oil and natural gas, unless a durable global economic recession or a serious crisis arises in one or more of the major emerging countries. China and India are now convinced that they will not pursue their industrialization and 'automobilization' by continuing to focus on oil as the main energy source. In addition, the use value of automobile vehicles is declining, while their total cost of ownership (TCO) and the cost of their externalities are rising steadily. Tolerance of congestion, pollution, accidents, the fragmentation of space and so on is falling sharply in developed countries. Finally, scientists say that global warming is continuing, and it has been shown that the efforts made to limit its increase are now insufficient.

The transfer of new technologies developed in other sectors and the race for innovations are the second condition for an automotive revolution. Now we can see that both are accelerating. There is so much public and private research taking place into renewable forms of energy and cleaner engines, and many announcements of new technical solutions to the problems introduced by these innovations. Renewable energies are entering a phase of mass industrialization. The third generation of agrofuels and batteries are being studied and some products are already being tested. Miniaturized electronic devices allow the optimization of both hybrid engines and batteries. New car architectures are being devised and new functionalities added. For example, the integration of electric motors within the wheels not only improves driving, but also makes possible the removal of the engine compartment. This creates new possibilities for car designs and uses.

The third condition for an automobile revolution is the formation of coalitions between public and private actors, government decisions, agreements between states to uphold and implement a specific solution, and in particular to make available in any place and any time the energy required by the type of engine adopted. The focus on the debate around the performance of the different powertrains often overlooks that these work in tandem with the energy they need to drive them. The use of the term 'automobile' to refer to existing vehicles is in fact a misnomer, because cars do not move by themselves, but by using an external source of energy that must be provided. This is essential to understanding how one fuel solution prevails over another. The oil-powered ICE quickly prevailed over electric motors in the early twentieth century, not because of its performance, which was poor, but because of the absence of an electrical grid deployed over the whole country. In contrast, oil could be transported and stored easily anywhere by private agents (Mballa, 1998). Utilities, governments and local governments were not at that time able to work together to invest and plan for the rapid electrification of urban and rural areas, despite the geopolitical dependency on oil that this inability created for more than a century. Today, agrofuels and gas are as easily available as oil, and they are able to use the existing oil distribution network. What is new today is the omnipresence of electricity. It has become even more readily available than oil, subject to the installation of adequate electric plug-in facilities in homes and public places. It can also be produced locally and partly stored. The vehicle itself can even produce a portion of the electricity it needs, thanks to photovoltaic panels and wind turbines, which are becoming increasingly common.

The fourth major condition for an automotive revolution is the existence of economic policy decisions that allow for the dissemination of new, cleaner cars. The most important decisions concern the policy of national income distribution, which determines the volume and structure of possible car demand; and environmental and energy policies, which determine the rapidity of adoption by buyers choosing cleaner cars. This fourth condition will be required only when one of the three possible transition scenarios outlined at the 18th International Conference of GERPISA in 2010 is imposed: the scenarios of diversity, of progressiveness, or of rupture (Freyssenet, 2011). Since the GERPISA conference, it has become possible to characterize these three scenarios more clearly, to present the first effects of their confrontation, and to evaluate their respective chances of succeeding.

The scenario of diversity

The scenario of diversity is already in progress, but paradoxically it has few supporters and is probably unlikely to be successful. The scenario is characterized by governments being able to impose on car manufacturers the type of energy available to them naturally or commercially in good conditions. In this case, carmakers would be required either to specialize in particular engine on specific markets, or to offer all types of engines if they wanted to be represented globally.

Indeed, in 2011, countries that are now the main automobile markets are distinguished by their energy preferences and their industrial ambitions. These countries are divided into five groups (note: some country names may appear twice as they have moved from one group to another). The first group is the countries that continue to focus on oil as a fuel, while expressing an interest in cleaner engines, including simple hybrid vehicles; the Middle Eastern countries and Mexico are typical of this group. The second group covers countries that have chosen agrofuels, such as Brazil and Sweden (Amatucci and Spers, 2010). The third group is composed of the countries preferring natural gas, such as Italy, Pakistan, Russia, Poland and Iran (Stocchetti and Volpato, 2010). In the fourth group, we find the countries that have chosen electricity, promoting plug-in hybrids and electric vehicles. These countries include the USA since the Obama Administration took power, China, India, South Korea since 2009, and all Western European countries apart from Italy and Germany. In the fifth and final group are countries that have not chosen to favour a particular technical solution. They consider that this choice is the concern only of the carmakers responsibility for production. So these countries just set standards for the reduction of fuel consumption and air pollution, according to technical and economic feasibility, and the evolution of energetic, climatic, political and social constraints. Among these countries were the USA before the Obama Administration, Japan, South Korea before 2009, Germany and the European Union (see Table 17.2).

Countries that defend the scenario of diversity – that is, the continuation of current trends – are countries that have a high-energy specialization or a strong economic ambition, but they know they cannot impose their choice on other countries. This is the case with Brazil, Sweden, the Middle East, Mexico, Russia, Italy, Pakistan, Poland, Iran and so on. Manufacturers, even if, for the moment, they follow national preferences, few among them have adopted a clear position in favour of the diversity scenario. Fiat and Volvo appear to be the only ones clearly to support this scenario. But their position

Table 17.2 Five groups of countries, according to their energy preferences for cleaner cars, with some recent changes in these preferences, as at 2011

Less polluting oil engines	Agrofuels	Natural gas	Plug-in hybrids and electric vehicles	Only objectives are pollutant reduction and fuel cell perspectives
Middle East countries Mexico	Brazil Sweden	Russia Italy Poland Iran Pakistan Iran	China, India France, Belgium, UK, Spain, Portugal, Denmark, Switzerland Israel, many small countries, many islands, etc.	Japan ← S. Korea ← Germany European Union
USA ⎤ Canada ⎦			→	

Note: Changes in preferences indicated by arrows.

has become increasingly blurred, after Fiat taking control of Chrysler, and since the Chinese manufacturer Geely bought Volvo. Producers of both agrofuels and oil are, of course, favourable to the scenario of diversity, as well as some local media and anti-globalization movements defending the independence of national energy. Supporters of diversity apparently do not attempt to form coalitions that could make their choice prevail across the world. Perhaps they are too confident in their local power and their traditional lobbying, or they are too divided among themselves.

But they should worry about the important changes that have been seen since 2009, as several countries, have joined the group in favour of plug-in hybrid and electric vehicles. This applies to both the USA and South Korea, which have clearly changed their positions, and Canada and Germany that are moving more cautiously. The German government is now a little more open to electric mobility, under pressure from Daimler and BMW, and from some regional administrations and political parties.

For the scenario of diversity to prevail, three conditions must be met. First, there should be in each group of countries at least one country with, on the one hand, a rapidly growing automotive demand and a potentially very wide market, and on the other, real geopolitical power. In this case, however, the price gap between the preferred engine and the engines favoured by other countries should not be too large. Otherwise the purchasing power of the population and the productivity of the country would be affected. Finally it would be necessary for the major global car manufacturers to accept a world market heterogeneity even higher than it is today.

The first of these conditions seems to have been satisfied. Who indeed can imagine changing, in the short term, the energy policy of the USA, China, India, Brazil, Russia and so on? It is currently unclear whether the second condition can be completed or not. Hybrid vehicles, which are more complex, should not see their total ownership cost fall below that of oil-powered vehicles. However, the total cost of ownership of electric-powered vehicles should decrease significantly in the medium and long term, for reasons we shall explain later. In this case, consumers could exert pressure on governments and manufacturers focusing on oil-powered engines to change their orientation. The third condition will be the result of the battle between car manufacturers: between those who will specialize in certain markets, the 'regional manufacturers', and 'global manufacturers', which will tend to reduce diversity.

If the scenario of diversity prevails, the consequences will therefore be a greater heterogenization of the world's automobile market; a greater complexity of the platforms; the necessity for more R&D investments; difficulties for carmakers pursuing a 'volume and diversity' profit strategy and a 'permanent reduction of costs' profit strategy, according to the typology we have proposed (Boyer and Freyssenet, 2000); and the formation of an oligopoly of carmakers specializing in a particular type of engine. Newcomers or 'born again' carmakers, as are some Russian manufacturers, could find market niches they need to protect from excessive competition. Finally, governments could maintain, or recover, a measure of control over both the manufacturers and the car market through their energy policies.

The scenario of progressiveness

Many people have imagined a gradual scenario, as it seems the only realistic and reasonable one, but it could turn into its exact opposite 'all at once'. The progression would be, via engines using agrofuels or natural gas, to hybrid engines, then to plug-in hybrids and finally electric motors. The transition would be at the rate of depreciation of investments, of the technical improvement of the different solutions, and of the natural renew of the world stock of cars. It appears that is this the only way to progress, because it is impossible that the whole world is able to change its type of vehicle overnight (Chanaron and Teske, 2007).

Supporters of this scenario are, above all, countries that merely set standards periodically to make the reduction of pollution even more stringent. They set the level of these standards and the frequency of change, taking into account the technical and financial resources of their national manufacturers. In fact, these were not among the more active protagonists in international negotiations to reduce greenhouse gas (GHG) emissions. They are, as we saw earlier (Table 17.2), Japan, South Korea, Germany and the European Union. But South Korea has clearly changed its mind in favour of electro-mobility since 2009. The German government has agreed to be more ambitious regarding the number of plug-in and electric-powered vehicles on the road (1 million by 2020). It refuses for the moment, however, to create new support for the purchase of these types of vehicles, despite the insistent demand of carmakers Daimler and BMW.

Among the supporters of the gradual scenario (see Table 17.3), there are logically almost all manufacturers pursuing a 'volume and diversity' profit strategy (Volkswagen, PSA, Ford, General Motors (GM), Hyundai–Kia); the manufacturers pursuing a 'permanent cost reduction at constant volume' strategy (Toyota); and a 'quality' strategy (Daimler, BMW and Porsche).

But the composition of this group is changing significantly: GM, Hyundai–Kia, Daimler and BMW now prefer the plug-in hybrid vehicles and electric vehicles. And Daimler and BMW also continue to explore fuel cell solutions.

Reasons for these surprising developments lie in the conditions of viability of a gradual scenario that, contrary to appearances, are extremely difficult to meet. For this scenario to be possible, several partly independent processes must be combined. There must be, simultaneously, a slow and controlled rise of oil prices; global warming that is less important than had been predicted; less strong pressure from governments, local authorities and

Table 17.3 Five (shifting) cleaner automobile strategies of some carmakers, giving priority to certain types of vehicle, August 2011

Less polluting fuels: gas, agrofuels	Hybrids versus plug-in hybrids	Hybrids versus all types of engine	Plug-in hybrids versus electric vehicles	Electric vehicles
Fiat Chrysler? Volvo Russian carmakers?	Toyota Honda Mazda Porsche	Ford PSA Volkswagen GM ⟶ Daimler ⟶ BMW ⟶ Hyundai–Kia ⟶	Mitsubishi BYD	Renault-Nissan, many Chinese and Indian carmakers, nearly all start-ups and others

Note: Changes in preferences indicated by arrows.

environmental movements for higher standards; a significant slowdown in the growth of large emerging countries, and their acceptance of the energy choices of established carmakers; the inability of manufacturers in China and India to produce, quickly and massively, a range of electric vehicles; and finally, technological barriers to substantially improved battery performance in the short to medium term.

These conditions are in fact difficult to meet and were certainly not fulfilled in 2011. So the manufacturers who chose the gradual scenario will be forced to launch, much earlier than expected, plug-in hybrids and electric models, while they are still trying to sell their first hybrid vehicles. The scenario of progressiveness could therefore be transformed into its exact opposite: the scenario that we have labelled 'all at once'. Rather than stagger investments and limit the technical and commercial risks, these car manufacturers seem to be pushed to do the reverse.

In the scenario of diversity, the global generalist carmakers are also driven to do everything, but they are assured of having markets corresponding to their different engines. This is not the case in the gradual scenario, which would become the scenario of 'all at once'. Manufacturers could see some of their engines quickly outdated technically or abandoned by customers, and not being able to depreciate their investments.

Yet the gradual scenario would have been ideal if the automotive industry had continued in its current structure. The economic and technical domination by the established carmakers could easily have been extended. The current balance of power between manufacturers and governments could continue, as would the different profit strategies of the carmakers. Newcomers would not have become threatening. The automotive supply chain, and the volume and structure of jobs would be able to evolve in a controlled manner.

However, if the gradual scenario became the 'all at once' scenario, and at the same time 'universal', the consequences for the carmakers favouring the progressive scenario would be a significant increase in both investment and risk, more power given to governments called on for help, a destabilization of the profit strategies of the firms concerned, increasing difficulties in competing with newcomers that would immediately privilege electric vehicles, and finally complications of coexistence and to control in the same group but with different value chains, and to manage the evolution of the volume and structure of labour.

The scenario of rupture

The third scenario is that of a rupture; that is, the immediate adoption of plug-in hybrid vehicles, or, even more radical, all-electric vehicles. This seems to be the riskiest scenario, but it is not the most improbable. The steps taken by the other engines in the progressive scenario are skipped here. This jump is a technical, industrial, environmental, stylistic, practical and symbolic rupture. The form of energy used and the engine type change completely, and the 'automobile system' is completely transformed. Various types of pollution generated by the ICE are for the most part eliminated, subject to the production of electricity by forms of renewable energy. The use of the automobile may be highly diversified (Beaume and Midler, 2009).

The two stages of the rupture

Ironically, the rupture takes place in two stages, but could succeed relatively quickly if certain conditions are met. During the first stage, only some uses of the car will be applicable to, and some users be interested in, electric vehicles. The second stage would see not only the generalization to other uses and users, but also the emergence of new applications and new consumers.

The lithium ion batteries, which are used to power the first plug-in hybrids and electric vehicles (called second-generation batteries), do not provide the same travelling distance as other engines before needing to be recharged (160 km as against 450 km) and their prices will remain high as long as they are not mass produced. The first step is for carmakers choosing this route to offer models for those engaged exclusively in travelling for short/medium distances. Cars that travel no more than 160 km per day is a very important criterion; for example, local delivery companies, rental cars, households' second car for commuting, cars moving about small countries and on islands and so on. This range of clients is sufficient to achieve in the near future the volume of production necessary to make the sale of electric vehicles profitable. At the same time, manufacturers have to work with other actors (electricity providers, public authorities and so on) to set up the infrastructure for electric charging or for the exchange of batteries in the areas concerned. Finally, they would have to make important investments in R&D to prepare a third generation of batteries that doubled their functioning distance and to conceive a new automobile architecture by exploiting the great potential of electric vehicles. All this would not be possible.

During the first stage, governments and local governments would have to offer grants for the purchase of electric vehicles to make them affordable to the earliest buyers, and in this way to break the vicious circle of high price and low demand. They would have to co-ordinate the setting of the battery recharge infrastructure, and to support R&D of third-generation batteries. Some of them might even want to penalize traditional cars by introducing carbon taxes or congestion charges.

Above a certain threshold of annual sales (100,000–200,000), economies of scale could quickly lower the price of batteries, therefore also the price of cars. The drop would be amplified by the lower cost of production and of the assembly of the vehicles themselves. The fully electric vehicle requires fewer components, and is easier and faster to produce and maintain.

The second stage of the rupture is the adoption of electric vehicles by the majority of drivers, once the third-generation batteries have been developed and if the price of oil has continued to rise. This conversion will be carried out more easily and quickly, as the manufacturers have taken advantage of the functional and symbolic potentialities offered by electric vehicles to make them into highly desirable and practical purchases. It seems that this is the scenario chosen by the countries that have adopted ambitious targets for plug-in hybrids and electric vehicles rolling on to the roads at different times (see Table 17.4).

Many active supporters to form coalitions

Supporters of the scenario of rupture are paradoxically numerous and active. They are working together to form coalitions, associations, consortia, 'initiatives' and so on. Among them (see Table 17.2), we find, first, as we have seen, countries already 'automobilized', such as the USA, Canada, Korea, Western Europe (including Germany, which appears to have abandoned the 'technological neutrality' it showed earlier, but for the time being with the exception of Italy), and countries in rapid 'automobilization' and with a strong industrial potential, such as China and India. Some of these countries are working together, notably to standardize the charging systems. This applies to the United States and China on one side, France and Germany on the other, etc. There are also many smaller countries (such as Israel, Switzerland, Georgia), and island countries (for example, Taiwan, Singapore, Hawaii, Iceland, New Zealand, Réunion) that have adopted this scenario.

But the most active supporters, who are pressuring manufacturers to speed up the launch of electric vehicles, are local authorities, car rental companies and companies with large fleets of vehicles. Regions and municipalities of large cities consider that electric vehicles could reduce both pollution

Table 17.4 Objective numbers of EVs (PHEVs + EVs) on the roads of the most ambitious countries, 2015–50

	2015	2020	2025	2030	2050
Switzerland		1 in 8 (500,000)			
Ireland		1 in 10 (200,000)			
China		1 in 12 (5,000,000)			
Spain	1 in 60 (250,000)				
France	1 in 62 (450,000)	1 in 16 (2,000,000)	1 in 9 (4,000,000)		
USA	1 in 136 (1,000,000)	1 in 20 (7,500,000)			
Canada		1 in 40 (2018) (500,000)			
Netherlands	1 in 450 (20,000)	1 in 45 (200,000)	1 in 9 (1,00,000)		
Germany		1 in 50 (1,000,000)		1 in 17 (6,000,000)	
European Union					1 in 4 (90,000,000)

and urban congestion, especially when the vehicles are rented or used for car-sharing. Among them are the following regions, provinces or prefecture administrations:

- California, Colorado, Tennessee and Oregon in the USA;
- Ontario and Quebec in Canada;
- Baden-Württemberg, Wallonia, eleven Spanish regions, Île-de-France, Brittany and the Vendée in Western Europe;
- West Midlands, Victoria and New South Wales in Australia;
- Saitama and Miyazaki in Japan;
- Guangdong and Wuhan in China, among others.

And the following cities or urban agglomerations:

- New York, Seattle, Houston, Washington DC, Chicago, Los Angeles, Vancouver Toronto and Mexico City in North America;
- São Paolo in South America;
- Guangzhou, Shenzhen, Shanghai and Hong Kong in China;
- London, Glasgow, Amsterdam, Rotterdam, Paris, Lyon, Rouen, La Rochelle, Nice, Strasbourg, Berlin, Munich, Hamburg, Hanover, Milan,

Rome, Florence, Madrid, Barcelona, Ankara, Stockholm, Copenhagen, Warsaw, Prague and Tel Aviv in Europe and the Middle East; and
- Tokyo, Yokohama, Seoul and Delhi in the Far East, which is not an exhaustive list.

Among the 40 cities in the countries most involved across the world in the fight against global warming (C40 Cities), the 13 largest have formed an EV Network to promote electric-powered vehicles and charging facilities.

Vehicles in company fleets generally travel short distances at low speeds on a daily basis. They are subject to usage restrictions in urban areas and represent a major expense to their firms. So the companies owning vehicle fleets consider that electric vehicles are already adapted to their needs and hope they will be more accepted in cities. They also know that electric vehicles are cheaper to use and maintain, and that the second generation of vehicles will be less expensive to buy. Among rental vehicle companies supporting electric vehicles are, in particular, Hertz, Avis, Nippon Rent-A-Car, Sixt, Athlon Car Lease, LeasePlan, ING Car Lease, Europcar and Arval. Among companies with large fleets of service vehicles are:

- Taxi companies, such as the Japanese companies Kashiwasaki, Matsuyama and Nihon Kotsu, the New York taxi companies, and the English company greentomatocars;
- utilities or telecommunications and mail companies, such as La Poste, TNT, DHL, UPS, FedEx, Austrian Post, Purolator, Bouygues Telecom, Orange and France Telecom;
- companies serving communities, businesses or individuals, such as GDF Suez, Turin Airport, San Francisco Airport, Domino's Pizza, Coca-Cola, Vinci Park and Veolia;
- electricity utilities, such as Tepco, EDF, EOS, San Diego Gas & Electric, ENDESA, Gaski Enerji, REW, A2A and Iberdrola;
- airlines or railway companies, as Air France, SNCF and Deutsche Bahn;
- urban transport networks such as RATP in France;
- the military, such as the Chinese Army and Navy;
- administrative services and institutions such as FedFleet, Yale, Rice and Stanford Universities, UCLA and NASA;
- industrial or commercial enterprises, such as General Electric (GE), Leclerc, Eiffage, SPIE, Areva, and the Japan Ryokan and Hotel Association.

Some of these companies have come together to launch joint calls for the purchase of electric vehicles. They assume a significant volume, as did some twenty French public and private companies: 100,000 electric vehicles within two years.

Car manufacturers that have opted for the rupture scenario were, first, Renault–Nissan, Mitsubishi, Chinese manufacturers (BAIC, Chana, Chery, Geely, Dongfeng, FAW and so on) and India (Tata, Mahindra and so on). Later, General Motors (GM) joined them, as along with Hyundai–Kia, and Daimler and BMW (see Table 17.3). We need to add to this list a myriad of smaller manufacturers, both old and new, who see the electric vehicle as an opportunity to create a niche, or to monetize their innovations, or even to compete with historical carmakers (see Table 17.5). Among these are:

- manufacturers of sports cars (Venturi, Lotus, Lightning, Malone TAZR and NLV);
- low-speed car companies, especially in China (Aixam, AEVCO, Biro, Bidewen, Chatenet, Eon Motors, GEM, Goupil, Kangda, Ligier-Matra and Shifeng);
- almost all of the many car start-ups that have appeared since the year 2000 (Tesla and Fisker for example);
- newcomers that may become major suppliers, such as Magna, Johnson Control and GKN;
- manufacturers of batteries, such as Bellier, Bolloré, BYD, Dassault, Electrovaya and Lynx;
- body-makers or assemblers, such as Micro-Vett, Alke, Balqon Corp., FAM, Oatmeal, Heuliez and Tazzari;
- manufacturers of electric motors, such as Electric Vehicles Europe and AC Propulsion;
- producers of logistics vehicles, such as INNOVEP and Motrec;
- engineering companies, such as Akka Technologies, Innovative Mobility and Zytek;
- electricity producers, such as ELCAT and the National Union Electric Company;
- importers, such as Miles Electric Vehicles; and
- manufacturers of bicycles and scooters, such as Ya Bao Jinan, Shandong Baoyi New Energy Vehicles, Piaggio, Scooter Street, Xero and Protanium.

Many people hope, particularly in countries and areas devoid of a domestic automotive industry or having lost their national carmakers over the previous decades, that these numerous players, new and different, could develop or relaunch automobile production (for example, the UK, Spain, Norway, Finland, Switzerland, Denmark, South Africa, Brazil, Hungary, the Netherlands and so on).

Suppliers, alone or in groups, are already making proposals for the provision of all or part of the charging required for electric powertrains and the infrastructure: Valeo, Michelin, Siemens, Schneider Electric, Bosch and so on. Each battery manufacturer is seeking to become the global leader: LG Chem, NEC, GS Yuasa, Hitachi, Panasonic, Toshiba, Saft, BYD and so on. All these have formed partnerships with carmakers. Fourteen US battery manufacturers and the Argonne National Laboratory have formed a

Table 17.5 Electric vehicle start-ups and newcomers in a range of countries (non-exhaustive census, June 2011)

	Start-ups, engineering companies	Suppliers, importers, bodybuilders	Carmakers, newcomers	Sport and low-speed carmakers, others
Australia	Eday Life			Alpha Lujo
Brazil				Obvio
Canada		Magna, Motrec, Nemo		BRP, Electro Wheels, Dynasty EC, Electroway, Zenn
China	Eagle, EuAuto, MyCar (Hong Kong), Yulon (Taiwan)		BAIC, BYD, Chana, Chery, Dongfeng, FAW, Foton, GAC, Geely, Hafei JAC, South East, Lifan, Zotye	Bidewen, Green Wheel, Jinan Baoya, Kangdi, Shandong Baoya, Shifeng, Wonder
Denmark				Protanium
Finland				ELCAT
France	Akka Technologies, Cocquelin, Courèges, Eco&Mobility, Electric-car, Fior-concept, Lumeneo, Muses, Volteis, Wattmobile	Bellier, Bolloré, Frappa, FAM, Gruau, Heuliez, SVE Dassault, Michelin		Aixam, Anruet, Chatenet, Eon-Motors, Goupil, Innovep, Ligier-Matra, Little Cars, Soffimat, Venturi
Germany	E-Wolf, Innovative Mobility	Ecocraft, Ruf		Smilies, Streetscooter
Hungary	Kenkuru Car			
India	Reva		Tata, Mahindra	
Italy	K-Way, Motus	Alke, Electric Vehicles Europe, Micro-Vett		Biro, Effedi, Piaggio, Tazzari, Pininfarina
Japan	Auto EV, SIM-Drive, Zero Sports			

Korea			CT&T
Norway	Think Global	Kewet ElbiNorge	
Russia	E-avtö		
Spain	Afaipada, Comarth		
South Africa	Optimal Energy, Peru ventures		
Sweden	Koenigsegg	EV adapt	
Switzerland	Mintset, Protoscar, Rinspeed		
UK	Liberty Elect, Lightning, Nice Car, Sinclair, Stevens Veh, Zytech	ECC, GKN, Smith Electric	Lotus, Malone TAZR, Modec, Murray, Xero
USA	AC Prop, Aptera, Commuter Cars, Azure Dynamic, Fisker, Phoenix, Segway, Tesla, Wrightspeed, Zap	Johnson Control, Balqon Corp, Envision, Miles Elect	AEVCO, Electrum, EZ-GO, Dynasty, GEM, Google, Myers, Shelby, Toro

Sources: Data from company websites and Wang and Kimble (2011).

National Alliance for Advanced Transportation Battery Cell Manufacture, so the USA is beginning to catch up in this area.

Producers and providers of electricity are working together to develop charging infrastructures and participate in the testing of electric vehicles and their management in real-life situations. Public research centres and universities are likely to be involved in joint programmes with industry, state and local governments in China, the USA, France, Germany and so on. Research is focused on batteries, infrastructure, materials, processes and so on. 'Mobility operators' are also emerging, such as Better Place, Mobile Tech, Mobility House TMH and Mobivia, offering complete and diversified solutions to local communities, businesses and individuals.

Sometimes some or all these different actors, governments, local manufacturers, suppliers, utilities, research centres and users come together in a consortium, as in the European consortium Green eMotion; the Electric Drive Transportation Association; the Electrification Coalition in the USA; Electric Mobility Canada; the Electric Vehicle Industry Alliance in China; the Electro-mobility Platform in Germany among others. They are becoming partners to test electric-powered vehicles in real-life conditions, as in the partnership between the Community of Agglomeration of Rouen, Elbeuf Austreberthe, and EDF, Schneider Electric, ERDF (Électricité Réseau Distribution France) and E. Leclerc in France.

Finally one can note numerous positions and actions in favour of electric vehicles. Some are expected, such as environmental movements or lobbying by citizens. Others are more unexpected, such as those of some churches. There is also the creation of dedicated media, trade shows and exhibitions. Might this flowering fade rapidly in the face of early difficulties, as has been the case several times in the past? (Fréry, 2000).

The conditions that make the rupture scenario possible are not technical, but geopolitical

The conditions for making this scenario possible are first, of course, widespread access to electricity, wherever it is necessary, at a very competitive price compared to other energy sources. Second, the price of oil must continue to rise rapidly and sharply, and the third generation of batteries must reach a similar autonomy to that of petrol-fuelled vehicles. Finally, the long-established carmakers must fear the Chinese and Indian car manufacturers skipping the stage of the ICE and becoming strong competitors in the electric car market (see Table 17.6).

Are these conditions being met? Electricity is available almost everywhere, at least as freely as petrol and at a lower price. It is not impossible that it could increasingly be produced locally, and even stored. The structural and economic conditions for a strong oil price growth already exist.

China and India are obliged to adopt the all-electric vehicle. They also have an interest in doing so as they have stated clearly that they wish to

Table 17.6 Number of cleaner car models in the world, by types of producer: on sale in 2011, and models to be launched by 2015 (non-exhaustive census, June 2011)

Producers	Hybrid	Plug-in hybrid	Electric
Established carmakers	38/19	2/22	12/30
BRIC carmakers		0/4	15/21
Start-ups, engineering		2/2	15/9
Suppliers, assemblers, others			42/4
Sport or low-speed carmakers			37/3

Note: BRIC = Brazil, Russia, India, China.

Table 17.7 Number of cleaner models, by type of vehicle: on sale in 2011, and models to be launched by 2015 (non-exhaustive census, June 2011)

Models	Hybrid	Plug-in hybrid	Electric	Fuel cell
Low-speed vehicle			37/4	
Minicar	1/0	0/2	27/19	
Small car	1/4	0/2	4/17	
Medium car	13/7	2/12	2/14	
Premium car	12/6	1/4	1/1	
Coupé	1/1		1/2	
SUV	5/4	0/3	6/2	
Commercial vehicle			23/2	
Sports car	3/2	1/6	6/3	0/1

develop a domestic automotive industry and become a powerful leaders in electro-mobility (Wang and Kimble, 2011). China already produces and sells millions of bicycles, scooters and low-speed electric cars (see Table 17.7). It seems that the foreseeable progress of battery autonomy will be sufficient to compete in the medium term with the ICE. The issues that the electric vehicle has to solve seem to be far less difficult than those that ICE vehicles had to overcome during their history (Mballa, 1998). Mass production of second- and third-generation batteries; the modularization of the vehicle that allows for the full electrification of the internal mechanism; the automation of assembly that allows for true modularization; and a drastic decrease in the number of parts and necessary employees should lead to a significant lowering of car prices. Only a radical change in the product can boost both profits and competition. A new profit strategy might even be invented, combining 'volume' and 'innovation' through the modularization and standardization of the powertrain. It could have the same importance as the combination of 'volume' and 'diversity' by General Motors during the interwar period thanks to the invention of platform sharing (Boyer and Freyssenet, 1999).

The mobilization of new players, the proliferation of initiatives and the launch of cleaner car models are finally creating a favourable movement that had no equivalent in the past.

The consequences of the rupture scenario, if it prevailed, would be considerable

The future of the automobile would be played out in China and India, whose manufacturers could become major players, though as yet they have to learn the skills that seem to be lacking in the domains of car safety and driving needs. The automotive supply chain would be completely changed, both in volume and the variety of its parts, making a greater need for more vocational education, and increased knowledge and skills. The control of the value chain could change hands. But the new 'core business' is still difficult to define: will it be the couple battery/electric motor, or the design and integration of new functionalities, or even a multi-modal mobility offer? It is clear to see, however, that the existing geography, economics and sociology of the automobile would be disrupted.

Two major problems, but not decisive in the ongoing transition, would not, however, be solved automatically: pollution and urban congestion.

Pollution from electric vehicles depends on the origin of the electricity used. The electric vehicle may be the best or the worst solution, in terms of CO_2 emissions, for example (Ungerer, 2009). But this is not what will hinder its adoption, if the real price of oil continues to grow, and if newcomers are pushing ahead of established manufacturers. However, it is likely that renewable energy, especially if we take into account the rapid progress of photovoltaic panels and wind turbines, can significantly replace fossil fuels in a reasonable time, which Germany is gambling on.

The congestion problem could increase. Indeed, electric vehicles could become more accessible, because their price could be lower than the price of oil-powered vehicles. They could also be easier to use because they could be more compact. Solutions, necessarily only partial, must be found. Among them, voluntary or compulsory car sharing and more comfortable, more frequent and more conveniently placed public transport. Some economists even imagine a ban on private car possession, making travel a public good available to all.

Conclusion

The indecision observed between the alternative forms of mobility could be the beginning of at least three scenarios. The consequences of these three scenarios on the geography, structure, economy and sociology of the world car industry are completely different. In the first scenario, each car producer will be able to find its own regional niche. In the second, only the most powerful carmakers will survive. In the third, the newcomers and the innovating enterprises will have the possibility of engaging in a rapid 'second automobile revolution'. The winning scenario will prevail after confused and unavoidably aggressive battles, not because of its technical superiority or it having the best environmental performance, but initially because of energy geopolitics and the profit strategies of firms. For these reasons, the third scenario, the scenario of electric vehicles, which appears today to be the most random, could prevail in the coming years, as the improbable petrol car scenario did a century ago (Freyssenet, 2011). This does not mean that carmakers who gamble on this scenario will necessarily be the winners. As is well known, the graveyard of car manufacturers is full of manufacturers have been technically innovative in their time.

It appears that the automotive industry is still very young, not only because highly populous continents began to buy cars in quantities never seen before, but in particular because this diffusion can only be completed by making radical changes to the means of propulsion, the architecture and the functionalities of motor vehicles.

References

Amatucci, M. and Spers, E. (2010) 'The Brazilian Bio Fuel Alternative', *International Journal of Automotive Technology and Management*, 10(1): 37–55.
Bardou, J.-P., Chanaron, J.-J., Fridenson, P. and Laux, J. (1982) *The Automobile Revolution. The Impact of an Industry*. Chapel Hill, NC: University of North Carolina Press.
Beaume, R. and Midler, C. (2009) 'From Technology Competition to Reinventing Individual Ecomobility: New Design Strategies for Electric Vehicles', *International Journal of Automotive Technology and Management*, 9(2): 174–90.
Boyer, R. and Freyssenet, M. (1999) 'Le monde qui a changé la machine. Essai d'interprétation d'un siècle d'histoire automobile. Quatorze textes préparatoires', Paris: GERPISA.
Boyer, R. and Freyssenet, M. (2002) *The Productive Models. The Conditions of Profitability*. London/New York: Palgrave, p. 126. First published in 2000 as *Les modèles productifs*. Paris: Repères La Découverte, p. 128.
Chanaron, J.-J. and Teske, J. (2007) 'The Hybrid Car: A Temporary Step', *International Journal of Automotive Technology and Management*, 7(4): 268–88.
Fréry, F. (2000) 'Un cas d'amnésie stratégique: l'éternelle émergence de la voiture électrique', *9th International Conference of Strategic Management*, Montpellier. Freyssenet, M. (2009a) 'The Second Automobile Revolution : Promises and Uncertainties', in M. Freyssenet (ed.), *The Second Automobile Revolution: Trajectories of the World Carmakers in the 21st Century*. Basingstoke, UK/ New York: Palgrave Macmillan, pp. 443–54.
Freyssenet, M. (2009b) 'Are We at the Outset of a Second Automobile Revolution? Inquiries Proposal', in B. Jullien (ed.), *Proceedings of 17th International Conference of GERPISA 'Sustainable Development in the Automobile Industry: Changing Landscapes and Actors'*. Paris. Available at: http://gerpisa.org: http://freyssenet.com/?q=node/1159.
Freyssenet, M. (2011) 'The Start of a Second Automobile Revolution. Corporate Strategies and Public Policies', *Economia e Politica Industriale*, 38(2): 69–84.

Jullien, B. (2008) 'A Framework to Enrich the Scientific, Political and Managerial Understanding of Sustainable Development Issues for the Automotive Industry: The GERPISA's 'Tradeoffs and Synergies' Approach', *International Journal of Automotive Technology and Management*, 8(4): 469–92.

Jullien, B. (ed.) (2010) *Proceedings of the 18th International Conference of GERPISA: 'The Greening of the Global Auto Industry in a Period of Crisis'*, Berlin. Available at: http://leblog.gerpisa.org.

Mballa, A. M. F. (1998) 'Historique d'une trajectoire technologique: le cas du système raffinage-automobile', Doctoral thesis in Economics, University of Grenoble.

Mom, G. (2004) *The Electric Vehicle: Technology and Expectations in the Automobile Age*. Baltimore, MD/London: Johns Hopkins University Press.

Stocchetti, A. and Volpato, G. (2010) 'Inquest for a Sustainable Motorisation: The CNG Opportunity', *International Journal of Automotive Technology and Management*, 10(1): 13–36.

Ungerer, P. (2009) 'Du moteur thermique au véhicule électrique: quels enjeux pour la recherche?', Slide presentation for the French Institute of Petroleum.

Wang, H. and Kimble, C. (2011) 'Business Model Innovation in the Chinese Electric Vehicle Industry', 19th International GERPISA Conference: 'Is the Second Automobile Revolution on the Way?' Paris, 8–10 June.

Appendix: The GERPISA International Network

GERPISA (the Permanent Group for the Study of the Automobile Industry and its Employees) started out as a network of French economics, management, history and sociology researchers interested in the automobile industry. Founded by Michel Freyssenet (CNRS sociologist) and Patrick Fridenson (EHESS historian), it was transformed into an international network in 1992 to carry out a research programme on 'The emergence of New Industrial Models'.

With Robert Boyer (CEPREMAP, CNRS, EHESS economist) and Michel Freyssenet supervising its scientific orientation and under the management of an international committee, the programme (which lasted from 1993 to 1996) made it possible, thanks to its study of the automobile firms' (and their transplants') trajectories, productive organization and employment relationships, to demonstrate that *lean production*, which, according to Womack *et al.*, the authors of *The Machine that Changed the World*, was supposed to become the industrial model of the twenty-first century, was in fact an inaccurate amalgamation of two completely different productive models – the 'Toyotian' and the 'Hondian'. Moreover, it showed that there are, have always been, and probably always will be several productive models that are capable of performing well at any given time. Shareholders, executives and employees are not only not obliged to adopt a *one best way*, they also have to devise a 'company governance compromise' covering the means that will allow them to implement one of the several profit strategies that are relevant to the economic and social environment in which they find themselves.

A second programme (running from 1997 to 1999) entitled 'The Automobile Industry, between Globalization and Regionalization', and co-ordinated by Michel Freyssenet and Yannick Lung (Montesquieu University – Bordeaux IV economist), tested the analytical framework that had been developed during the first programme in an attempt to better understand the new wave of car manufacturer and component-maker internationalization that had been observed over the previous decade. The outcome was that the viability of the choices being made depends primarily on the chosen profit strategies' compatibility with the growth modes in the areas in which the investments are being made.

The third programme (2000–2) has been developed under Yannick Lung's co-ordination. It focuses on the issues at stake in the 'Co-ordination of Knowledge and Competencies in Regional Automotive Systems'. Supplementing existing studies of forms of regionalization in the automobile industry, the programme analysed the sector's new contours as well as the development of new relational and co-operative modes among its actors.

The fourth programme (2003–6), entitled 'Variety of Capitalism and Diversity of Productive Models' and co-ordinated by Bruno Amable (CNRS economist) and Yannick Lung, discussed the thesis of national productive models. It showed the diversity of productive models in each type of capitalism by exploring the interaction between micro, meso and macro levels. It emphasized the importance of political compromises at the meso and macro levels to manage competition and employment conditions.

The fifth programme (2007–10), developed under the direction of Bernard Jullien (Bordeaux IV economist), focused on 'Sustainable Development and the Automobile Industry'. It analysed the emergence of this theme and compared the discourses and practices of automobile entreprises. The impact of the 2008–9 crisis has allowed the broad issue of sustainability to be placed at the crossroads of ecological, economical and social sustainability, and to identify the synergies and trade-offs between these different dimensions.

In 2011, GERPISA included 350 members from 27 different countries. Since 2010, it has become a Groupement d'Intérêt Scientifique based at the École Normale Supérieure de Cachan. As such, it is supported by the French Ministries of Industry and of the Environment, by the professional associations of French car companies (CCFA), of French suppliers (FIEV) and of French dealers and repair shops (CNPA); also by the École des Hautes Études en Sciences Sociales (EHESS) and Montesquieu University – Bordeaux IV.

The international steering committee comprises the following members: Robert Boyer (CNRS-EHESS, Paris), Jorge Carrillo (Colegio de la Frontera Norte, Mexico), Elsie Charron (CNRS, Paris), Dan Coffery (University of Leeds, UK), Michel Freyssenet (CNRS, Paris), Patrick Fridenson (EHESS, Paris), Takahiro Fujimoto (University of Tokyo, Japan), Bernard Jullien (Université de Bordeaux IV), Bruno Jetin (Université Paris XIII), Ulrich Jurgens (WZB, Berlin, Germany), Yveline Lecler (MRASH/IAO, Lyon), Yannick Lung (Université de Bordeaux IV), Tommaso Pardi (EHESS, Paris), Sigfrido Ramirez (Institut européen, Florence, Italy), Mario Sergio Salerno (University of São Paolo, Brazil), Koichi Shimizu (University of Okayama, Japan), Koichi Shimokawa (Hosei University, Tokyo, Japan), Carole Thornley (Keele University, UK).

References

Womack, J. P., Jones, D. T. and Roos, D. (1990) *The Machine That Changed the World: How Lean Production Revolutionized the Global Car Wars*. New York: Rawson Associates.

GERPISA's publications

GERPISA edits, in English and French, a quarterly review entitled *Actes du GERPISA* and a monthly newsletter, *La Lettre du GERPISA*. The review combines the writings the network's members have presented on a specific topic in various work meetings. The newsletter comments on news from the automotive world and provides up-to-date information on what is happening in the network. Findings from the first and second programmes have been published in a series of books:

Programme 'Emergence of New Industrial Models'

Freyssenet, M., Mair, A., Shimizu, K. and Volpato, G. (eds) (1998) *One Best Way? Trajectories and Industrial Models of the World's Automobile Producers*. Oxford/New York: Oxford University Press. French translation: *Quel modèle productif? Trajectoires et modèles industriels des constructeurs automobiles mondiaux*. Paris: La Découverte, 2000.

Boyer, R., Charron, E., Jürgens, U. and Tolliday, S. (eds) (1998) *Between Imitation and Innovation: The Transfer and Hybridization of Productive Models in the International Automobile Industry*. Oxford/New York: Oxford University Press.

Durand, J.-P., Stewart, P. and Castillo, J.-J. (eds) (1999) *Teamwork in the Automobile Industry: Radical Change or Passing Fashion?* London: Macmillan. First edition in French: *L'avenir du travail à la chaîne*. Paris: La Découverte, 1998.

Lung, Y., Chanaron, J.-J., Fujimoto T. and Raff, D. (eds) (1999) *Coping with Variety: Product Variety and Production Organization in the World Automobile Industry*. Aldershot: Ashgate.

Shimizu, K. (1999) *Le Toyotisme*. Paris: La Découverte.

Boyer, R. and Freyssenet, M. (2002) *The Productive Models: The Conditions of Profitability*. London/New York: Palgrave/Macmillan. First edition in French: *Les modèles productifs*. Paris: La Découverte, 2000. Other editions: *Los modelos productivos*. Buenos Aires/Mexico: Lumen Humanitas, 2001. *Produktionmodelle, Ein e Typologie am Beispiel der Automobilindustrie*, Berlin: Edition Sigma, 2003. *Los modelos productivos*. Editorial Fundamentos, Madrid, 2003. *Oltre Toyota. I nuovi modelli produttivi*. EGEA, Università Bocconi Editore, Milan, 2005.

Boyer, R. and Freyssenet, M. (1999/2006) *Le monde qui a changé la machine, un siècle d'histoire automobile*, 1999, 2006. Available at: http://freyssenet.com.

Programme 'The Automobile Industry, between Globalization and Regionalization'

Humphrey, J., Lecler, Y. and Salerno, M. (eds) (2000) *Global Strategies and Local Realities: The Auto Industry in Emerging Markets*. London/New York: Macmillan/St Martin's Press.

Freyssenet, M., Shimizu, K. and Volpato, G. (eds) (2003) *Globalization or Regionalization of the American and Asian Car Industry?* Basingstoke: Palgrave Macmillan.

Freyssenet, M., Shmizu, K. and Volpato, G. (eds) (2003) *Globalization or Regionalization of the European Car Industry?* Basingstoke: Palgrave Macmillan.

Charron, E. and Stewart, P. (eds) (2004) *Work and Employment Relations in the Automobile Industry*, Basingstoke: Palgrave Macmillan.

Carillo, J., Lung, Y. and van Tulder, R. (eds) (2004) *Cars, Carriers of Regionalism?* Basingstoke: Palgrave Macmillan.

Programme 'Co-ordination of Knowledge and Competencies in Regional Automotive Systems'

Lung, Y. and Volpato. G. (eds) (2002) 'Reconfiguring the Auto Industry', *International Journal of Automotive Technology and Management*, 2(1).

Froud, F., Johal, S. and Williams. K. (eds) (2002) 'The Tyranny of Finance? New Agendas for Auto Research', *Competition & Change*, 6(½) (double issue).

Lung, Y. (ed.) (2002) 'The Changing Geography of the Automobile Industry, Symposium', *International Journal of Urban and Regional Research*, 26(4).

Calabrese, G. and Lung, Y. (eds) (2003) Designing Organisations to Manage Knowledge Creation and Co-ordination, *International Journal of Automotive Technology and Management*, 1/2 (special issue).

Programme 'Variety of Capitalism and Diversity of Productive Models'

Hirt, O. (ed.) (2005) 'Variety of Capitalism and Diversity of Productive Models', *Actes du GERPISA*, 38.

Pardi, T. (ed.) (2006) 'State and Politics in the Automobile Industry', *Actes du GERPISA*, 40.

Lung, Y., (2008) 'Modèles de firmes et formes de capitalismes', *Revue de la régulation*, 2.

Freyssenet, M., (2008) 'Stratégies et modèles nationaux de croissance. Proposition d'une démarche et esquisse d'iun schéma d'analyse', *Revue de la régulation*, 3.

Jullien, B. and Smith, A., (2008) *Industries and Globalization: The Political Causality of Difference*, Basingstoke: Palgrave/Macmillan, 2008.

Programme 'Sustainable Development and the Automobile Industry'

Jullien, B. (2007) 'A Framework of Sustainable Development Issues for the Automobile Industry', *International Journal of Automotive Industry and Management*, 1(1): 1–19.

Freyssenet, M. (ed.) (2009) *The Second Automobile Revolution: Trajectories of the World Carmakers in the Twenty-first Century*. Basingstoke: Palgrave Macmillan, 2009.

Jullien, B. and Lung, Y. (2011) *Industrie automobile: La croisée des chemins*. Paris: La Documentation Française.

Information on GERPISA's activities can be obtained by contacting:
GIS GERPISA, École normale supérieure de Cachan, Bât Laplace, 61, avenue du Président Wilson, 94235 Cachan cedex.
Telephone: +33 1 47 40 68 53
E-mail: gerpisa@ens-cachan.fr
Website: http://gerpisa.org/

Index

Notes:
Bold = extended discussion or term highlighted in text;
f = figure, n = endnote/footnote, t = table.

AC/DC converter   219, 226, **227**
  see also   electric motors
acceleration   **292**, **294**, 295f, 296n, **298**
ACEA (European Automobile
  Manufacturers' Association)   110n,
  289n
Achtnicht, M.   287, 302
Active Wheel (Michelin)   58
ADEME (Environment and Energy
  Management Agency, France)   98
Adzel joint venture   234
AEA Technology   29(n2)
aero-engines   122
  see also   alternative engines
aesthetics   3, 50–1, 53–5
affordability   25, 142, 243, 248, 254,
  270, 273t, 277, 278f, 294, 299, 312
  see also   price
Africa   149–50t, 150, 169, 181, 243
Agassi, S.   58–9
Aggeri, F.   91–2, 101
agreements and joint ventures in EV
  industry   7–8, 23, **225–39**
  case studies   226, **233–7**, 238
  disruptive innovations   **226–9**
  economic theory   226, **229–33**
  growth strategies   **229–31**
  implications for policy-makers   8,
    226, **238**
  and innovation   **232–3**
  literature   234
  some concluding remarks   **237–8**
agribusiness   6, 165, 166, 172, 174, 181t
agrofuels   28, 305–9, 310t
Ahlvik, P.   161(n2), 161
air pollution   39, 43, 60, 140, 274t,
  276f, 279, 307
air quality   162, 271, 279, 294
air-conditioning   216, 244

Ajzen, I.   126, 138
Alaska   246
alcohol   169, 171, 181t
  see also   ethanol
Algeria   149t, 155t
Allegrini, I.,   145–6n, 146, 148n, 162
Alston, L. J.   166, 182
alternative engines   29(n1), 135t, 286
  versus alternative fuels   301(n1)
  versus conventional engines (mental
    perspectives)   287, **292–4**, 295f,
    301(n7), 302
  see also   car engines
alternative fuels   29(n1), 162, **266–7**,
  280, 284, 291t, 304
  forecasting framework for
    evaluating   9, **263–85**
  research   278, 278f, 279
  scenarios   2, 9, 263, 267, **272**,
    276–84
Alternative Motor Fuels Act (California,
  1988)   176
alternative vehicles (AVs)   2, 13, 16, 25,
  29(n1), 240, 284
  automotive industry race   14–15,
    19–22, 30(n9–11)
  consumer attitudes   9, **286–303**
  context   7
  definition   301(n1)
  development   7
  development (surrounding
    conditions)   **7–10, 205–322**
  motivation for purchase   26
  production plans   25
  technological trajectories   **4–7**, 23,
    87–204
  see also   battery electric vehicles
aluminium   54, 59, 192, 197
Amable, B.   324

328 Index

Amatucci, M. **xxi, 6, 164–84**
analytic hierarchy process (AHP)
  advantages and disadvantages **267**
  application to problem 263, **271–2,
    273–5t**
  construction of hierarchies **271–2**
  criteria to find 'best motor vehicle
    fuel' **271–2, 273–5t,** 276f
  'forecasting tool' 266–7
  levels 9, 263
  pair-wise comparisons 271, 281–2f
  result-synthesis 271
  sensitivity analysis 271, 272, 277
analytic hierarchy process: 'best motor
    vehicle fuel' quest 9, 263
  'culture' category 271–2, **273–4t,**
    276f, 276–7, 281f, 282
  'development time' category 272,
    275t, 276f, 281f, 283
  'economics' category 271–2, **273**t,
    276f, 276–7, 281f, 283
  'environment' category 271–2,
    274–5t, 276f, 276–7, 281f, 282, 283
  'long-term sustainability'
    category 271–2, **275t,** 276f, 281f,
    282, 283
  scenarios **2, 272, 276–84**
Anderson, J. 243, 252
Annual Energy Outlook 2007 (EIA) 219
architectural innovation **117,** 196f
'architectures' 36, 200
  comparative advantage 36
  'compatibility' 36
  constraints and **35–6**
  evolution 38–9
  integral type 34, 35, 36, 42
  modular (mix and match) type 34,
    35, 39
  products and processes 38, 38f
Argentina 152t, 155t
Argonne National Laboratory 315,
  318
Armenia 155t
artefacts (products) 35, 36
  definition 40
  design information 33
  design, production,
    consumption **40–2**
  design, production, usage
    conditions 46

'growing complexity' 46
integral-type 'architecture' 39
'no specific architecture' 38
structure, function, operation 41f
'such as prototypes' 91
see also new products
ASEAN **37**
Asensio, A. 57–8, 67
Asia ii, 2, 37, 40, 48, 149–50t
Asia-Pacific 111f, 122, 150
*Asociación española de fabricantes
  de automóviles y camiones*
  (ANFAC) 140, 162
asset development **96**
*Associação Nacional dos Fabricantes de
  Veículos Automores* (ANFAVEA) xiii,
  166, 168f, 175, 182(n1, n3)
Audi 116, 134, 135–6t, 138(n6), 300,
  301t, 301(n8)
Audi A6 299
Austin (Texas) 29
Austin, H. 56
Austin Seven (1922) 56
Australia 150, 155t, 235, 313, 316t
Austria 27t, 151t, 153f, 156–7t, 159
Autolib Paris 29, 50, 58, **62–5**
  fees 65
Automobile Club d'Italia (ACI) 146–7n,
  147, 148n, 161
automobile design
  Loewy **55**
automobile industry
  C21 challenge 47
  classical 256
  crisis **46–7**
  future **46–7**
  mainstream 246
  non-mainstream 8
  traditional versus new players 23
  see also automotive industry
Automobile Industry Executive Group
  (GEIA), Brazil 174
automobiles 41–2
  architecture 46, 312
  bodywork 54
  'commodification' (developed
    nations) **43–4**
  design evolution 39
  development cycles 104
  devices (various) 18

versus digital commodities (definitive difference)   43
'dominant mode of transport'   54
eco-friendly   14
integral versus modular type   45
low-cost   45, 46
'misnomer'   305–6
'oversized means of transport'   50
period of transition   66
small, popular   56
see also   cable cars
automotive chain   168f
innovation   **174–9**
path dependence   **174**
automotive consumption practice   **287–9**, 301(n2–4)
automotive industry   25, 101, 277, 278
'best example of industrial rationality'   257
competition between 'new' and 'old' technologies in race for sustainable solutions   5, **103–23**
dominant design challenged by environmental issues   **70–1**
financial difficulties   65
hybrid manufacturing model wanting   257
innovative design   **3–4**, **69–85**
innovative design and sustainable development   **13–31**
institutional and technological innovation   **175–8**
new business model required   257, 260
over-production and market segmentation   **258–9**, 261–2(n1)
R&D processes for sustainable development   **3–4**, **69–85**
race for AVs   14–15, **19–22**, 30(n9–11)
restructuring (in Europe)   260
'sailing-ship' effects   5, **103–23**
'still very young'   321
strategic crisis   256, 257, **258–9**
territorial organization   24
urban mobility   8–9, **254–62**
value-added versus oil import bill (Europe)   142
see also   carmakers

automotive industry: greening   **1–10**, 19
learning from experiment   70, **78–81**, 81–2
new R&D processes   70, **78–81**
urban mobility   8–9, 14, **254–62**
see also   second automotive revolution
Automotive News   106, 112, 119, **122**
automotive sector   303
capital accumulation model (social pressure to transform)   3, 49
commercial innovations   167, **178–9**
profit model   49
automotive telematics services   102
Avis-RATP-SNCF-Vinci Park consortium   63

Baden-Württemberg   313
bagasse (biofuel)   100
Baker, R. A.   144n, 162
Bangalore   37
Bangkok   242
Bangladesh   152t, 155t
Banque National de Paris   237
Bascap (Bolloré subsidiary)   65
batteries   16, 17, 21, 24, 30, 51, 53, 93, 95–6, 105, 107–8, 112, 120, 161, 193, 216, 223, 229, 236, 261, 305
bulk, weight, power, safety   22
capacity   119
characteristics   225
charging time   22, 210, 227
costs   26, 45
energy and power density   234
essential materials   28
improved performance   19
limited storage capacity   268
new   230, 237
new (partnerships between firms)   234
new chemicals   233
rechargeable   23, 219
safety criterion   17–18
second-generation   311, 319
secondary   **227**
third-generation   312, 318, 319
see also   energy storage
battery autonomy   318, 319

battery charging/recharging 7, 59, 61, 94, 100, 158, 217, 225–6, **227**, 228, 234–5, 244, 245, 268, 304
   cost-efficient 213
   daily patterns 207
   en-route versus at-home **235**, 306
   new forms of organization 233
   overnight 211, 248, 250–1
   at parking areas **235**
   'smart' versus 'non-smart' **211–12**
   technical constraints 211, 212
   techniques 210
   see also charging
battery component producers 23
battery control systems 112, 118
battery electric vehicles (BEVs) 209, 210, 218, 220, 292, 294, 298, 304
   global projection (2020) 215
   refuelling 210
   see also CNG vehicles
battery energy 209
battery exchange 211, **235**, 311
battery maintenance 18
battery manufacturers 22, 65, 120, 315, 318
   competition with carmakers 23–4
battery materials 44–5
   'high cost' 45
battery packs 209, 237
battery performance 310
battery prices 25
battery recycling 237, 248
battery rental 97
battery sharing 77
battery technology 99, 210, 279, 280
BB1 (concept car) 58
Beaume, R. xxiv, 93, 102
behaviour **127**
behavioural model 126, 138
Belarus 152t, 155t
Belgium 27t, 151t, 153f, 155t, 157t, 307t
Berggren, C. **xxi, 5, 6–7, 103–23, 185–204**
Berlin 29, 126
Bernard, J. 2
Bernardes, R. C. 175, 182
Bertone (Italy) 58
Bertone, G. 56

Better Place 29, 58–9, 66, 94, 226, 235–6, 238, 318
bicycles 53–4, 209, 244–5, 250, 257, 260–1, 302, 315, 319
Bijker, W. B. 298, 302
biodiesel 17, 164, 265, 271, 276f, 282f, 282, 283
   B20 biodiesel 268–9
   pollution and safety issues **268–9**
biofuels 4, 14, 16–17, 21, 28, 29(n1), 159, 163–4, 169–71, 301(n1)
   EU directive (2003) 25
biogas 6, 143–4, 148, 154, 159–60, 161(n7), 163
biomass 17, 21, 66, 160, 170
biomethane 17, 159, 160, 163
biotechnology 169, 279
bipolar lead-acid battery 227
Bluecar (Autolib project, 2005–) 50, **62–5**; 64f
Bluecar (Pininfarina-Bolloré, 2009–) 64
BMW 22, 30(n8), 113n, 117, 237, 308, 309, 310t, 315
   average CO_2 emissions of cars sold in Europe (1997–2010) 141t
   diesel engine patents (1990–2007) 115f
'Boeing 787' 208
Bolivia 155t
Bolloré Group 24, 66, 66(n3), 315, 316t
Bolloré-Pininfarina joint venture 63–4
Bompard, E. **xxi, 207–24**
Bordeaux IV 324
Borloo, J.-L. 100
Bosch/Robert Bosch 104, 118, 175–8, 184
   partnership with Daimler 230, 234, 237
   partnership with PSA Group 233, 237
   product development executives 182(n1)
Bosnia and Herzegovina 151t
Boston Consulting Group 26, 30, 261–2(n1)
Bower, J. L. 22, 30, **241**, 252
Boyer, R. ii, xxiii, 323, 325
BP 150, 162
BP Statistical Review of World Energy (2010) 262(n2)

Index   331

braking energy   189, 199
see also   regenerative braking
Brandberg, A.   161(n2), 161
brands   40, 60, 75, 191, 291t, 299, 300
  brand design   37
  brand identity   57, 58
  brand image   22, 58
  brand loyalty   288, 301(n3), 302
Brazil   xxi, xxvi, 28, 155t, 307t, 307–8, 313, 315, 316t
  car sales (1980s)   180
  economic growth (1972–9)   173t
  external debt (1972–9)   172, 173t
  flex-fuel evolution (2003–12)   178f
  GDP (1972–9)   173t
  oil production   172
  origin of name   169
  Portuguese arrival   166–7, 168
  sugar exports   169
  sustainable development strategies   6
  vehicles powered exclusively by ethanol   172
Brazil: ethanol and automotive industries   6, 17, 21, **164–84**
  automotive industry: institutional and technological innovation   **175–8**
  automotive sector: commercial innovations   167, **178–9**
  economic growth, external debt, oil account (1972–9)   173t
  ethanol production: commercialization innovation   166, **171–4**, 175
  ethanol production: institutional and technological innovation   **169–71**, 182(n7)
  ethanol: phases in use as fuel   **179–81**
  flex-fuel technology (number of cars/trucks, 2003–12)   178f
  innovation in automotive chain   **174–9**
  innovation within ethanol chain   164, **167–74**
  institutional changes and governance structure   167f
  lessons   6, 165, **179–81**
  model   167, 167f
  path dependence in automotive chain   **174**
  path-dependence and innovation in sugar cane sector   164, **168–9**, 182(n6)
  system innovation and institutional theoretical approaches   164, **165–7**, 168f, 182(n2–5)
  technological innovation (market adoption)   167f
Brazil: Law 17,917 (1931) and Law 25,174 (1948)   171
Brazil: Ministry of Environment (MMA)   178n, 183
Brazil: protocol of 19 September 1979   172, 175
Brazil: regions
  centre   171
  north-east   165, 169
  south   170, 171
  south-east   165, 170
Brazilian Agricultural Research Organization (EMBRAPA)   xiv, 166, 168f, 169, 182(n7)
  website   182(n5)
Brazilian VW   177, 179, 182(n1)
Bremmer, I.   265–6
BRIC   xiii, 234, 243, 258
  carmakers   319t
  registered cars (estimate for 2014)   262(n1)
Brittany   313
Brown, R. C.   144, 162
Brown, T.   53–4, 67
Brundtland Commission   51, 67
Buick Y-Job model   55
Bulgaria   155t, 157t
buses   4, 27, 76, 77, 157t, 159, 189, 255, 261
  CNGVs   154
  electric   209
  hybrid   185, 187f, 187, 198–200, 202t
  parallel hybrid versus series hybrid   193, 201–2t
  see also   heavy hybrid vehicles
business   18, 21, 49
  innovative strategy   29
  see also   'environment/business'
business model innovation   2, 322
  characterisation   **241–2**, 252–3
  emerging markets   **242–3**, 252

business model innovation: Shifeng group (case study) 8, 244, **246–50**
  background **247**
  building business model for LSEVs **247–8**
  building market for LSEVs **248–50**
  methodology **247**
business model innovation and development of EV industry in China 7, 28, **240–53**
  business models **240–4**
  defining low-speed EV 8, **245–6**
  e-mobility in China **244–6**
  emerging markets 8, **242–3**
  implications for Western carmakers 8, **250–2**
business models 4, 29, 94, **240–4**, 252, 253
  automotive industry 257, 260
  carmakers **13–14**
  'defined in various ways' 241
  differentiation 241
  disruptive 244
  originating in emerging markets **243–4**
business opportunities **134**
BYD Automobile Company 24, 231, 237, 310t, 315, 316t

cable cars 257, 261
  see also cars
Calabrese, G. ii, xix, **xxi–xxii**, **1–10**, **13–31**, 49, 56, 67, 326
Calanz 243
California 25, 116, 126, 177, 235, 242, 313
California: Air Resources Board 124, 176
California: Fuel Cell Partnership 125
California Institute of Arts 66(n1)
Calkins, M. 251, 252
Camry programme (Toyota, 2011–) 119
Canada 149–50t, 156t, 236, 307t, 308, 312–13, 313t, 316t, 318
car body-makers 315, **316–17t**
car consumers
  stated preferences 287, **289–91**, 301(n5-6)
  sub-frames 294

car emissions/vehicle emissions 9, 161(n5), 290, **291t**
  decarbonization 259
  electrification 'by no means permanent solution' 22
car engines **292**
  versus car preferences 301(n5)
  consumer perspectives (existing literature) 286–7
  firm perspectives **125–6**
  see also diesel engines
car mass and fuel consumption (trends 1980–2005) 289f
car ownership 251, 260
car parks 61, 238
car prices 13, 29(n2), 319
  'vehicle price' 19
car purchases/sales
  current trends 9
  information sources 288, 296
  'two-stage process' 288
car rental 58, 260, 312
car-focused mobility crisis **254–5**
car-sharing 54, 61, 62, 66, 77, 94, 313
  see also Autolib
Car2go 29
carbon dioxide (CO_2) 31, 124, 144, 146, 207, 269
  greenhouse effect (versus methane) 143
carbon dioxide emissions 1, 13, 14, 15, 17, 22, 29(n2), 30(n8), 71, 77, 90, 95, 98, 101, 101(n1), 147, 158, 259, 269, 279, 287, 291, 302
  data 141, 161(n1)
  EU target (2020) 25
  proportion derived from road transport 140
  sequestration 19
carbon dioxide emissions-reduction 6
  by substitution and by 'methanization' 146, 146t
  investment 19
carbon monoxide (CO) 141, 268, 269–70
carbon taxes 312
carmakers ii, 6, 14, 168f, 177, 232, 233, 235, 237–8, 259, 279, 280, 304, 306–8, 309, 311, 312, **316–17t**, 318, 319t, 323

agreements and joint ventures with
    suppliers 7
attitudes 4
changes in preferences 310t
cleaner automobile strategies
    (2011) 310t
competition with battery
    manufacturers 23–4
design and manufacturing
    platforms 58
diesel engine patents (1990–2007) 115f
hydrogen technology: behavioural
    intention 133
hydrogen technology: perspectives 5,
    124–39
hydrogen technology: perspectives
    (three groups) **125**
international competition 57
Japanese competition intensified
    (1980s) 56
measures of innovativeness **104–6**
most powerful 320
newcomers 310–11, 320–1
option for rupture scenario
    (listed) **315**
patents related to hybrids (1992–2007)
    113f
'permanent reduction of costs' profit
    strategy 308, 309
perspectives on engines **125–6**
profit strategies 310, 311, 319, 321,
    323
'quality' strategy 309
'regional' versus 'global' 308
strategies (role of industrial
    designers) 3
strong relationships with suppliers
    (EV components) 8
traditional 230, 231
trajectories 13
'vehicle makers': new forms **6–7,
    185–204**
'volume and diversity' profit
    strategy 308, 309
Western (implications of Chinese
    LSEV production) **250–2**
see also automobile industry
cars 13, 52, 164, 255, 257, 264, 309
age 142, 145–6, 160
age (Europe/USA averages) 140

'by-products of mobility'
    (Garibaldo) 8, 254, 256
cleaner models 319t, 319–20
CNGVs 154
common parts 39
dual-fuelling (CNG and petrol) 143
electronic control 39
flex-fuel evolution (Brazil, 2003–12)
    178f
integral-type architecture 32
luxury-level 288–9
new 146
'new architectures' 305
new ones sold versus old ones
    scrapped (no correspondence) 142
older (equipped with CNG
    device) 145–6
perceptions 66, 255
performance criteria 77
petrol-CNG hybrid 161(n4)
premium sector 259
'private and collective' **77**
scrapping 142, 154, 160; *see
    also* End-of-Life Vehicle Directive
second-hand 301(n3)
size 288–9
small 32, 104, 319t
'that facilitate use of public
    transport' 78
trend towards comfort 288–9, 303
see also city car
Carvalho, R. Q. xxvi
case studies 37, 49, 50, **60–5**
    agreements and joint ventures in EV
        industry 226, **233–7**, 238
    AutoX 4, **69–85**
    electric vehicles **93–6**
    innovative design **3–4, 69–85**
    Shifeng Group 8, **246–50**
CCFA (professional association of
    French car companies) 324
CENARGEN (Genetic Resources and
    Biotechnology Centre for Plants,
    Brazil) xiii, 169, 182(n7)
Central America 149–50t, 169
Central Europe 26
Centro de Tecnologia Canavieira
    (CTC) xiii, 166, 168f, 169, 182(n4)
*Centro Esperienze Costruzione Modelli
    Prototipi* (Cecomp) 64

334 *Index*

CEPREMAP   xiii, 323
Ceschin, F.   29, 30
CHAdeMO consortium (Japan)   236
Chanaron, J.-J.   ii, 20, 30, 107, 121, 325
charging/recharging   215, 311
  facilities   314
  fast   211
  infrastructure   27, 90, 94, 98, 236–8, 315, 318
  methods   90
  profiles   218
  scenarios (EVs)   **216–19**
  timing 'not unimportant'   **216–17**
  smart devices   219
  technical features   211
  see also   refuelling
charging/recharging stations   29, 58, 64–5, 95–7, 100–1, 218, 251
  recharging distance   311
charging/recharging systems   17, 18, 99, 213, 259
  smart   221
  standards   25, 236
Charue-Duboc, F.   **xxii, 4–5, 89–102**
Chee, W.   266, 284
Chesbrough, H. W.   240, 241, 252
Chevrolet Volt (GM, USA, 2011–)   109, 119
Chile   152t, 155t
China   xxiii, xxvii, 24, 38, 40, 44–5, 168, 171, 218, 228, **231**, 235, 237, 259, 278, 305, 307t, 307–8, 310, 312, 313, 315, 316t, 318, 322
  adoption of all-electric vehicle   318–19
  business model innovation   **7, 240–53**
  car production and sales   28
  carmakers   310t
  cars   264
  CNG pump prices (versus petrol and diesel, 2011)   152t
  CNG reserves (2008)   149t
  CNGVs and CNG filling stations   155t
  coastal areas   37
  definition of LSEV   **244**
  development of EV industry   **7, 240–53**
  economic opening up (1980s–)   248
  EV development   7
  EVs: target number   313t
  growth in demand for oil   264
  household income (rural versus urban)   248, 252, 253
  industrial architectural geopolitics   **37**
  leapfrogging strategy   28
  locally-designed vehicles   32
  production of two-wheeled e-vehicles   244
  provincial or local governments   245, 249, 250
  see also   BRIC
China: General Administration of Quality Supervision, Inspection, and Quarantine (AQSIQ)   249
China: Ministry of Industry and Telecommunications   245, 249
China: Ministry of Technology   249
China: National People's Congress   249
China: State Council   249
China: State Council: Development Research Centre   249
China Electric Engineering Technology Association   247
China Statistical Yearbook (2010)   248, 252
choice of fuel   282, 283
  see also   analytic hierarchy process
Christensen, C. M.   22, 30, 120–1, **241–2**, 246, 243, **252**
Chrysler   56, 246, 310t
Cicero   164
cities   13, 29, 45, 52–3, 76, 95, 112, 119, 143, 160, 189, 193, 199, 247–8, 260, 312
  central areas   27, 260
  coastal   280, 283
  listed   **313–14**
  'mega cities'   258
  'metropolises'   256
  small cars   104
  'vertical-shaped'   257
  see also   urban areas
Citroën 2 CV (1948)   56
Clapaud, A.   67
Clark, K. B.   117, 121
Clean Air Act Amendments (USA, 1990–)   176
'clean diesel' engine   104, **108–17**, 120
Clermont Ferrand   62

Clio 183
Cloodt, M. 105, 121
Club of Rome 51, 67
clutch 190f, 192
CNG (compressed natural gas) 4, 16, 301(n1)
  average cost-saving (versus petrol and diesel) **151–2t**, 153
  average emissions in urban situation (Euro standards) 145t, 146
  average pollutant emissions 145t
  conversion incentives 160
  ecological benefits 143, 144
  economic and environmental benefits 140
  features (as fuel) 142
  octane index and heat of combustion comparison 144t
  'possible intermediate stage' 141, 142
  price (versus petrol and diesel) 150–3
  see also gas
CNG cars/engines 158, 207
  environmental performance 144, 161(n4)
  number on road (Italy) 147t
CNG cars: sustainable mobility **5–6**, 16, 21, 28, **140–63**
  assumptions 142
  crucial stage in race towards ZEVs **141–3**, 161(n1)
  data deficiencies 146
  economic perspective: CNG availability and cost **148–54**, 161(n6)
  future technological opportunities for development of CNGVs **158–60**, 161(n7)
  literature 162
  payback time 153, 153f
  policy implications **160–1**
  reducing emissions through conversion to CNG (Italian example) **143–8**, 161(n2–5)
  see also compact cars
CNG cylinders 161(n4)
CNG dual-fuel-powered products 143
CNG filling (home network) 28–9
CNG fuel injection systems 161(n4)
CNG refuelling stations 154, 159, 160, 161

CNG technology
  'can be applied to existing cars' 140–1
CNG vehicles (CNGVs) 21, 22, 28, 207
  bi-fuel 6
  diffusion **154–7**
  future technological opportunities **158–60**, 161(n7)
  'not ultimate solution' to pollution 158
  price and performance 16
  see also commercial vehicles
CNPA 324
CNR-Ceris (National Research Council: Institute for Research on Firms and Growth), Moncalieri iii, xxi, xxv, xxvii, 2
CNRS xiii, xxii–iii, 323, 324
Cobasys 234
Cold War 33
Coles, C. B. 241, 242, 251, 252–3
Colombia 155t
Colorado 313
commercial vehicles 44, 319t
  CNGVs 154
  see also consumer attitudes towards alternative vehicles
compact cars 43, 46, 51
  integral-type versus modular-type 35, 38
  see also concept cars
companies/firms xix, 1, 5, 23, 25, 28, 33, 36, 40, 46, 52, 82, 165, 175, 180, 238, 250
  aggregate number of patents 106
  capability-building ability 38, 38f
  Chinese 35
  engagement with hydrogen technology (three groups) 132–3
  EV industrial sector 7–8
  'governance compromise' 323
  high-technology 233, 237–8
  industrial 89, 314
  internal capabilities 230
  innovation behaviour (structural model) 127
  inventive performance **105**
  Japanese 35, 42, 43, 47
  Korean 47
  large fleets of service vehicles 314
  most dynamic 232

companies/firms – *continued*
  partnership agreements (case studies) **233–4**
  partnerships with public institutions (case studies) 233, **234–5**
  perspectives: car engines **125–6**
  perspectives: hydrogen technology **5, 124–39**
  small 49, 233
  specialized 238
company fleets 66
  listed **314**
components 26, 44, 54–6, 59, 104, 108, 110, 114, 117–20, 175, 196–7, 202t, 233, 236, 242, 312, 323
  costs 105
  design and manufacture (new players) 23
  most important (for EVs) **226–8**
  new 8, 19
compressed biogas (CBG) 17
compressed natural gas *see* CNG
concept cars 53, 56–8, 64, 66, 104
  electric 57
  two-seater 58
  see also dual-fuel cars
concepts-knowledge (C-K) theoretical framework 4, 70, 74, 80, 91
congestion 76, 217–18, 254, 258, 305, 313
congestion charges 312
Connecticut 246
consumer attitudes 7, 128
consumer attitudes towards alternative vehicles **9, 14, 286–303**
  analysis of engine choice **299**, 300f
  car mass and fuel consumption (trends 1980–2005) 289f
  interviews 301t
  mental perspectives of conventional and alternative engines 287, **292–4**, 295f, 301(n7)
  stakeholder consultation **300**
  stated preferences of car consumers 287, **289–91**, 301(n5–6)
  summary and conclusion **296, 298–9**
  trends in automotive consumption practice **287–9**, 301(n2–4)

  see also diesel vehicles
consumer 'framing' 2, 9, 286–7, 298
consumer movements 52
consumer pressure 308
consumer segments 287, **294–6, 297f**, 301(n8–9)
  see also market segmentation
consumers 15, 29, 53, 57, 65, 78, 90, 111, 134, 140, 154, 165, 167–8f, 170, 176–8, 180–1, 213, 235, 236, 243, 251, 273t, 277, 279, 280
  car choice: environmental factors 'do not play major role' 9, 14, 289, 298
  empowerment (C21) 288
  'internal search' versus 'external search' 288
  see also customers
consumption 41–3, 45, 51, 222
  artefacts **40–2**
  Fordist revolution 50, 55
  impact on oil price **264**
consumption mode
  challenged by eco-design 52
context 3, 7, 8, 92, 212, 213
  product usage 4
  technological trajectories in AVs 4
control systems 105, 187, 201, 203t, 226
  hybrid and electric vehicle applications 200
  patents (1990–2009) 188f
  software 233
Cooper, A. C. 107, 121
corn: ethanol yield per acre 269
cost/s 9, 15–19, 29(n4), 59, 63, 65, 72, 79, 93, 150, 175, 263, 312
  initial 36
  material versus variable 45
  switching to alternative technology 23
  see also running costs
cost-benefit analysis 92
cost-competitiveness 58, 59
cost-constraints 57
cost-effectiveness 186, 197
cost-reduction 13, 23, 26, 43, 58, 59, 63, 78, 105, 108, 112, 116, 119, 126, 178, 186, 228, 232
'cost-saving green car' 76

cost-sensitivity   105
Coulomb Technologies   237
Creswick, B. P.   267, 277, 278n, 284
Cristal transport system   62, 63f
Croatia   151t
Crowther, S.   55, 67
cryogenic fuels   270
CTC   *see*   Sugar Cane Technology Centre
Cybergo (eight-passenger shuttle)   60–2; 61f
cylinder deactivation   19, 29(n5)
Czech Republic   27t, 151t, 156–7t

DAF Trucks   136t
Dahlman, C. J.   175, 183
Daim, T. U.   xxii, 9, **263–85**
Daimler   24, 113n, 135t, 138(n6), 185, 236, 308–9, 310t, 315
  agreement with Bosch   230, 234, 237
  agreement with BYD (2010)   231, 237
  agreement with Evonik Industries   234, 237
  agreement with Tesla   234, 237
  diesel engine patents (1990–2007)   115f
  emissions of cars sold in Europe (1997–2010)   141t
  fuel cell R&D   126, 138(n3)
Danaher Motion   191
Dassault   24, 57–8, 315, 316t
David, P. A.   164, 166, 183
dealers/dealerships   26, 324
dedicated switch stations (batteries)   211
Delhi   162, 163, 314
delivery   186, 192, 197
delivery vans   210
Delphi (components)   175
demand   13, 28, 137t, 141, 277, 280, 306, 308, 312
  electricity   216
  'more than simply sales levels'   9, 286
Denmark   25, 27t, 59, 157t, 235, 307t, 315, 316t
Depuis, D.   136t
design   23, 35, 45
  artefacts   **40–2**
  computer-assisted   56
  concept-knowledge theory   4, 70

  environmental responsibilities versus corporate strategies   60
  evaluation and validation   73
  integral-type   3
  meaning   40–1
  modular-type   3, 39, 42
  rule-based   2, 4, 73, 78, 80
  Sloanian model versus design thinking model   60
  validation   81
  see also   industrial design
'design' (engineering principle)   35
design companies   56
design costs   59
design information   **32–3**, 34, 40–1, 43–4, 48
  evolutionary pathway   35
  formal aspect   34
design location   2
  comparative advantage   3, **34–5**, 36
design process   91
  'increasingly integrated'   57
design quality   47
design reasoning   74
design research
  human-centred perspective   **53–4**
  lack of continuity   58
design spaces
  new   **75–6**
design structure constraints   39
design theory   70
design thinking   3, **49**, 60
  new approach to mobility   **50–4**
design-driven approach   2, 15
designers
  'social responsibility'   51
Designers Associés   64–5
Detroit   47, 111
Deutsche Bank   20, 20n, 30, 214, 223
Diedre Design   50, **60–2**
diesel   17, 19, 150
  average emissions in urban situation (Euro standards)   145t, 146
  average pollutant emissions   145t
  market battlefield (USA)   **116–17**
  octane index and heat of combustion comparison   144t
  price (versus CNG)   150–3
diesel car sales   106
  Europe and USA (2005–10)   110f

diesel engine/s   20, 44, 122, 124, 125, 138(n5), 140, 144, 158, 161(n4), 191, 193, 268, 292, 299, 300f
  average emissions (Italy)   148t
  consumer 'frames'   294, 298
  cost advantages and disadvantages   116
  evolution (European success story)   **109–11**, 115f
  failure in USA   111
  'modular innovation'   117–18
  negative reputation (USA)   111, 116
  number of cars on road (Italy)   147t
  patents   106, 115f
  small   29(n4)
  see also   electric engines
diesel fuel/s   144
  low-sulphur (lack of availability in USA)   116
diesel hybrids   120
diesel price   171
diesel technology   5, 103–4
  contest with hybrid vehicles   104
  patent activity (1992–2007)   112, 113f
  patents relating to emissions reduction (1990–2007)   114f, 114
diesel vehicles
  advanced   26
  relative costs   120
  see also   E85 vehicles
Dijk, M.   **xxii**, 5, 9, 14, 19, 30, **124–39**, **286–303**
direct fuel injection   19
direct injection (DI) diesel systems   298
  consumer 'frames'   294, 295f
disruptive innovations   2, 241, **242**, 252
  'better theory' needed   242, 252
  EV industry   **226–9**
disruptive searchlights   82
disruptive technologies   7, 22, 30, 126, 138(n8), **241**, 244, 246, 251–2
  see also   technology
distance   18, 22, 50, 95, 96, 100, 311, 312
  long   13, 237, 251
  medium   311
  short   66, 246, 258, 311, 314
  see also   commuters
distributed power generation   221

distribution grid
  EV-supporting   7, 215, **219–21**
  see also   electricity grids
distribution mode
  challenged by eco-design   52
distribution networks   16, 18, 28
  constraints (EVs)   7, **207–8**
DIWAhybrid (Voith Turbo)   193
Dominican Republic   156t
door-to-door transport system
  solution to crisis   304–5
Dow Chemical   126, 234
downstream area (EV sector)   **236, 237**
drive-train components   117
dual clutch transmission   19
dual fuel cars   21, 177
  average emissions (Italy)   148t, 148
  first generation   176
  versus flex fuel   176
  petrol/CNG   148t, 148, 154
  petrol-LPG   154
  see also   electric cars
Dublin University   236
Dunning, J. H.   231, 239

E85 (ethanol) vehicles   269, 270
  see also   electric vehicles
Earl, H.   55
East Asia   243
Eastern Europe   26, 151–2t, 153
EASYBAT project   236
Eco and Eco car   4, **69–85**
  AutoX 'did not know what was valuable'   79
  connection to other transport modes 'critical'   77
  'that supports use of public transport'   80–1
'Eco and Eco Innovation'   74
eco-design   3, **50–4**
eco-innovation   16, 69, 81
École Normale Supérieure de Cachan (Paris)   iii, xx, 324
'ecological' (notion)   **76**
ecology   324
'ecology of the artificial'   52
Economic Commission for Europe
  blue corridor report (2003)   159, 162
'ecosystem' (product support)   4–5, 97, 99
  micro-level   **95–6**

'ecosystem' formation **94–6**
EDF 97, 236, 237, 318
Edgerton, D. 107, 121
Edison project (manufacturing platform) 59
Edström, K. 203t
efficiency 26, 33, 62, 170, 179, 192, 212, 227–8, 232, 236
  criterion to find best motor vehicle fuel 273t, 276f
'Efficient Dynamics' concept (BMW) 117
EFTA countries 163
EFTE (European Federation for Transport and Environment) 13, 29(n2), 31
Egardt, B. 203t
Egypt 149t, 152t, 155t, 168
EHESS xiv, 38n, 323, 324
Eisenhardt, K. M. 247, 252
Elbeuf Austreberthe 318
electric batteries 16, 29(n1), 259
electric cars (1899–) 19, 30, 54, 58, 71, 244, 266, 271, 276f, 279, 286
  equipped with CNG-powered range extender 158, 160
  mass-produced 59
  models on sale (2011 and by 2015) 305t
  see also ethanol car
electric/al drive systems 193
  components 187
  technological dynamics 186, **187–8**
Electric Drive Transportation Association 318
electric engines 14, 90, 259
  see also engines
Electric and Hydrogen Vehicle Symposium (EVS 21) 134–5, 138(n8)
'Electric Mobility Canada' 318
electric motor development for heavy hybrid vehicles **6–7**, **185–204**
  from broad search to focused development **197–200**
  comparative analysis of development of electric motor applications 186, **194–7**
  data collection **200–1**
  development of electric motors for hybrid vehicles at Volvo and Scania 186, **189–94**
  documents 201
  hybrid technology and electric motor technology **189–90**
  hybrid vehicle development processes at Volvo and Scania **186–7**
  'important challenge' 186
  interviews 186, **201**, 201–3t
  interviews at Scania 201–2t
  interviews at Voith 203t
  interviews at Volvo 202t
  interviews with professors 203t
  knowledge bases (comparative analysis) **194**
  organizing effective collaborations (comparative analysis) **196–7**
  patents **200**
  Scania: problems of developing two technologies in parallel **198–9**
  Scania-Voith collaboration 186, **193–4**
  Scania-Voith electric motor development **192–4**
  sourcing and knowledge development **190–1**, **192–3**
  technological dynamics of electrical drive systems 186, **187–8**
  technology and application novelty (comparative analysis) **194–5**, 196f
  technology as search process with spillover effects **200**
  'three essential issues' 7, **197–8**
  Volvo: advantage of focus **199–200**
  Volvo-Kollmorgen collaboration 186, **191–2**
  Volvo-Kollmorgen electric motor development **190–2**
electric motor efficiency 225
electric motor suppliers 185
electric motors 118, 120, 187–8, 191, **227**, 230, 234, 236–7, 244, 246, 304, 306, 309, 315
  AC (asynchronous versus synchronous) 189
  AC versus DC 189
  asynchronous 193
  axial **189–90**
  classification by operating principle 189–90
  control system 191
  in-house production 185
  new materials 233

electric motors – *continued*
  patents (1990–2009)   188f
  radial   **189–90**
  three-phase   190
  transverse flux   **189–90**
  Volvo patent (2005)   191
electric plug-in facilities   306
electric power   222, 236, 259
electric power systems   7
  'electronic power' systems   22
electric vehicle (EV) components
  strong relationships (carmakers and suppliers)   8
electric vehicle firms
  agreements and joint ventures   7–8, 225–39
  China   **244–6**
  China (business model innovation)   7, **240–53**
  clusters   238
  corporate flexibility required   226
  corporate growth strategies   7–8
  downstream   8
  new players   225–6
  partnerships with electricity producers   233, **235–6**
  structure   225–6
Electric Vehicle Network   314
electric vehicle sector: agreements and joint ventures   **236–7**
  core business   **236**, 237
  downstream area   **236**, 237
  upstream area   **236**
Electric Vehicle Industry Alliance (China)   318
Electric Vehicle Technology for Réunion (VERT)   **100–1**
electric vehicles (EVs)   16–18, 24, 26, 28–9, 30(n7, n9–10), 30, 32, 59, 65, 66(n3), 104, 117, 119, 122, 124, 161, 249, 251, 261, 268, 296, 304, 307t, 307–8, 310t, 310–12, 319t, 321
  affordability   21
  battery-run   21–2
  breakthrough strategies   90
  classification   7, **208–10**
  cost and benefit analysis for year 2020   20, 20t
  diffusion   7, 234
  diffusion scenarios   213, 224
  dissemination 'not easy'   **44–5**
  distance able to travel   100
  environmental breakthrough innovations   4
  'eternally emerging' technology (Fréry)   19, 31
  'first steps in life-cycle'   225, 237
  fleet renewal   29
  impact on grid   7, **218–19**
  incentives   25, **216–19**
  innovation race   101
  issues to be solved   319
  lack of continuity   58
  maintenance issues   96
  'massive deployment' (by 2020)   214
  models   20
  modular   3, 44, 45, 63
  'most likely scenario'   237
  number (2015–50 targets)   313t
  performance structure (comparison with ICEs)   22
  potential diffusion   210
  price gaps   30(n11)
  product support   99
  profitability   311
  pure versus hybrid   225
  recharging   95
  refuelling   7, **210–12**
  role in reducing electricity supply and demand mismatch   221
  running costs   314
  sales   27, 27t, 207
  second-generation   314
  small urban   54
  start-ups and newcomers (2011): listed   **316–17t**
  subsidies   228
  supporting distribution grid   7, 215, **219–21**
  technological progress   208
  testing   318
  total energy requirement (forecast for 2050)   215
  turnover   207
  urban   228
  usage cost (versus ICEs)   27
  see also flexible-fuel vehicles
electric vehicles: balancing strategic intent and experimental approach   **4–5, 89–102**

case study analysis 93–6, 101(n3)
characteristics of overall strategy 93–4
conclusions 99
conducting parallel pilot testing to support formation of 'ecosystem' 94–6
data-gathering 93
discussion 96–9
environmental innovation 91–3
further research required 99
initiatives related to environment 91–3
literature 89, 90–3, 97, 98
managing projects with high levels of uncertainty 90–1
pilot project on Réunion 95–8, 100–1
pilot projects as means of developing assets and generating support 96
pilot projects as opportunity for collective learning 96
pilot projects as tool for building micro-level 'ecosystem' 95–6, 97
research questions 89, 91, 93
role of pilot testing in Renault's EV strategy 95
ZEV initiative 93–4
electric vehicles: efficient use of electricity (charging scenarios and economic incentives) 216–19
ex-ante versus real-time mechanisms 218–19
load profiles (Italy) 216, 217f
market mechanisms versus top-down control 217–18
rational use of electricity 217–18
when to recharge is not unimportant 216–17
electric vehicles: impact on electric energy systems 7, 208, 213–19
charging scenarios and economic incentives 216–19
efficient use of electricity 214, 216–19
scale 213–14
electric vehicles: technical impacts on system
macro perspective 214–15
micro perspective 215
problems 215–16

electric vehicles and power grids: challenges and opportunities 7, 22, 207–24
classification of electric vehicles 7, 208–10
EV impact on electric energy systems 7, 208, 213–19
refuelling EVs 7, 210–12
smart grids 7, 208, 212–13
'time for joint research projects' 222
V2G: EV supporting distribution grid 7, 215, 219–21
electric-petrol hybrids (Toyota takes initiative) 108–9
electrical appliances 46
electrical components 120, 189
electrical energy system (EES) 221
'electric energy systems' 207, 213–19
electrical engineering 105, 118, 119, 202t
electricity 14, 22, 27, 30(n12), 66, 100, 119, 234, 306–7
access 318
bi-directional flow 220, 222
cost 222
distributed storage 222
ex-ante versus real-time mechanisms 218–19
generation, transmission, and distribution devices 208
global consumption 219
global production (2008) versus EV requirement (2050 forecast) 215
home outlet 211
import capacity 216
'just-in-time' production 216
market mechanisms versus top-down control 217–18
non-polluting sources 213–14
peak-demand hours 215–22
pollution and safety issues 268
production (decoupled from consumption) 222
production and distribution cycle 207–8
production-transmission system (centrally-operated) 221
rational use 217–18
use scenarios 218–19
electricity consumption 95, 225

electricity demand   217, **235–6**
electricity distribution   214, 218, 238
electricity grids   7, 101, 207, **208**, 210, 211, 214, 235, 268, 283, 306
  hardware   215
  impact of ground transport   209
  low-voltage (LV) versus medium-voltage (MV)   215, 220
  performance   215
  see also   smart grids
electricity load profiles   216, 218
electricity prices   28, 31, 90, 219
electricity storage   212
electricity supply   207, 213, 218, 219
  diffusion of EVs   7
  smart grids (emerging paradigm)   **212–13**
electricity taxes   27t, 28
electricity utilities   24, 311, 315, 318
  large fleets of service vehicles   314
  partnerships with EC firms   233, **235–6**
electricity-generating plants   **208**
  de-carbonization   22
  dispersed   214
  fuels   266
  global capacity (2008)   215
  inefficiency   216–17
  polluting   217
  primary input prices   213
  see also   energy companies
electrification   19, 120
electro-technical industry   186
electromagnetic grid   259
electromagnetic valve train   29(n5)
electronic:
  commands   53
  control systems   108
  control units   236
  devices   46
  injection systems/devices   176, 177
  management systems   18
  mapping   260
electronics   226, 236
  revolution in cars   23
Elektromotive   236
Elmquist, M.   **xxii–iii, 3–4, 69–85**
ElvoDrive (Voith Turbo)   193
EM-Motive joint venture   234
Embraer   175

EMBRAPA   *see*   Brazilian Agricultural Research Organization
emissions   xix, xx, 3, 22, 25, 41, 52, 76, 89, 100, 101(n1), 110, 116, 124, 137(n1), 176, 268, 283
  cars sold in Europe (1997–2010)   141, 141t
  see also   pollution
emissions legislation   110, 120, 125, 134
  see also   European emission standards
emissions testing
  distinct cycles   25
emissions-reduction   117
  conversion to CNG (Italy)   **143–8**
Empresa Brasileira de Pesquisa Agropecuária (EMBRAPA)   xiv, 166, 168f, 169, 182(n7)
  website   182(n5)
End-of-Life Vehicle (ELV) Directive (wef 2007)   24–5
Enel   236, 237
Enerdata   142, 162
energy   28, 32, 35, 45, 50, 51, 107, 194, 214
  alternative sources of improvement (impact on oil price)   **265**
  clean   257
  cost comparisons   21
  new green   225
  price gap   27
  scare resources   35, 39
energy companies   23, 94, 95, 100–1
  'energy producers'   98, 235
  see also   power plants
energy consumption   18, 35–6, 58, 76, 79, 94, 210
  regulations   39
energy converters   190f, 191
energy density   19, 234
energy efficiency   141, 208, 240, **264**, 286
Energy Efficiency and New-Energy Automotive Industry Planning 2011–2020 (Chinese policy document)   249
energy policies   306, 309
energy production   100, 260
Energy Research Institute of Beijing   264

Index 343

energy security   240
energy storage   112, 185, 190–3, 201, 210
  see also lithium
energy supply   256
  political risk   265–6
  and time horizon (portfolio of solutions)   259, 262(n2–3)
energy-saving/conservation   43, 46, 47, 212
engenho   169
engenhoca   169
engine capacity (horsepower)   41, 89, 179, 269, 293f, **294**, 295f, 296n, **298**, 299
engine choice   **299**, 300f
engine design   94
engine efficiency   144, 161(n3–4), 268, 301(n8)
engine structure   295f
engine systems   19
engine timing (advances)   19
engine types (diversification)   44
engine volume   293f, 295f, **298**
engine weight   293f, 294, 299
engines
  'alternative' versus 'conventional'   287, **292–4**, 295f, 301(n7), 302
  endothermic   259
  hybrid electric   287
  low-pollution   53
  oil-powered   308
  thermal efficiency   161(n2)
  traditional   21
  see also flex-fuel engines
environment (general)
  AHP category   271–2
  business   20, 25, 242
  day-to-day   51
  economic   175–6, 323
  industrial   54
  institutional   165, 166, 168f, 181
  macroeconomic   xix
  organizational   168f
  political   20, 175–6
  social   54, 128, 323
  urban versus non-urban   256
environment (natural)   1, 4, 6, 18, 22, 26, 28, 45, 53, 72, 127, 136, 137t, 217, 222, 228, 240, 243, 251, 261, 271–2
  see also 'perceived environmental risk (EV)'
environmental concerns   **277, 278f, 278–81**
environmental constraints   32, 35, 39, 46, 47
Environmental Grenelle on Réunion: Making Innovation Work (2007–)   100
environmental impact   7, 30, 50, 90, **207**, 214, 290, **291t**, 291, **292**, 293f, 296, 296t, 297f, **298**, 301(n7)
environmental initiatives   4, 90, **91–3**, 97
environmental innovation   89, **90–3**
environmental issues   52, 69
  challenge to dominant automotive design   **70–1**
  impact on oil price   **265**
environmental movements   309, 318
environmental performance   xix, 212, 320–1
environmental policies   139, 259, 306
environmental regulations   9, 44, 127, 290
environmental studies   72
environmental taxes   89
ERDF   318
Espace Autolib stations   65
Estonia   151t, 157t
ethanol   4, 17, 159, 271, 276f, 279, 282f, 283
  for aircraft   180
  alcohol-based   269
  from cellulose   181t
  consumer price   174
  future (as fuel)   180, 181t
  'GHG neutral'   269
  new sources   180
  pollution and safety issues   **269–70**
  'second-generation'   170
  system innovation (mature market economy)   166
  use as fuel (common elements of three phases)   **180–1**
  use as fuel (three phases)   **180**, 181t
  yield per acre of corn   269

ethanol and automotive industries (Brazil)   6, **164–84**
  economic growth, external debt, oil account (Brazil, 1972–9)   173t
  transactions (TA 1–2, TE 1–2)   167, 168f, 168n
ethanol car/engine   6, 166, 174–6
  'reconversion' to petrol   179
  see also   hybrid cars
ethanol chain   168f
  innovation   164, **164–74**
ethanol industrialization   168f
ethanol market
  establishment and deregulation (1999)   169
ethanol price   171
ethanol production   179
  commercialization innovation   166, **171–4**
  governance structure   170
  institutional and technological innovation   **169–71**, 182(n7)
ethanol production organizations   166, 182(n2)
ethanol retail distribution   182(n1)
ethanol sugar cane industry   6, 166
ethanol value chain   164
ethanol-only engine   172, 175–9
  second phase   170–1, 179–80, **180**, **180–1**, 181t
  see also   gasohol
ethanol-petrol mixtures   180
ethyl alcohol (ethanol)   269
'EU FP7'   236
Eurasia   149–50t
Eurasia Group   265–6
'Euro NCAP tests'   24
Europe   ii, xxiii, 1, 2, 8, 20–1, 24, 26, 30, 44, 46, 104, 106, 114, 142, 149, 235, 243–4, 250–1, 303, 313–14
  automotive market (size)   258, 261(n1)
  biogas use 'increasing'   159
  'clean diesel' engine   104, **108–16**, 120
  definition of LSEV   244, **245**
  diesel car sales (2005–10)   110f
  EV deployment (2020 scenario)   214
  hybrid car sales (2006–9)   111f

industrial architectural geopolitics   **37**
LNG consumption   150
natural gas production (2009)   150t
reserves of CNG (2008)   149t
sourcing strategies (specialization)   **118–19**
Europe: Western   110, 151t, 153, 153f, 312, 313
Europe 2020 strategy   29(n4)
Europe's Energy Portal   27, 27n, 31
European Automobile Manufacturers' Association (ACEA)   110n, 289n
European Centre for Mobility Documentation (ECMD), Netherlands   289n
European Climate Foundation   259, 262(n3)
European Commission   111–12, 147
European Community (EC)   52
European Council   162
European emission standards
  Euro 0 standard   146–7, 146–7t
  Euro 0-3 standards   144, 160
  Euro 0-4 standards   145t, 146–7, 147t
  Euro 1-6 standards   14, 141
  Euro 3 standard   140
  Euro 4 standard   24
  Euro 5 standard (2009)   24, 145–6t, 146
  Euro 6 standard (2015)   24, 120
  see also   emissions legislation
European Environment Agency   101(n1), 102
European Federation for Transport and Environment (T&E)   141n, 163
European Fuel Cells and Hydrogen Joint Undertaking   259
European Parliament   141, 162
European Patent Office (EPO)   **105–6**, 113n, 114f, 115n, 193, 195n, 200, 204
European Union   14, 25, 26, 60, 137t, 140–1, 228, 234, 236, 307, 307t, 309
  data source   145–6n, 148n, 162
  directive 92/61/EEC   246
  directive 2002/24/EEC   246

electric mobility (transnational development projects)   25
EVs (target number)   313t
'political implosion'   258
political leadership 'missing'   25
Regulation EC 715 (2007)   162
European Union: EU-15   101(n1)
European Union: EU-27   89, 163
 diffusions of CNGVs   154
 natural gas vehicles (latest figures)   157t
Eurostat   27n, 31
Evonik Industries   234, 237
excise duties   16, 25, 27t, 28, 153
experimental approach versus strategic intent (EVs)   **4–5, 89–102**
Exxon   150, 162

FAW   231, 237, 315, 316t
FCC (Spanish company)   160
FDI   xiv, 230
Federal Motor Vehicle Security Standards (USA)   245
Ferrara   260
Ferrato, E.   xxvi
FIAT   19, 22, 112, 113n, 174–5, 181t, 258–9, 307–8, 310t
 average CO2 emissions of cars sold in Europe (1997–2010)   141t
 diesel engine patents (1990–2007)   115f
FIAT 500 model   52
FIAT 500 'Topolino' (1936)   56
FIAT 600 (1956)   56
FIAT Powertrain (FTP)   182(n1)
FIAT-Chrysler alliance   59, 235, 236
'Field of Invention' (USPTO)   106
FIEV   324
Figueiredo, J. B. de O.   175
filling stations   16, 29, 178, 248, 268
 see also   petrol stations
Finizio, G.   53, 67
Finland   151t, 153f, 155t, 157t, 315, 316t
Fischer, F.   30(n6)
Fisker   226, 315, 317t
fleet renewal   26, 29
flex-fuel   168f
 versus dual-fuel   176
flex-fuel car   181t

flex-fuel engines   164, 165, 170, 174, 177
 see also   hybrid engines
flex-fuel technology   6, 166, 168, 175–9, 182(n1), 184
 'Phase III'   **180, 180–1**, 181t
flexible fuel vehicles (FFVs)   25, 269
 see also   fuel-cell vehicles
Ford, H.   50, 54–5, 67
Ford   109, 174, 175, 309, 310t
 emissions   141t
 fuel cell R&D   126
 patents   112, 113f, 115f
Ford C-Max   300f
Ford Europe   112
Ford Explorer (SUV)   109
Ford Fiesta   300f
Ford Focus   299, 300f
Ford Model T   55
Ford Nederland   301t
Fordist industrial revolution   50, 55
forecasting framework and scenarios: AV fuels   **9**, 25, **263–85**
 AHP model   **9**, 263, 266–7, **267**, 276f
 AHP model: application to problem   263, **271–2, 273–5t**
 alternatives   263, **268–71**
 challenge (growing oil demand versus shrinking supply)   **263–7**
 conclusion   **284**
 criteria to find best motor vehicle fuel   9, 263, **273–5t**
 meaning of oil price   263, **264–7**
 pollution and safety issues   **268–71**
 results and discussion   **281–3**
 scenario 1: 'status quo'   9, 278f, **278–9**
 scenario 2: 'environmental challenge'   9, 278f, **279**
 scenario 3: 'economic challenge'   9, 278f, **280**
 scenario 4: 'catastrophe'   9, 278f, **280–1**
 scenarios   9, 263, 267, **272**, **276–84**
 see also   alternative vehicles
fossil fuels   6, 25, 28, 50, 66, 103, 166, 208, 259, 263, 266, 279, 282–3
 reserves   60
Foster, R.   108, 121
Francastel, P.   54, 67

France xxii–iii, xxv, xxvii, 25, 27–8, 30(n12), 54, 60, 96, 100, 109, 116, 135, 236, 246, 287, 307t, 314, 316t, 318
  AV market 28
  CNG: payback period 153f
  CNG pump prices (versus petrol and diesel, 2011) 151t
  CNGVs and CNG filling stations 156t
  diesel technology 112
  distance travelled between home and workplace 50
  dual-fuel cars 154
  EVs 27t, 313t
  natural gas vehicles (latest figures) 157t
  night-time recharging 98
France: Ministry of Environment 324
France: Ministry of Industry 324
Frankfurt 57, 58, 120, 260
Fréry, F. 19, 31
Freyssenet, M. ii, xxiii, 9–10, 304–22, 325–6
  co-founder of GERPISA 323
Fridenson, P. 323
Fritschtak, C. R. 175, 183
fuel consumption 18, 21, 41, 65, 69, 71, 94, 209, 289f, 289–90, **291t**
fuel cost 273t, 275t, 276f
fuel economy 14–16, 21, 43, 105, 109, 110–12, 117, 120, 158, 186, 198–9, 251, 265, 287, **290**, 290t, 292, 294, 301(n8), 303
fuel efficiency 9, 273t, 290, 291, **292**, 293f, 294, 295f, 296, 298, **298**, 299
fuel gauges 55
fuel mix 119, 172
fuel prices 5, 25, 103, 120, 111, 180, 207, 213, 287, 292
fuel use **292**
fuel use lifetime 275t, 276f
fuel-cell electric vehicles (FCEVs) 209
fuel-cell vehicles (FCVs) 4, 5, 14, 16, 21–2, 28, 29(n3), 107, 126, 137t, 137–8(n2), 209, 271, 319t
  R&D investment 138(n3)
  see also fully-electric vehicles
fuel-cells 19, 21–2, 71, 103, 259, 261, 265, 303–4, 307t, 309

  see also hydrogen
fuelling infrastructure development time 275t, 276f
fuelling pipeline of future projects **81**
fuels
  distribution network 21
  'less polluting' 310t
  new 22
Fujimoto, T. ii, **xxiii**, **3**, **32–48**, 325
fully-electric vehicles 124, 287, 298, 312
  'all-electric vehicles' 10, 28, 311, 318–19
  'pure EVs' 214, 237, 240
  see also green vehicles
FUPET consortium (Japan) 236
future 4, 5, 7, 38, 44, **46–7**, 50–1, 53, 66–7, 69, 70, 78, **81**, 82, 96, 107, 119–20, 125, 129, 131–2, 134, 153–4, 180, 181t, 212, 225, 231, 250, 259, 260, 266, 272, 276, 277, 284
  best motor fuel 9, 263
  city size 258
  CNGVs **158–60**
  see also Knightian uncertainty

Galicia 29
Garibaldo, F. **xxiii**, **8–9**, **254–62**
Garnero, M. 173n, 183
gas 16, 21, 259, 262(n2), 306, 310t
gas carrier ships 17, 149, 160
  see also LNG
gasohol (fuel) 172, 174–5, 179, **180–1**, 181t
  see also ethanol-only engine
gasoline *see* petrol
Gastaldi, L. 91, 102
GDP 173t
GEA (urban planning consultancy) 62
Geels, F. W. 165, 183
Geely (Chinese owner of Volvo) 308, 315, 316t
Geisel, E. 172, 183
General Command for Aerospace Technology (CTA, Brazil, 1947) 168f, 181t, 175
General Electric (GE) 234, 314
General Electric Motorcars (GEM, 1998–) 246, 317t

Index   347

General Motors (GM)   20, 57–8, 109, 111, 174–5, 178, 182, 234, 258, 309, 310t, 315, 319
  agreement with LG Chem   230
  agreement with SAIC   231, 237
  EV   242
  fuel cell R&D   126
  patenting   112
  patents: diesel engines (1990–2007)   115f
  patents: hybrids (1992–2007)   113f
  plug-in hybrids   119
  style department (1927)   55
General Motors Ampera   30(n10)
General Motors Volt   21, 30(n10), 189
generation capacity (electricity)   7, **207**
generators   192, 192f, 193
  patents (1990–2009)   188f
Genetic Resources and Biotechnology Centre for Plants (CENARGEN)   xiii, 169, 182(n7)
genetically-modified crops   169
  sugar cane   180
Geneva   57, 63, 112
Georgia (country)   155t, 312
geothermal plants   18
Germany   xxii, 25–8, 30(n12), 109, 116, 126, 135, 236, 287–8, 302, 307–9, 312, 316t, 318
  biomethane use   159
  changing position (type of vehicle)   307t, 308
  CNG pump prices (versus petrol and diesel, 2011)   151t
  CNG system: payback period   153f
  CNGVs   154
  CNGVs and CNG filling stations   155t
  EVs (target number)   313t
  natural gas vehicles (latest figures)   157t
  regional administrations   308
GERRI project (2007–)   100
Geurts, F.   301t
GGE (energy content in gallon of petrol equivalent)   273t
Ghia, C.   56
Ghosn, C.   101(n3)
Giacosa, D.   56
Gifford, J. D.   144, 162

Gilfillan, S. C.   107, 121
Giolito, R.   52–3, 67
global warming concerns   13, 15, 35, 39, 43, 240, 274t, 276f, 277–80, 305, 309, 314
Gothenburg: Chalmers University of Technology   203t
Goyal, S. P.   143, 162
GPS   260
grain alcohol (ethanol)   269
Greece   151t, 153f, 155t, 157t
'green' consumers   296t, 297f, 299
Green eMotion project   236, 318
'green innovation approach'   225
green vehicles
  architecture   3, **32–48**
  see also heavy duty vehicles
greenhouse gases (GHG)   52, 101(n1), 102, 140, 159, 208, 221
  along entire fuel cycle   274t
  emission-reduction   104, 146, 213–14, 309
  emission-reduction: international efforts (impact on oil price)   **265**
  emissions   89, 103, 112, 222, 271, 274t
  'illusory abatement'   119
  methane   143–4
Greening European transportation infrastructure project   236
Grid Point   237
GS Yuasa   234, 315
GSM   234
Guangdong   313
Guédon, P.   63
Gulf of Aden   266
Gulf of Mexico   265

Hagedoorn, J.   105, 121
Hao, Z.   248, 253
Hart, S. L.   242, 243, 252
Harvard Business School   xxiii
Hatchuel, A.   70, 73, **82**, **83–4**, 92, 101–2
Hawaii   312
Hayek, N.   57, 58–9
heat of combustion   144t, 161(n2)
heavy duty vehicles (HDVs)   110, 154, 157t, 159
  use of CNG 'highly recommended'   160

348 *Index*

heavy hybrid vehicles   6–7, **185–204**
  electric motor applications
    (comparative analysis)   186, **194–7**
heavy hybrid vehicles: electric motor
    development   **192–4**
  knowledge bases (comparative
    analysis)   194, 195t, 195f
  sourcing and knowledge
    development   **190–1, 192–3**
  technology and application novelty
    (comparative analysis)   **194–5**,
    196f
  Volvo-Kollmorgen collaboration   186,
    191–2
  see also   hybrid electric vehicles
Henderson, R.   117, 121
Henrique, D.   168
High Graph Architecture   65
high-voltage (HV) systems   105, 193
  distribution and transmission
    failure   222
  feeder of distribution grid   221
Holmén, M.   105, 121
Honda   22, 104, 122, 135t, 138(n6),
    243, 310t, 323
  fuel cell R&D   126
  hybrid technology (US market
    battlefield)   116
  patents related to hybrids (1992–2007)
    113f
  sourcing strategy (via media)   119
  US patenting   112
Honda FCX Clarity   21
Honda Insight   109
Hong Kong   313, 316t
Hoover, H.   257
horsepower   *see*   engine capacity
household electrical appliances
  commodification   40, 45
Hungary   151t, 157t, 315, 316t
hybrid buses   187f, 187, 193, 198–200
hybrid car sales   106
  USA (2000–10)   109f
  USA/Japan versus Europe (huge
    difference)   **111–12**
hybrid cars   28, 158, 188, 271, 276f,
    279–80, 319t
  models on sale (2011 and by
    2015)   305t
  see also   hydrogen cars

hybrid diesel-electric propulsion
    systems   208–9
hybrid driving systems   22, 194, 195f
  Volvo patent (1997)   191
  Volvo patent application (2009)   191
hybrid electric traction systems   209
hybrid electric vehicles (HEVs)   3, 4, 16,
    20, 112, 117, 124–5, 138(n5), 185,
    203t, 210, 240, 298–9
  'charge-sustaining' versus 'charge-
    depleting'   **209**
  consumer attitudes (frames)   286–7,
    292–4, 298
  cost and benefit analysis for year
    2020   20, 20t
  frames (2000, 2005)   293f
  market share   294, 296, 301(n9)
  micro-hybrid, mild-hybrid, power-
    split full-hybrid   209, 214
  purchase motivation   291, 303
  sales worldwide (by 2011)   286
  see also   hybrid vehicles
hybrid electric-petrol cars
  versus clean diesels   104, **108–12**
hybrid engines   21, 50, 144, 305, 309
  cost disadvantage versus diesel   116
  petrol-electricity   5, 22
  see also   hydrogen engines
hybrid fuel   282f, 282, 283
hybrid parallel powertrains   200
hybrid power control unit (HPCU)   190
hybrid powertrains   104–5, 118, 119,
    185, 190–1, 193–4, 196f, 199
  classification   189
  innovation (architectural and
    modular)   118
hybrid technology   69, 126, 135
  and electric motor technology
    189–90
  major challenges   112
hybrid vehicles   13, 18, 32, 44, 45, 90,
    107, 112, 117, 185, 121–2, 201, 227,
    283, 304, 307–8, 310t
  CNG-electric   161
  development processes (Scania and
    Volvo)   **186–7**
  patents   105–6
  'temporary step'   20–1, 30
  see also   light-duty vehicles
hydrocarbons   141, 270

Index

hydrogen 14, 17, 30(n8), 66, 103, 144, 160, 270, 276f, 279, 282f, 282–3
  mixing with CNG 158
  production and storage 126
  see also fuel-cells
hydrogen cars/vehicles 19, 21, **129**, 130f, 161, 287
  scenario 18
  'without fuel cells' 16
  see also passenger cars
hydrogen engines 5, 22, 137–8(n2)
  see also internal combustion engine
hydrogen fuelling systems 267, 284
hydrogen infrastructure 280
hydrogen sulphide 143
hydrogen technology: firm perspectives 5, 14, 16, 18, **124–39**
  'central question' 124
  date of survey 138(n8)
  discussion and conclusion 125, **133–4**
  engagement in developing a hydrogen vehicle **129**, 130f
  firm perspectives on car engines **125–6**
  'major question' for business 134
  methodology 125, **126–8**, 133
  possible sources of bias 128
  potential drivers of engagement **129–33**
  questionnaire 127, **128–37**, 138(n5, n9)
  questionnaire survey respondents **134–5**
  results: engagement and beliefs regarding hydrogen technology 125, **128–33**, 135, 137–8(n2)
  stakeholder consultation **134–6**
  stakeholder interviews **135**
  stakeholder interviews: format **135–6**
Hyundai-Kia 309, 310t, 315

Iceland 151t
IDEO (design firm) **53–4**
IER Group (subsidiary of Bolloré Group) 65
Île-de-France 313
Independent Paris Transport Authority (RATP) xvi, 74, 83(n3), 314

India 44, 45, 59, 168, 278, 305, 307t, 307–8, 310, 312, 315, 316t
  adoption of all-electric vehicle 318–19
  carmakers 310t
  cars 264
  CNG pump prices (versus petrol and diesel, 2011) 152t
  CNGVs and CNG filling stations 155t
  growth in demand for oil 264
  industrial architectural geopolitics 37
  low-cost vehicles (integral nature) 32
  see also BRIC
Individual and Public Urban Transport Programme (TULIP) 58
Indonesia xxvi, 149t, 152t, 155t
Induct (company, 2004–) 50, **60–2**
industrial architectural geopolitics 37
industrial design (in-house) 56, 57
  see also innovative design
industrial designers and challenges of sustainable development 3, 15, **49–68**
  Bluecar for Autolib project 64f
  Bolloré-Pininfarina, Bluecar, and 'Autolib' in Paris 50, **62–5**
  case studies 3, 49, 50, **60–5**
  conclusion **65–6**
  Cristal transport system 63f
  Cybergo and Modulgo Induct projects with Diedre Design **60–2**
  design thinking as new approach to mobility and eco-design theory **50–4**
  Diedre Design sketches for Cybergo vehicle 61f, 62
  LOHR Industrie group and Cristal project 50, **62**, 63f
  mobility as product/service: challenge from new players 50, **60–5**
  mobility as service offer: from style to design 50, **54–60**
industrial development 244
industrial instruments 46
industrial locations 255
industrial perspective
  new paradigm implied **257**
industrial policy 238, 259, **260–1**
  horizontal versus vertical measures 25

industrial rationality   257
industrial renewal   260
industrial sector   158
industrial standards   225, 232, 236
industrial transition   260–1
industrialization   48, 50, 51, 56, 174, 305
  competence/capability   104, **106**
industry   170
  capability-building competitiveness   38, 38f
  non-polluting complexes   51
industry groups   41
industry structure   34–5
  effect on company strategies   **231**
information technology   15, 255
  ICT   212, 226, 257, 259
infrastructure   6, 8, 18, 25, 28–9, 89, 92, 95, 126, 142–3, 160, 228, 241, 255, 257, 259, **260**, 270, 311
  CNG-hydrogen mixture   158–9
  cost   271
  cost-benefit ratio (criterion to find best motor vehicle fuel)   273t, 276f
  designed for mobility   256
  urban   261
innovation   ii, xxvii, 3, 49, 67, 82, 89–90, 102, 104, 138(n5), 213, 230, 253, 294
  agreements and joint ventures in EV industry   **232–3**
  agricultural   165
  alternative approach   54
  architectural versus modular   5, 108, **117–18**, 120
  attacker's advantage   108, 121
  attitudes   **127**, 129, 130
  automotive chain   **174–9**
  changing nature   73
  commercial   181t
  commercial (automotive sector) **178–9**
  as cultural project (Manzini)   51
  'design-driven process'   **49**, 53
  discontinuous (and technological trajectories)   14, **15–19**, 29–30(n3–8)
  environmental breakthrough   4
  within ethanol chain   164, **167–74**
  FFE and FE   72
  human-centred approach   53
  incremental   73
  incremental (versus 'exploratory' projects)   90
  institutional   181t
  institutional, technological, commercial (Brazil)   **6**, **164–84**
  institutional and technological (automotive industry)   **175–8**
  literature   15, 78, 79
  market adoption   167f
  potential sources   243, 252
  sugar cane sector   **168–9**
  social   180
  technological   180, 181t
  typology (Henderson and Clark, 1990)   117, 121
innovative design
  for alternative vehicles   **3–4**, **11–85**
  and sustainable development in automotive industry   **2**, **13–31**
innovative design   3–4, 16, 69–85
  addressing sustainable development through innovative design   70, **74–8**
  'C' phase: introducing disruptions   **76–7**, **82**
  capturing value   **79–80**
  case studies: AutoX   4, **70–3**
  characteristics   73
  complementing R&D processes   **71–3**
  conclusion   **81–2**
  creating breakthroughs 'taking limited risks'   **80–1**
  discontinuous innovation and technological trajectories   14, **15–19**, 29–30(n3–8)
  dominant design challenged by environmental issues   **70–1**
  empirical studies   70, 79, 82
  framework   73
  fuelling pipeline of future projects   **81**
  industrial policy implications   15, **24–9**, 30(n11–12)
  innovative and sustainable strategies (way of developing)   **80–1**
  'K' phase: creating new design spaces   **75–6**, **82**
  KCP method   4, 70, **74–8**, 82, 83(n3–4), **83–4**

learning from experiment **78–81**
literature 70, 72, 78, 79
'missing knowledge' (identification method) **78–9**
new potential strategic spaces (way of crafting) **81**
new research areas (identification) 82
new strategic players and relationships 15, **23–4**
'P' phase: proposing path through sustainable strategies **77–8**, **82**
producing new knowledge **78–9**
quest for breakthrough innovation at AutoX **71–3**
race for AVs in automotive industry 14–15, **19–22**, 30(n9–11)
research methodology 75
short-term initiatives and long-term strategies 80
sustainable development in automotive industry **70–3**
towards new R&D processes in greening of automotive industry 70, **78–81**
valuation criteria **79–80**
innovative and sustainable strategies
path through **77–8**, **82**
way of developing 78, **80–1**
Institut National de Recherche en Informatique et en Automatique (INRI) 60
institutions
supranational 181, 181t
see also public institutions
integral-type products
competitive advantage 36
integrated systems
urban mobility **261**
'integrated-type manufacturing' 33
internal combustion engine (ICE) 8, 14–22, 45, 49–50, 59, 71, 89–90, 94, 100–1, 104, 107, 117, 124, 130, 208–9, 233, 240, 244, 246, 271, 286–7, 294, 311, 318–19
versus EVs (usage cost) 27
environmental performance 22
fossil-fuel-driven 19
hydrogen-powered 22, 30(n8), 137(n2)
improvement 19
improvement (incremental) 103
oil-powered 304–5, 306
penalties for use 228
performance 16
performance structure (comparison with EVs) 22
petrol and diesel 4
spark-ignition 161(n3)
transition to FCV 28
wider scenario 17
see also petrol engines
International Energy Agency (IEA) 264
inverters 191, 226, **227**
investment 6, 13, 16, 25, 30(n9), 55, 57, 60, 90, 92, 94, 99, 101, 111, 125, 132, 141, 143, 160, 166, 169–70, 180, 185, 200, 208, 214, 215, 221, 226, 228–9, 235, 238, 255, 257, 264, 277, 279, 309, 310–12, 324
greenfield 230
infrastructural 18, 28
internal 232, 233
lock-in (technological and economic) 256
in production capacity (impact on oil price) **265**
public 18, 259, 261
quasi-irreversibility **167**
reluctance 213
types 230
Iran 149t, 152t, 155t, 174, 264–6, 307, 307t
Iraq 149t
Ireland 27t, 153f, 313t, 157t
Israel 25, 59, 94, 235, 307t, 312
Italy xxi–iii, xxv–vii, 6, 26–8, 30(n12), 58, 162, 168, 236, 260, 307, 307t, 312, 316t
AV market 28
average CO2 emissions of cars (standard, fuel type, class) 148t
cars on road, 2009 (class, fuel type, Euro standard) 147t
CNG filling stations 155t
CNG pump prices (versus petrol and diesel, 2011) 151t
CNG system: average payback period 153f
CNGVs 154, 155t

Italy – *continued*
  electric vehicles (2010–11 statistics) 27t
  electricity load profiles 216, 217f
  electricity statistics 224
  emission-reduction through conversion to CNG 143–8, 161(n2–5)
  natural gas vehicles (latest figures) 157t
  peak power 215
  pipelines 149

Jacobsson, S. 105, 121
Jakarta 242
Janatzy, C. 54
Japan 24, 28, **32–48**, 103–4, 110, 114, 120, 236, 307, 307t, 309, 313, 316t
  CNG pump prices (versus petrol and diesel, 2011) 152t
  CNGVs and CNG filling stations 156t
  commodification and loss of competitiveness **40**
  comparative advantage **34–5**
  design cost (comparative advantage) 34
  fuel cell R&D 126
  hybrid car sales (2006–9) 111f
  industrial architectural geopolitics 37
  in-house sourcing **118–19**
  manufacturing capability ('integration' type) 34
  over-engineering, under-engineering problems 47
  patent protection 105
  product design, quality, function 'excessive' 40
Jato Dynamics 27n, 31
Johnson, P. 254, 262
Joiun 65
Josefson, L. 203t
Juice Technologies 226
Jullien, B. ii, **xix–xx**, 1, 324, 326

Katz, B. 54, 67
Kazakhstan 149t
Kazazian, T. 53, 67
KCP method 4, 70, **74–5**, 82, **83–4**
  K (knowledge-sharing) phase: creating new design spaces 74, **75–6**, 82
  C (concept) phase: introducing disruptions 74, 75, **76–7**, 82
  P (proposal) phase: proposing path through innovative and sustainable strategies 74, **77–8**, 82
KCP process 78, 80–1
  'complementary activity' to R&D 81
  'cross-disciplinary approach' 82
  'describes new value spaces' 79–80
  helps 'to generate new valuation criteria' 80
  technique to decide which ideas should be explored' **79**
Kemp, R. 19, 30
Kempton, W. 220, 222
Khan, A. R. 248, 252
Kimble, C. **xxiii–iv**, **7**, **240–53**, 317n, 319n, 322
'Knightian uncertainty' (known unknowns) 73, 79, **90–1**
  see also scenarios
knowledge 74, 83(n4), 138(n4)
  external 72, 82, 132, 133
  incomplete 291
  missing areas (identification) 75
  new 79, 82, 198
  new (identification) 81
  producing new **78–9**
  research-based 24
knowledge accumulation 92, 106
knowledge base/s 23, 82, 190, 195t, 195f, 202t
  architectural 194, 195, 198, 200
  electric motor development for heavy hybrid vehicles (comparative analysis) **194**
  internal 7, 72, 82, 197, 198
knowledge development
  Scania **192–3**
knowledge gaps 79
knowledge models 2
knowledge networks 24
knowledge transfer 29
knowledge-based clusters 121
Kollmorgen 186, **191–2**, 194–5
  architectural knowledge 195t
  brands 191
  component knowledge 195t

data source   191, 204
  subsidiary of Danaher Motion   191
Korea   155t, 312, 317t
Kroon, A.   187
Kubic, M.   290n, 302
Kubitschek, J.   174
Kuwait   149t
Kyoto standards   147
Kyrgyzstan   152t

La Jamais Contente (electric car, 1899)   54
labour-retaining capability (ASEAN)   37
Lacerda, A. C.   173n, 183
Lambert-Pandraud, R.   301(n3), 302
Landi Renzo company   161(n4)
Lane, B.   **290–1**, 303
Latvia   151t, 155t, 157t
Lausanne   62
Law of Road and Transportation Security (China)   245
Le Masson, P.   73, 80, **84**, 91, 102
Le Quément, P.   56
lead-acid battery   210, 244, 246
Leclerc, E.   318
LED/OLED   xv, xvi, 19
Lenfle, S.   91, 102
Lévy, R.   56
Lexus LX400h   116
Liaocheng City: Gaotang County   249
Liechtenstein   152t
life-cycle emissions   25
light trucks   *see*   trucks
light-duty vehicles   210
  battery-charging systems   210–11
  see also   low-speed electric vehicles
Lille   159
Limits of Growth (Meadows *et al.*, 1972)   51, 67
Lindqvist, S.   107, 121
Linköping University   xxi, 203t
liquefied natural gas (LNG)   149, 162
  global trade (2006)   150
  pollution and safety issues   **270**
  refuelling stations   159
  source of supply   150
  storage   270
  tankers   150
  vehicles   159–60
  see also   natural gas

liquefied petroleum gas (propane)   16, 28, 144, 154, 271, 276f, 279, 282f, 283, 304
  average pollutant emissions   145t
  pollution and safety issues   **270–1**
  LPG drivers   294
  LPG vehicles   21
liquid hydrogen   19, 30(n8)
lithium cobalt oxide (LCO)   227–8
lithium iron phosphate (LFP)   228
lithium manganese oxide (LMO)   228
lithium nickel cobalt aluminium oxide (NCA)   228
lithium nickel manganese cobalt oxide (NMC)   228
lithium polymer batteries   60, 210
  'polymer battery'   19
lithium titanate (LTO)   228
lithium-ion (Li-Ion) batteries   17–21, 24, 63, 93, 210, 225, 227–8, 230, 311
  see also   batteries
lithium-metal-polymer (LMP) batteries   63
lithium-sulphur   227
Lithuania   151t, 157t
Liu Cheng Qiang   247
Liu Yifa   247, 249
Loch, C. H.   91, 102
Loewy, R.   **55**, 67
LOHR Industrie group   50, **62**, 63f
London   29, 254, 313
London, T.   242, 252
Lotti, G.   52, 60, 67
Lotus   24, 315, 317t
low-speed electric vehicles (LSEVs)   8, **316–17t**, 319
  advantages   251
  business model   **247–8**
  business model: dual structure of Chinese economy 'key factor'   248
  Chinese exports   250
  commercialization (wanting)   246, 251
  consumer acceptability   251
  definition (international consensus lacking)   **244–5**
  main market   247
  market   **248–50**

low-speed electric vehicles (LSEVs) – *continued*
  models on sale (2011 and by 2015) 305t
  'not seen as road vehicles' 248, 249
  potential advantages for urban consumers 248
  prospects outside China 251
  running costs 'low' 248
  scenarios 250
  Shifeng involvement (2008–) 247
  see also 'new-energy vehicles'
Lucassen, J. 301t
Lung, Y. ii, 323–6
Luxembourg 151t, 153f, 156–7t

Maastricht 301t
Machine that Changed the World (Womack *et al.*, 1990) 323, 325
Madeira 168
Madlener, R. xxiv, 9, 14, **286–303**
Madrid 163
Magneti Marelli 175–9, 182(n1)
Magnusson, T. xxiv, 5, 6–7, **103–23**, **185–204**
Magretta, J. 241, 252
Mahla, S. K. 144n, 163
Malaysia 149t, 152t, 155t
MAN 185
Mandart, D. 62
manufacturing 23, 25, 191, 201, 257, 267
  costs 18
  design information view 48
  Fordist 50, 55
  'integrated' type 36
manufacturing capability
  'division of labour' type 34
  'integrated' type 34
manufacturing capability and architecture of green vehicles **3**, 15, 18, **32–48**
  basic forms of production (structure, function, and operation of artefacts) 41f
  commodification (developed nations) **43–4**
  commodification and loss of competitiveness (Japan) **40**
  comparative advantage of design location **34–5**
  competitive advantage in integral-type products **36**
  conditions of commodification **42–3**
  constraints and architectures **35–6**
  crisis in automobile industry and future **46–7**
  design, production, and consumption of artefacts **40–2**
  design-based comparative advantage 38f
  dissemination of EVs 'not easy' **44–5**
  increasing complexity versus commodification **45–6**
  industrial architectural geopolitics **37**
  'level-headed discussion' required 45
  manufacturing management (broad concept) **32–4**
  organizational capability and architecture **34, 38–9**
Manzini, E. 51, 53, 67
Marchionne, S. 258–9
Maritz 287, 303
market adoption 181
market context 213
market distortion 26
market dynamics 140, 259
market economy 242
market entry mode 231
market experience 119
market failures 225, 231, 234, 238
market forces 261
market imperfections 229, 230, 231
market intelligence department 75
market launch 200
  versus economic success 105
market leadership loss 241
market maturity 27
market mechanisms
  electricity (rational use) **217–18**
market needs 33, 36
market niches/niche markets 14, 53, 89, **92**, 92–3, 97, 104, 112, 199, 241, 251–2, 259, 286, 309
market pressure (MP) 29, 128–31, 137t
  see also perceived capabilities
market pull 15
market push 49, 54
market requirements 38, 38f, 39

Index 355

market segmentation 3, 51, 56, 58, 66, 101, 256, **258–9**, 260, 299, 301(n3)
  see also consumer segments
market share 40, 119–20, 244, 258
market size 308
market transactions strategies (corporate growth) 229
market trends 72, 135
marketing 72–3, 75, 116, 267, 287, 301t, 301(n7)
  competence/capability 104, **106**
  'green' 52
markets 24–5, 36, 106, 165, 175, 177, 207, 211, 309
  'base-of-pyramid' 242–3, 252, 253
  domestic 244
  external 172
  global 40, 45, 47, 134, 181t
  high-income 243
  internal 261
  international 171
  mainstream 241
  mature 258
  new 15, 50, 65, 233
  potential 199
  pure competitive 178
  selected 200
  world heterogeneity 308
Markides, C. 242, 243, 252
Marx, K. H. 256
Matra 56, 63, 316t
  concept car (2005) 64
maximum speed 293f, 295f, 296n
  see also speed
Maxwell (supplier) 192
Mazda 310t
Meadows, D. H. 51, 67
Megane 59
Melville, K. **59**
mentality 278f, 278–9
  conventional versus alternative engines 287, **292–4**, 295f, 301(n7), 302
Mercedes 57
Mercedes S-class 299
Mercedes-Benz 116
Mercosur 165
methane 142, 150, 159, 160, 270
  greenhouse effect 143–4
  pipeline networks 6, 143

thermal efficiency 161(n2)
methodology
  business model innovation (case study) **247**
  firm perspectives on hydrogen technology **126–8**
Mexico 152t, 155t, 307, 307t
Michelin 24, 56, 58, 316t
micro hybridization 117, 120
Middle East 149–50t, 150, 307, 307t
  instability (impact on oil price) **266**
Midler, C. **xxiv**, xxvii, **4–5**, **89–102**
Milan 29
Mines ParisTech 74
Minguet, C. xxvii
minicars 30(n10), 319t
Mississippi 246
Mitchell, D. W. 241, 242, 251, 252–3
Mitsubishi 119, 234, 310t, 315
  electric vehicles 44
  partnership with PSA Group 237
Miyazaki (Japan) 313
MMA (*Ministério do Meio Ambiente*) xv, 178n, 183
mobility 7, 54, 72, 240
  'actual product' **256**
  car-focused (crisis) **254–5**
  case studies 3
  'collective as well as personal right' 256
  design thinking as new approach **50–4**
  individual right to be safeguarded (Garibaldo) 8, 255
  'low-cost package' 76
  'network dimension' 255
  new 49, 58
  'primary drive needing to be elaborated' 257
  as product/service: challenge from new players 50, **60–5**
  'real product' 256
  as service 52, 53
  as service offer: from style to design 50, **54–60**
  short-distance 66
  social approach 66
  as social asset (Garibaldo) 8, 255
  social organization 49
  subscription system 66

Mobility House TMH   318
mobility operators   23, 318
mobility platforms   261
mobility system/s   257
  networked social dimension   8, 261
'mobilization capability' (China)   37
Mobivia   318
Modulgo (small EV)   **60–2**
Moldova   151t, 155t
Monaco   134
Montalvo, C.   **xxiv–v**, 5, **124–39**
Montana   246
Montbéliard   62
Montesquieu University – Bordeaux IV   324
Morris, M.   241, 253
Morvannou, P.   xxiii
motor cycles   243, 261
motor drive system   191
  see also   electric drive systems
motor power   209
motor shows   57, 58, 63, 90, 120, 134
motor vehicle architecture   321
motorbus   255
motorcycles   164, 244–5, 248
multi-mobility   54
Mytelka, L. K.   287, 289–90, 290n, 301(n5), 303

National Agency for Petroleum, Natural Gas and Biofuels (ANP Brazil)   xiii, 168f, 171, 182
National Alliance for Advanced Transportation Battery Cell Manufacture   318
National Association of Motor Vehicle Manufacturers (ANFAVEA, Brazil)   xiii, 166, 168f, 175, 182(n1)
  website   182(n3)
National Highway Traffic Safety Administration (USA)   246
National Institute for Research in Computer Science and Control (INRI)   60
National Research Council: Institute for Research on Firms and Growth (CNR-Ceris, Moncalieri)   iii, xxi, xxv, xxvii, 2

natural gas (NG)   4, 16, 21, 28, 29(n1), 138(n5), **142–3**, 154, 271, 276f, 279, 282f, 283, 304–5, 307, 307t, 309
  efficiency of WTW supply path   144
  geographical distribution   148, 149t
  known reserves   6, 143
  pipelines   160
  production (2009)   150t
  refining process   148
  reserves   17, 148, 161(n6), 162
  technological advances   149
  see also   compressed natural gas
NEC   230, 234, 237, 315
Netherlands   xxii, xxiv, 27t, 151t, 153f, 155t, 157t, 169, 315
  AVs (consumer attitudes)   **286–303**
  EVs (target number)   313t
  new-car prices   289, 301(n4)
New European Driving Cycle (NEDC)   100, 144, 145t, 146, 161(n5)
new product development (NPD)   72, 89
  literature   97
  NPD process   191
  NPD projects   197
  spillover effects   200
New South Wales   313
New York   266, 314
New York ISO (Independent System Operator)   220, 223
New York Stock Exchange   171
New Zealand   155t, 312
NGVA Europe   152–3n, 154, 156–7n, 163
niche markets   *see*   market niches
nickel-hydride battery technology   19
nickel-metal hydrate battery   210
nickel-metal hydride (NiMH) batteries   20, 108
Nielsen, L.   203t
Nigeria   149t, 265, 266
Nijhuis, J.   **xxv**, 9, 14, **286–303**
Nissan   22, 59, 90, 135t, 138(n6)
Nissan Leaf (EV)   119, 228
  see also   Renault
nitrogen oxide   30(n8)
  see also   oxides of nitrogen
nitrous oxide (N_2O)   159
noise   26, 76, 110, 160, 293f, 295f

non-methane hydrocarbons (NMHC) 145t
North, D. C. 165–6, 182, 183
North America 2, 21, 149–50t, 150, 243
North American Reliability Corporation (NERC) 219
Norway 27t, 149t, 151t, 153f, 155t, 315, 317
NOx *see* oxides of nitrogen
nuclear energy 28, 100, 259, 261, 275t

Oak Ridge National Laboratory (Tennessee) 216, 219
Obama administration 307, 307t
Obasanjo, O. 265
Oceania 149–50t
octane 144, 144t, 158, 269
Ogden, J. M. 143, 162
oil 154, 162, 214, 259, 262(n2), 291, 308
　less-polluting engines 306–7, 307t
　new sources (difficult locations): impact on oil price 265
　percentage devoted to road transport 142
　potential replacements 264
　unavailability 283
oil companies 14, 100, 265, 278, 280
oil imports 142, 172, 173t, 174, 268–9, 271, 274t, 283
'oil peak' 264
oil price 15, 25–6, 162, 179, **277–81**, 282, 299, 305, 309, 312, 318
　fluctuations 290
　meaning 263, **264–7**
　peak (potential triggering factors) **264–6**
　uncertainty 9, 263, 277
　upper limit 280
oil reserves 17, 148, 278, 279
oil shocks
　(1970s) 56, 103, 111, 165–6, 171–2, 174, 176, 180, 265–6, 287, 289–90
　('late 1980s') 111
oil supply 258, 280
Oliveira, E. S. de 175
Olson, M. 166, 183
omnibus (horse-drawn) 254, 255, 256
on-board chargers 210, 211

One North East 235
Ontario 313
OPEC behaviour/impact on oil price **265**
Opel 141t, 258, 300, 301t
Opel Ampera 21
Oregon 313
Our Common Future (Brundtland Commission, 1987) 51, 67
oxides of nitrogen (NOx) 106, 110, 141, 145t, 146, 269, 271, 282
　NOx Trap 114f
Ozaki, R. 291, 303

Pacific Northwest National Laboratory (PNNL) 216, 219
Padua 62
Pakistan 155t, 307, 307t
Palo Alto (California) 29
Panasonic 315
Panasonic/Sanyo (Toyota Group) 118, 229
Papanek, V. 51, 66(n1), 68
parallel hybrid configuration (Volvo) 190, 190f
parallel hybrid electric drive
　Scania patent (2001) 193
'parallel hybrid system' 186–7, 201, 201t
parallel hybrid technology 200
parallel pilot testing
　'ecosystem' formation support **94–6**
parallel structures versus series layout (range-extender) 21
parallel systems (hybrid powertrains) **189**
parallelism (in development processes) 98
'parallelism and selection' 91
Pardi, T. 2, 326
Paris xxi, xxiii, xxv, xxvii, 29, 50, 57, **62–5**, 90, 254
Parkeon 234
particulate filter 111
particulate matter (PM) 141, 160, 269
　smaller than 10 micrometres in diameter (PM10) 145t, 146
passenger cars/vehicles 4, 26, 185, 188, 250–1
　see also rental cars

Patel, P. 105, 122
patent applications 200
  Scania (2010) 193
  Volvo (2008–9) 191
patent data **105–6**
  literature 114, 116
patents/patenting 79, 104, 118, 135, 194, 197
  Danaher Motion 191
  diesel technology (1893–) 109
  electric motor (Volvo, 2005) 191
  electric motor development **200**
  electric motor development (hybrid vehicles, 1990–2009) 188f
  hybrid drive system (Volvo, 1997) 191
  parallel hybrid electric drive (Scania, 2001) 193
  Voith 193
  weak performance of European producers **112–16**
path dependence/path-dependency 6, 32, 165, 167
  automotive chain **174**
  definition (David) 167
  and innovation in sugar cane sector 164, **168–9**
  sources **13–14**
  theory 2
Paulus, I. 136t
Pavitt, K. 105, 122
Pelata, P. 101(n3)
Pennsylvania 246
Pennsylvania Railroad Company 55
Perrin, J. 101(n3)
Persian Gulf instability
  impact on oil price **266**
Persson, M. 161(n7), 163
Peru 155t
Perugia 260
Petrobras 172
petrol 17, 19, 27, 44, 53, 144, 154, 164, 177, 269–71, 276f, 282f
  average emissions in urban situation (Euro standards) 145t, 146
  combination with ethanol 176
  competitive disadvantage (rural China) 248
  octane index and heat of combustion comparison 144t

  pollution and safety issues **268**
  premium 'blue' 174
  price (versus CNG) 150–3
  rationing 172
  retail price 20
  taxes 27t
  thermal efficiency 161(n2)
  unleaded 27t, 30(n12)
petrol distribution monopolies **172**
petrol engines/petrol ICE 6, 16, 20, 110, 124–5, 138(n5), 140, 143–4, 158, 161(n4), 176, 179, 227, 274t, 280–1, 299, 300f, 318, 321
  average emissions (Italy) 148t
  cost advantage versus diesel 116
  evolution 119–20
  number of cars on road (Italy) 147t
  patent activity (1992–2007) 112, 113f
  see also aero-engines
petrol hybrid vehicles 103, 112, 120
  market battlefield (USA) **116–17**
  patent activity (1992–2007) 112, 113f
  see also plug-in hybrid electric vehicles
petrol price/s 28, 60, 65, 109, 170, 171, 174, 278–80, 283
petrol stations 165, 172, 174
  see also filling stations
Peugeot (Peugeot-Citroën/PSA Group/PSA Peugeot Citroën) 22, 56–8, 104, 110–11, 113n, 114, 117, 119, 122, 134, 135–6t, 138(n6), 236, 300, 301t, 309, 310t
  average CO2 emissions of cars sold in Europe (1997–2010) 141t
  diesel engine patents (1990–2007) 115f
  diesel hybrid (2011) 120
  partnership with Bosch 233, 237
  partnership with Mitsubishi 237
Peugeot '3008 Hybrid 4' 233
Peugeot Automotive Design Network (2004) 57
Peugeot Ion 63
photovoltaic panels/systems 19, 100, 101, 306
Phylla solar car 235
Piedmont 30, 235
pilot projects 4–5, 29, 90–3, 99
  characteristics 92
  ecosystem formation support **94–6**

Index 359

long-term framework agreements 98
means of developing assets and generating support 96
opportunity for collective learning 96
parallel 94–6
role in Renault's EV strategy 95
simultaneous 98
source of learning 89
strong local component 97–8
tool for building micro-level 'ecosystem' 95–6, 97
Pininfarina, B. 56
Pininfarina Group 63–4, 66, 66(n3), 316t
pipelines 6, 17, 143, 149, 150, 154, 160, 270
Pizzinatto, N. K. xxvi
plug-in hybrid electric vehicles (PHEVs) 10, 28–9, 109, 112, 117, **119**, 120, 214, 216, 220, 222, 223–5, 234, 237, 304, 307–12, 313t, 319t
blended plug-in; extended plug-in; green-zone plug-in 209
cost-benefit analysis (for 2020) 20, 20t
deployment (impact on US power grid) 219
global projection (2020) 215
models on sale (2011 and by 2015) 305t
refuelling 210
see also smart vehicles
Poland 152t, 155t, 157t, 307, 307t
Polaris Industries 246
policy implications 2
agreements and joint ventures in EV industry 8, 226, **238**
business model innovation (China) 8, **250–2**
CNG cars **160–1**
innovative design 15, **24–9**, 30(n11–12)
Polk 110–11n, 122
pollutants 4, 69, 141, 143, 146
trans-border 35, 39
pollution 6, 9, 16, 26, 28, 76, 158, 164, 207–8, 215, 217, 240, 251, 258, 263, 282–3, 305, 311
responsibility 'shared by designers' (Papanek) 51
see also emissions

pollution issues **268–71**
pollution-reduction 228, 307, 307t, 309, 312
Porsche 309, 310t
Porter, M. E. 175, 183
Porter's diamond 175
Portugal 25, 27t, 151t, 153f, 156–7t, 166–9, 182(n6), 307t
power control unit **226**
power electronic converters 219, 220
power electronics 16, 105, 118, 120, 187, 188, 190, 200
patents (1990–2009) 188f
power flows
bi-directional 215–16
'power for convenience' consumers 296t, 297f
power grids
electric vehicles: challenges and opportunities **7**, **207–24**
power industry 25
Power Information Network 110n
power plant capacity 215
power plant emissions 268
power plant inefficiency 221
power plants 212
coal-burning 119, 268
fossil-fuel 22
oil-fired 221
thermal and nuclear-thermal 222
thermo-electric 214
see also electricity-generating plants
power profile 211–12
daily (smoothing) 222
power quality 219, 220
PowerStream 236
powertrain architecture 209
powertrain control systems 16
powertrain systems 187
powertrain technologies 103
powertrains 7, 13, 71, 76, 104, 106, 107, 110, 117, 256, 304, 305
clean 9, 260
electric 315
future evolution 120
hybrid 116, 118
hybrid or purely electric 209
modularization and standardization 319
parallel hybrid 199

PPG   234
Prencipe, A.   104, 122
Prius   *see*   Toyota Prius
PROALCOOL programme (1975–)   **166**, 168f, 172, 179, 181t
  law No 75,593 (Brazil, 1975)   169
  processes   3, 32–5, 38, 38f, 41, 89–90, 98, 272
  high-involvement   288
  hybrid vehicle development   **186–7**
  industrial   256
  new   91, 105, 256
  see also   R&D processes
product appearance   55
product architecture   33, 35, 45, 242, 246, 253
  integral versus modular   46
product categories   35, 38
product concepts   81
product cycle hypothesis   40
product design   15, 39, 42
product development   15, 56, 58, 80, 99, 175
product development collaboration   7
  adaptation to knowledge contributions of organizations involved   197
product development competence   104, **106**
product differentiation   79, 80, 81
product failure   241
product function   33–4, 39
product launch technology frame   **134**
product life-cycle   53, 70, 72, 76
product life-cycle management (PLM)   57–8
product strategy   52, 53
product structure   33–4, 39
'product support'   99
product-process architecture   3, **33–4**
  'integral type'   34
  'modular (mix and match) type'   34
production   1, 2, 15, 21, 26, 32, 45, 46, 51, 105, 181, 222, 265
  artefacts   **40–2**
  basic forms   41f
  constraints   42
  internationalization   58
  methods   19
  over-capacity   65

see also   mass production
production costs   19, 59, 312
production goods   41
production mode
  challenged by eco-design   52
production processes   3, 33–4
productivity   36, 56, 170, 181, 308
products   30, 32, 36, 41, 51–2, 78
  cultural meanings   49
  electronic   3, 15, 45
  high-involvement   288
  integral-type   37, 47
  modular-type   43
  modular-type (capital-intensive)   37
  modular-type (knowledge-intensive)   37
  modular-type (labour-intensive)   37
  modularization and commodification   **42–3**
  new meanings   3, 49
  see also   artefacts
propane   *see*   LPG
prototype electrified lanes   211
prototypes   57, 58, 80, 95–6, 99, 125, 167f, 175, 178, 181t, 187f, 187, 190, 193, 195, 197–200, 235, 305t
  reduction in number   50
  tool for validation and learning   97
PSA Group/PSA Peugeot-Citroën
  see Peugeot

Qatar   149t
quadricycles   58, 246
Quebec   313

radial flux technology   193–5, 196f, 200
Ragazzi, E.   **xxv, 207–24**
Ranis, G.   248, 253
rapid-charging stations   **235**
rare earths   225
recharging   *see*   charging
recycling   3, 24, 52, 61, 237
reduction gear   **227**
refineries   266
  accidents (impact on oil price)   **265**
refuelling   207, 274t, 276f, 294
  see also   battery charging
regasification   150, 154, 160, 162
regenerative braking   18, 19, 89, **227**, 228
  see also   braking energy

Régie Autonome des Transports Parisiens (RATP)   74, 83(n3)
Renault   4, 22, 57, **59**, 66, 83(n2), 110, 135t, 138(n6), 175
  average CO2 emissions of cars sold in Europe (1997–2010)   141t
  EV: commercial launch   90, 101(n2)
  EV strategy: breakthrough   **93–4**
  EV strategy: case study   **93–6**, 101(n3)
  EV strategy: pilot testing   **95**
  four electric car projects   59, 94
  'style centre' (later Department of Industrial Design)   56–7
Renault: Electric Vehicle Business Development (BDVE, 2009–)   93–5, 101(n2)
Renault Fluence   59
Renault Kangoo Express Z.E.   59
Renault Logan   59
Renault Technocentre (1998)   57
Renault Twingo   57, 59
Renault Twizy   59, 66(n2)
Renault ZEV initiative   **93–4**
  battery technology   93
  business model   94
  optimization for electric engines   94
Renault-Daimler alliance   59
Renault-Nissan   19–20, 30(n9), 114, 119, 310t, 315
  agreement with NEC   230, 234, 237
  agreement with One North East   235
  agreements with municipalities   231, 235, 237
  diesel engine patents (1990–2007)   115f
  patents related to hybrids (1992–2007)   113f
  see also Nissan
renewability
  criterion to find best motor vehicle fuel   275t, 276f
renewable energy   17, 148, 164, 170, 220–2, 268, 271, 274–5t, 276f, 279, 305, 311
  distributed generators   214
  small-scale distributed generators   212
rental cars   311, 313
  companies listed   **314**
  see also saloon cars

research and development   14, 30, 56–8, 60, 66, **71–3**, 103, 111, 120, 124–6, 161, 177, 191, 201, **201–2t** 228–30, 259, 312
  conventional methods 'no longer sufficient'   81
  high levels of cost   232
  in-house   126
  new processes in greening of automotive industry   70, **78–81**
  personal interviews   118, 123
  role   **261**
  rule-based   79
  stage-gate processes   73
  technology frame   **134**
research and development:
  agreements   237
  centres   232
  departments   105
  expenditure   134
  investment   138(n3), 232, 236, 308
  managers   105
  networks   233
  phases   16
  processes   **3–4**, 16, **69–85**
  subsidies   234
research octane number (RON)   144t, 161(n3)
Réseau Francilien de Recherche sur le Développement Soutenable (R2D2)   82(n1)
retailing   97, 172, 174, 211, 212
Reunion pilot testing   95, 97, 98, 312
  EV technology   **100–1**
Reva   316t
Reva: G-Wiz EV   24
Rheims Auscher   54
Ricart, J.   243, 253
Rifkin, J.   66, 68
Riley, R. Q.   266, 284
Riskin, C.   248, 252
Rolfo, S.   2
Romania   27t
Roordink, R.   301t
Rosenbloom, R. S.   241, 252
Rothschild   54
Rouen: Community of Agglomeration   318
4R Energy Corporation   237

Russian Federation 28, 155t, 264, 307t, 307–9, 317t
  carmakers 310t
  CNG pump prices (versus petrol and diesel, 2011) 152t
  natural gas production (2009) 150t
  reserves of CNG (2009) 149t
  'Soviet Union' 37
  see also BRIC
RWE 236, 237

Saab 135t, 138(n6)
Sadarangani, C. 203t
safety 17–18, 24–6, 28, 32, 35–6, 39, 41–4, 46–7, 54, 58, 60, 71, **268–71**, 288, 290t
  heaviness versus lightness bias 55
Sagem 74
SAIC 231, 237
'sailing ship effect'
  origin **106–8**, 121
'sailing ship effects' in global automotive industry 5, 19, **103–23**
  conclusion and discussion 104, **119–21**
  diesel car sales (Europe and USA, 2005–10) 110f
  electric-petrol hybrids (Toyota takes initiative) **108–9**
  evolution of diesel engines (European success story) **109–11**
  hybrid car sales (Europe versus Japan/USA: explanation for huge difference) **111–12**
  hybrid car sales (USA 2000–10) 109f
  hybrid car sales (USA, Asia-Pacific, Europe, 2006–9) 111f
  'hybrid electric-petrol cars versus clean diesels 104, **108–12**
  'key questions/issues' 119, 120–1
  literature 103, 106–7, 114, 116
  measures of car manufacturers' innovativeness **104–6**
  measuring technological competence **105–6**
  modular versus architectural innovation' 104, **117–18**
  origin of 'sailing ship effect' **106–8**
  plug-in hybrids 119

product development, industrialization, marketing competence **106**
  selection bias 108
  sourcing strategies (in-housing sourcing in Japan, specialization in Europe) 104, **117–19**
  technological activities (weak patent performance of European producers) **112–16**
  technologies and market performance 104, **108–12**
  US market (diesel versus petrol hybrids) **116–17**
Saitama (Japan) 313
Sánchez, P. 243, 253
São José dos Campos (São Paulo) 175
São Paulo 170, 313
São Paulo state 170, 171
São Paulo University: Technological Research Institute (IPT) 168f, 177
SAP (software company) 58
Saudi Arabia 149t
Scania 185, 194, 200, 204
  architectural knowledge 195t
  component knowledge 195t
  data source 192n, 204
  heating system (patent application, 2010) 193
  hybrid vehicle development processes **187**
  in-house development 192
  interviews 201, **201–2**
  parallel hybrid system 199
  patent for parallel hybrid electric drive (2001) 193
  permanent-magnet synchronous motors (transverse flux) 190
  problems of developing two technologies in parallel **198–9**, 200
  series hybrid application 199
  series hybrid powertrain 192, 192f
  sourcing and knowledge development **192–3**
  technology and application innovations of electric motors relative to suppliers 196f
  timeline 187f
Scania-Voith collaboration 186, **192–4**, 195–8, 200

technology innovations relative to vehicle manufacturers 196f
scenario of diversity 10, 306–9
conditions required for success 308
scenario of progressiveness 10, 309–11
scenario of rupture 10, 311–20
  cleaner car models (by types of producer) 319t
  consequences 320
  electric vehicle start-ups and newcomers (2011) 316–17t
  many active supporters to form coalitions 312–18
  number of cleaner models (by type of vehicle) 319t
  technical versus geopolitical conditions 318–20
  two stages 311–12, 313t
scenarios 7–8, 24, 212, 259
  advantages 276–7
  AV fuels 9, 263
  in confrontation (second automotive revolution) 9–10, 304–22
  definition 272
  non-quantitative factors 277
  use 276
  see also uncertainty
scenarios: 'best motor vehicle fuel' 272, 276–84
  assumptions 277, 282
  catastrophe scenario 277–8, 278f, 280–1, 281–2f, 283, 284
  choice of fuel results 282f
  economic challenge scenario 277, 278f, 280, 281–2f, 282, 284
  environment challenge scenario 277, 278f, 279, 281–2f, 282–3
  environmental concerns 277
  external influences 277
  first-tier criteria comparison results 282f
  mentalities 278f, 278–81
  oil price 277
  players/actors 277
  status quo scenario 277, 278f, 278–9, 281–2, 283
Schendel, D. 107, 121
Schneider Electric 98, 234, 318
scooters 244–5, 248, 315, 319

second automotive revolution: scenarios in confrontation ii, xx, 9–10, 28, 304–22
  'door-to-door' transport system (solution to crisis) 9, 304–5
  economic policy decisions 9–10, 306
  formation of coalitions between public and private actors 9, 305–6
  'gradual' versus 'all at once' scenario 309–11
  scenario of diversity 10, 306–9, 320–1
  scenario of progressiveness 10, 309–11, 320–1
  scenario of rupture 10, 311–20, 320–1
  three of four conditions about to be fulfilled 9, 304–6
  transfer of new technologies developed in other sectors 9, 305
Segrestin, B. xxv–vi, 3–4, 69–85
Seoul 314
Serbia 152t, 155t
series hybrid technology 187, 189, 192, 192f, 194, 200
services 52, 258
  after-sale 14
  cultural meanings 49
  new 242
  new meanings 3, 49
  use-oriented and result oriented 29
Sevastyanova, K. 291, 303
Shandong Province 247
Shanghai 62, 242
Shanghai Kanleqiu Science and Technology Company 247
Shifeng (Group) Company Ltd (1993–) 316t
  background 247
  business model innovation (case study) 246–50
  location 249
  production of LSEVs (2007–) 248–9
Shimano (bicycle parts manufacturer) 53
Shimizu, K. ii, 325
Shove, E. 289, 303
Shrestha, R. M. 143, 162
Siciliano, L. de B. 175
Siemens 118, 237

Silicon Valley (California)   24
Sina-Newchance New Energy
    Technology Co. Ltd.   247
Singapore   152t, 155t, 236, 312
Sloan, A.   55
Sloanian model   **60**
Slovakia   152t, 155t, 157t
Slovenia   152t, 157t
Smart car (petrol-engined)   57
Smart EV   24, 59, 63
    usage cost   27–8m 30(n12)
Smart Fortwo   66(n2)
smart grids   xxi, 2, 7, 207–8, **212–13**, 215, 218, 221–2, 225, 235–7
    feasibility   213, 223
smart meters   218, 219
smart vehicles   257, 260
    see also   sport utility vehicles
Smith, A.   ii, 89, 92, 97, 102, 326
sodium nickel chloride   227
software fuel sensor (SFS)   176–7
soil   51, 170, 268, 269
solar cars   19, 64
solar energy   18, 71, 76, 98, 101, 268, 274–5t, 279
Solar Print   236
Somalia   266
Sorgenia (Italy)   236
sourcing   5, 108
    electric motor development   **192–3**
    in-house sourcing (Japan) versus specialization (Europe)   104, **117–19**, 120–1
    Scania   **192–3**
South Africa   315, 317t
South America   2, 149–50t, 313
South Asia   243
South Korea   37, 47, 307t, 307–9
Southeast Asia   243
Southeast Electric Reliability Council   216
Spain   25, 27–8, 30(n12), 151t, 153f, 155t, 157t, 182n6), 307t, 313t, 313, 315, 317t
speed   55, 248, 251, 258, 314
    see also   LSEVs
Spers, E. E.   xxi, **xxvi**, 6, **164–84**
split-axle hybrid   233, 237
sport utility vehicles (SUVs)   46, 71, 109, 210, 319t

    see also   ultra low-emission vehicles
sports car   315, **316–17t**, 319t
    see also   used cars
stakeholders   92, 95–6, 99, 127, **300**
    firm perspectives on hydrogen technology   **134–6**
Steg, L.   287, 303
Stern Report (2007)   108, 122
Stocchetti, A.   **xxvi**, **5–6**, **140–63**
Stockholm   187, 314
Stockholm: SL (public transport provider)   197
Stockholm: Royal Institute of Technology (KTH)   203t
stop-start features   19, 89, 117
storage   271
storage capacity (oil)   266
strategic intent versus experimental approach (EVs)   **4–5**, **89–102**
Studebaker models   55
Stumpf, U. E.   175
sugar cane (*Saccharum officinarum*)   6, 100, 159, 165–7, 168f, 180–1, 181t, 269
    path-dependence and innovation   **168–9**, 182(n6)
Sugar Cane and Ethanol Industry Association (UNICA)   xvi, 166, 168f, 170, 181, 182(n2)
Sugar Cane Technology Centre (CTC), Brazil   xiii, 166, 168f, 169, 182(n4)
sugar price   169–70, 171, 180
sulphur oxides (SOx)   282
Sumitomo   237
suppliers   57, 58, 82, 125, 128, 137t, 167, 168f, 175, 177–9, 233, **316–17t**, 319t, 324
    agreement and joint ventures with carmakers   7
    complementary technical knowledge   197, **198**
    independence   **6–7**, **185–204**
    late versus early involvement   196–7, 198
    listed   **315**
    new   226
    strong relationships with carmakers (EV components)   8
    tier-one   23
supply chains   5, 150, 263, 310

supply and demand   xxii, 23, 142, 154, 165, 171, 178, 258
  market-clearing price   217
  mismatch (role of EVs in reducing)   221
  oil   **264**
  smart grids   212
Sushandoyo, D.   **xxvi, 5, 6–7, 103–23, 185–204**
sustainability   170, 180, 181t, 183, 261, 278f, 279, 284, 290, 324
  social urban model   256
  transitions   102
  see also analytic hierarchy process
sustainable development   8, 14, 29, 92, 98, 240
  addressed through innovative design   70, **74–8**
  aesthetic form   51
  Brundtland definition   51
  concept   2
  industrial designers and challenges of   3, **49–68**
  innovation design   **3–4**
  innovative design in automotive industry   **13–31**
  literature   xix
  new R&D processes in automotive industry   **3–4, 69–85**
  requirements and constraints   3
  scope   3, 15
  strategies (Brazil)   6
sustainable mobility   **5–6, 140–63**
Sustainable Mobility Institute   101(n2)
sustainable solutions
  'new' versus 'old' technologies (global automotive industry)   **5, 103–23**
Sweden   xxi, xxiii, 25, 27t, 111, 151t, 153f, 155t, 157t, 160, 307, 307t, 317t
  biogas production   159
Swedish Hybrid Vehicle Centre (SHVC)   203t
Switzerland   27t, 151t, 155t, 159, 160, 307t, 312, 313t, 315, 317t
synchronous AC generator   **227**
synthetic fuels   17, 21
Syria   168

Taiwan   **37**, 40, 312, 316t
Tajikistan   152t
Tata Motors   315, 316t
Tata Motors: Nano   45
tax incentives   90, 97–9, 116, 172, 177, 179, 181t, 273t, 293f
tax on industrial products (IPI, Brazil)   172
taxation   1, 27t, 28, 89, 110–12, 228, 245, 249, 291t, 291, 296, 301(n4, n9), 312
taxis   45, 77, 314
technological innovation/s   xix–xx, 6, 92, 139, 164–5, 167, 181t, 197, 199, 228, 241, 292, 298, 302
  Brazilian ethanol production   **170–1**
  electric motors relative to suppliers   196f
  impact on oil price   **265**
  relative to vehicle manufacturers   196f
technological trajectories   2, 8, 14, **15–19**, 29–30(n3–8), 104
  alternative vehicles   **4–7, 87–204**
technology/technologies   3, 5, 36, 49, 52, 70, 168f, 226, 272
  adoption 'involves more than producing technologically-elegant solution'   8, 240
  alternative   125
  CNGVs   **158–60**
  co-existence   108
  'commercialization via business model'   240, 243
  design process   89
  and market performance   104, **108–12**
  mature   19
  new   8, 226, 233–4, 237–8, 259, 305
  'no single objective value' (Chesbrough)   240
  'old' versus 'new' (race for sustainable solutions in automotive industry)   5, **103–23**
  'old' versus 'new' ('sailing-ship effect')   **107**
  'old' versus 'new' (synthesis)   120
  put in motion   181
  relationship between frame, willingness, and belief structure   133f

366 Index

technology/technologies – *continued*
  'search process with spillover effect's **200**
  second-generation 176–7
  state-of-art 232
  successful adoption 8
  supply-side and demand-side co-evolution 165
  system-dependent and capital-intensive 108
  unpredictability 8, 226
  see also disruptive technologies
Technology Service Centre for EVs 247
Teece, D. J. 241, 253
Tenconi, A. **xxvi, 207–24**
Tennenbaum, M. 101(n3)
Tennessee 313
Terna 216n, 224
Teske, J. 20, 30, 107, 121
Tesla Motors 24, 226, 238, 315, 317t
  agreement with Daimler 234, 237
Thailand 152t, 155t
Thalès 74
Tholen, J. xxiii
Tianjin 62
TK Advanced Battery 234
TNO 29(n2)
'TNS-Emnid/AutoScout24' 288, 303
Tokyo 314
Tomić, J. 220, 222
top-down control
  electricity (rational use) **217–18**
torque 108, 110, 158, 192, 227, **294**, 295f, **298**
torque-to-inertia ratio 189
Toyota 19, 21–2, 90, 104, 117–23, 185, 229, 234, 236–7, 309, 310t, 323
  average CO_2 emissions of cars sold in Europe (1997–2010) 141t
  challenge 47
  construction of US truck factory 'overly slow' 46
  diesel engine patents (1990–2007) 115f
  electric-petrol hybrids **108–9**
  fuel cell R&D 126
  hybrid evolution 114
  hybrid technology (US market battlefield) 116
  in-house sourcing **118**, 119, 121
  manufacturing system (evolution) 47
  over-reliance on US market **46–7**
  patents related to hybrids (1992–2007) 113f
  'pioneered patents in hybrids' 112
  plug-in hybrids 119
  quality problems 46
  recall problem 32
Toyota Hybrid System (THS) 118
Toyota Prius (1997–) 16, 58, **108–9**, 119, 122
  driving comfort 291
  launched in Netherlands (2000) **292–4**, 301(n7)
  post-purchase justification versus pre-purchase motivation 291
  UK buyers (questionnaire) 291
Toyota Production System 33
trade-offs xix, 1, 18, 267, 324
  'battery leasing' versus 'fast charging' 18
traffic congestion *see* congestion
tramcar system 255
tramways 62, 256, 257
transformers 215, 216
Transitec (Swiss consultancy) 62
transmission (transferring power to wheels) 117, 118, 120, 190, 190f, 191–2
transmission grids (electricity) 212
transport 54, 65, 94, 100, 150, 174, 240, 261
  emissions 89, 101(n1)
  inter-nodal 255
  long-distance 13, 237, 251
  revolution (C19) 254
Transport Urbain Libre Individuel et Public (TULIP) 58
transverse flux machine/motor (TFM) 193, 195, 196f, 197
Tropsch, H. 30(n6)
trucks 4, 135, 136t, 185, 199
  flex-fuel evolution (Brazil, 2003–12) 178f
  light 178f, 247
  parallel hybrid 187, 201–2t
  three-wheeled 247
  see also heavy hybrid vehicles
turbochargers 19, 110, 158
Turin xxi, xxv–vii, 56, 64, 314

Turkey 155t
Turkmenistan 149t
tyres 19, 24

UAE 149t, 156t
UBS 117, 122
Ukraine 152t, 155t
Ulm 29
ultra low-emission vehicles
  (ULEVs) 124, 125
  see also vehicles
underground railways 209, 255, 256,
  261
União da Indústria de Cana de Açúcar
  (UNICA, Brazil) xvi, 166, 168f,
  170, 181, 182(n2)
United Kingdom xxiii, 24, 27–8,
  30(n12), 58, 235, 287, 307t, 315,
  317t
  CNG pump prices (versus petrol and
    diesel, 2011) 151t
  CNG system (payback period) 153f
  CNGVs and CNG filling
    stations 156t
  electric vehicles (2010–11
    statistics) 27t
  natural gas vehicles (latest
    figures) 157t
United Nations 159, 162
  UNFCCC 102
United States ii, xxvii, 20, 24, 28,
  40, 44, 112, 114, 120, 149, 177,
  194, 228, 235, 237, 243, 250, 257,
  270–1, 290, 299, 307, 312–13, 317t,
  318
  'America' 8, 244, 251
  anhydrite ethanol 176
  automobile crisis 32
  boom and bust 46
  cars 264
  changing position (new preference
    for plug-in hybrid and EVs) 307t,
    308
  CNG reserves (2008) 149t
  CNGVs and CNG filling
    stations 156t
  coal-fired power plants 268
  definition of LSEV **244–5**
  diesel car sales (2005–10) 110f
  electronic fuel injection 176

EV deployment (2020 scenario) 214
EVs (target number) 313t
hybrid car sales (2000–10) 109f
hybrid car sales (2006–9) 111f
hybrid and diesel sales (2007–12
  advance estimate) 117
industrial architectural geopolitics
  37
'low-level of diesel patenting' by
  European car-makers 116
low-sulphur diesel fuel lacking 116
manufacturing capability
  ('division-of-labour' type) 34
market battlefield (diesel versus petrol
  hybrids) **116–17**
natural gas production (2009)
  150t
'playing a catch-up game' 103–4
Prius sales 108–9
tariff on ethanol imports 171
US Congress 109, 171
US Department of Defence 266
US Department of Energy
  (DOE) 126, 219, 268, 269, 285
US dollar decline **266**, 283
US Energy Information
  Administration (EIA) 148,
  149–50n, 161(n6), 162, 219
US Environment Protection Agency
  (EPA) 176
US Patent and Trademark Office
  (USPTO) **105–6**, 113n, 114f,
  115n, 188, 188n, 191, 193, 195n,
  200
  see also North America
University of California
  (Berkeley) 213–14, 223
Upadhyay, J. D. xxvi, 9, **263–85**
Uppsala University 203t
urban areas 18, 27, 210, 237, 249, 306,
  313
  planning 251, 255, 260
  pollution 21
  subscription services 66
  transport networks (service
    vehicles) 314
urban mobility 3, 15, 52, 54, 62, 66
  electric systems 29
  product of systemic change **8–9**,
    254–62

urban mobility: greening of automotive industry 8–9, 14, **254–62**
  crisis of car-focused mobility 8, 254–5
  energy supply and time horizon (portfolio of solutions) 259, 260, 261, 262(n2–3)
  forma urbis and infrastructures 260
  industrial policies 260–1
  integrated systems and role of R&D 261
  mobility as actual product 256
  over-production and market segmentation 258–9, 261–2(n1)
  shaping the network 255–7
  systemic change 259–61
  utility value as vector for change 8, 256–7, 261
urbanization 255
used cars 142, 179
  see also automobiles
Uzbekistan 149t, 152t, 155t

Valeo 104, 118
Vallourec 74
Van den Burg, S. xxv
Van der Cluijs, Mr 301t
van der Heijden, P. 136t
VAT 301(n4)
Vauxhall 141t
vehicle cost
  criterion to find best motor vehicle fuel 273t, 276f
  vehicle development time 275t, 276f
vehicle maintenance cost
  criterion to find best motor vehicle fuel 273t, 276f
vehicle maker: new forms 6–7, **185–204**
vehicle modularization 319
vehicle ranges 56, 57, 58
vehicle-to-grid (V2G) 222
  EV-supporting distribution grid 7, 215, **219–21**, 222
vehicles 7, 19, 100
  eco-efficient 9, 286, 299
  energy-efficiency 7
  energy-saving 53
  high-fuel-economy 65
  lighter 53
  new fuels 266
  obsolete (most polluting) 141–2
  public 77
  range-extender 13
  'specific-purpose' 260
  two-seater 30(n10)
  see also ZEVs
Velib system (bicycle-sharing) 65
Vendée 313
Venezuela 149t, 155t
VERT (EV Technology for Réunion) **100–1**
Vervaeke, M. xxvii, 3, 15, **49–68**
Vezzoli, C. 29, 30
Viarisio, E. 2
vibration 191–2, 209
Victoria (Australia) 313
Virginia and Carolinas (VACAR) 216
Vitali, G. **xxvii, 225–39**
Voith/Voith Group 186, **193–4**, 195
  architectural knowledge 195t
  component and architectural knowledge (electric motors/hybrid drive systems, 1990–2009) 194, 195f
  component knowledge 195t
  divisions 193
  interviews 201, 203t
  patents 193, 197
  transverse flux motor 192
  Turbo division 193
volatile organic compounds (VOC) 145t
Volkswagen (VW) 104, 110, 112, 113n, 117, 122–3 135, 174–5, 301(n8), 309, 310t
  agreement with FAW 231, 237
  average CO_2 emissions of cars sold in Europe (1997–2010) 141t
  diesel engine patents (1990–2007) 115f
  diesel engines 116
  Golf platform (sixth generation, 2008–) 117
  strategic approach 21
Volkswagen (Brazil): Gol Power 1.6 Total Flex (2003–) 177, 179
Volkswagen Beetle (1946) 56
Volkswagen BlueMotion 117, 301(n8)
Volkswagen Jetta 116

Volkswagen Touareg Hybrid   189
Volpato, G.   ii, xxiii, **xxvii**, 5–6, **140–63**, 325–6
voltage-boosting converter   226, **227**
Volvo   123, 194, 196, 204, 307–8, 310t
  advantage of focus   **199–200**
  architectural knowledge   195t
  competences   190–1
  component knowledge   195t
  data source   190n, 204
  in-house development   190, 191–2
  interviews   201, **202t**
  parallel hybrid configuration   190, 190f
  parallel hybrid powertrain   199–200
  patent applications (2008–9)   191
  patent for electric motors   195
  patents   191
  permanent-magnet synchronous motors (radial)   190
  sourcing and knowledge development   **190–1, 192–3**
  technology and application innovations of electric motors relative to suppliers   196f
Volvo: business package team (BPT)   191, 201, **202t**
Volvo-Kollmorgen collaboration   197, 198
  electric motor development   **190–2**
  technology innovations relative to vehicle manufacturers   196f
Volvo Trucks   185, 200
  hybrid vehicle development processes   **186–7**
  timeline   187f
Volvo V60 model   21
VTLIB (Véolia urban transport)   63
VU Log (urban mobility information systems)   62

Wallonia   313
Wang, H.   xxiv, **xxvii, 7, 240–53**, 317n, 319n, 322
Washington DC   126
waste   76, 144, 159
watch design   57, 58
weight/weight-reduction   19, 59, 60, 112, 227, 225, 234, 246, 271, 289f
Weil, B.   73, **83–4**, 85
well-to-wheel (WTW)   19, 144, 161, 207
Well-to-Wheels Report (European Commission, 2007)   153–4, 162
West Indies   169
West Midlands (Australia)   313
Williamson, O. E.   166, 184
wind energy   18, 76, 220, 222, 268, 274–5t, 279, 306
Winebrake, J. J.   267, 277, 278n, 284
Womack, J. P.   323, 325
Wuhan   231, 313

Xerox Corporation   241, 252

Yang, D. H.   248, 253
Yedla, S.   143, 162
Yin R. K.   247, 253
Yokohama   314
Yu, A. S. O.   176, 184

Zero Emission Vehicles (ZEVs)   13, 25, 100, 119, 140, 142, 260
  see also   alternative vehicles
ZF (component maker)   118
Zhang, C.-H.   144, 163
Zhao, J.   240, 253
zinc-air battery   227
Zirpoli, F.   xxiii
Zoe (Renault)   59